Shale Gas, the Environment and Energy Security – A New Framework for Energy Regulation

NEW HORIZONS IN ENVIRONMENTAL AND ENERGY LAW

Series Editors: Kurt Deketelaere, *Professor of Law, University of Leuven, Belgium and University of Dundee, Scotland* and Zen Makuch, *Reader in Law, Barrister, Imperial College, London, UK*

Environmental law – including the pressing considerations of energy law and climate change – is an increasingly important area of legal research and practice. Given the growing interdependence of global society and the significant steps being made towards environmental protection and energy efficiency, there are few people untouched by environmental and energy lawmaking processes.

At the same time, environmental and energy law is at a crossroads. The command and control methodology that evolved in the 1960s and 1970s for air, land and water protection may have reached the limit of its environmental protection achievements. New life needs to be injected into our environmental protection regimes – perhaps through the concept of sustainability in its environmental, economic and social forms. The same goes for energy policy and law, where liberalization, environmental protection and security of supply are at the centre of attention. This important series seeks to press forward the boundaries of environmental and energy law through innovative research into environmental and energy law, doctrine and case law. Adopting a wide interpretation of environmental and energy law, it includes contributions from both leading and emerging international scholars.

Titles in the series include:

Renewable Energy Law in the EU
Legal Perspectives on Bottom-up Approaches
Edited by Marjan Peeters and Thomas Schomerus

Environmental Enforcement Networks
Concepts, Implementation and Effectiveness
Edited by Michael Faure, Peter De Smedt and An Stas

Regulation of the Upstream Petroleum Sector
A Comparative Study of Licensing and Concession Systems
Edited by Tina Hunter

Regional Environmental Law
Transregional Comparative Lessons in Pursuit of Sustainable Development
Edited by Werner Scholtz and Jonathan Verschuuren

Earth Governance
Trusteeship of the Global Commons
Klaus Bosselmann

Waste Policy
International Regulation, Comparative and Contextual Perspectives
Alexander Gillespie

Shale Gas and the Future of Energy
Law and Policy for Sustainability
Edited by John C. Dernbach and James R. May

The Privatisation of Biodiversity?
New Approaches to Conservation Law
Colin T. Reid and Walters Nsoh

Aquaculture Law and Policy
Global, Regional and National Perspectives
Edited by David L. VanderZwaag, Nigel Bankes and Irene Dahl

Charting the Water Regulatory Future
Issues, Challenges and Directions
Edited by Julien Chaisse

Shale Gas, the Environment and Energy Security
A New Framework for Energy Regulation
Ruven Fleming

Shale Gas, the Environment and Energy Security

A New Framework for Energy Regulation

Ruven Fleming

Groningen Centre of Energy Law, University of Groningen, the Netherlands

NEW HORIZONS IN ENVIRONMENTAL AND ENERGY LAW

Edward Elgar
PUBLISHING

Cheltenham, UK • Northampton, MA, USA

Published by
Edward Elgar Publishing Limited
The Lypiatts
15 Lansdown Road
Cheltenham
Glos GL50 2JA
UK

Edward Elgar Publishing, Inc.
William Pratt House
9 Dewey Court
Northampton
Massachusetts 01060
USA

A catalogue record for this book
is available from the British Library

Library of Congress Control Number: 2017936582

This book is available electronically in the **Elgar**online
Law subject collection
DOI 10.4337/9781786433176

MIX
Paper from
responsible sources
FSC® C013604
FSC
www.fsc.org

ISBN 978 1 78643 316 9 (cased)
ISBN 978 1 78643 317 6 (eBook)

Typeset by Servis Filmsetting Ltd, Stockport, Cheshire
Printed and bound by CPI Group (UK) Ltd, Croydon, CR0 4YY

To those with an idea.
Keep pushing.

Contents

Acknowledgements

This book has been written to prove a point – or rather two points – as will become clear from the introduction. This approach has been inspired by some of the greatest legal writers of the present, past and ancient past – Arie Trouwborst and Nicolas de Sadeleer; Konrad Hesse, Robert Alexy and Ronald Dworkin; Friedrich Carl von Savigny. Their writings on topics as diverse as environmental law, constitutional law and legal theory are among the best and most inspiring out there.

Moreover, this book has been influenced by the waves, the sand and the sea of the city of Aberdeen and the thinking that prevails at its university. Many people I met in Aberdeen contributed to the book and that starts with the fine library staff who were always there to help, with whom I could exchange thoughts and who embraced me with warmth as if I were a member of their extended family. Actually, I became one. The same, of course, goes for all my Aberdonian friends, but I wish to mention three people in particular who provided a helping hand and/or lent their expertise to this project. These are Dr Piti Eiamchamroonlarp, Dr Tanatthep Tianprasit and Ms Alexandra Ungureanu, LLM.

There are two people who stood out in influencing this book, Prof. John Paterson (who is now Vice-Principal for Internationalisation at the University of Aberdeen) and Prof. Anatole Boute (who is now teaching at the Chinese University of Hong Kong). John always maintained an upbeat mood and a positive attitude, was incredibly supportive and backed me in all situations throughout the project. His kind nature as well as his profound knowledge on energy law make him a genuine role-model for students. Anatole took a more critical approach. It is fair to say that this book would not exist without his training, his leadership, his relentless obsession with detail and his drive for scholarly perfection. By giving his best to this project he not only took the book to a whole new level, but also channelled my thinking and, actually, made me a legal scholar. I owe these two gentlemen a debt that I will probably be unable to pay back.

The final version of this book was written under the impression of my transfer to the University of Groningen. Discussions with the great people at the Groningen Centre of Energy Law and beyond helped to shape this book. I am particularly grateful to Prof. Martha Roggenkamp, who provided me with unlimited support and the precious freedom and time to finalize this book. Martha is a truly inspiring person, whose energetic nature enabled her to create a buzzing, close-knit community of energy law scholars in Groningen, which makes this an extraordinarily pleasant place to be. There simply is 'never a dull moment in energy law', as she would put it.

Furthermore, I wish to acknowledge the contribution of two German universities. The first is the University of Marburg, which contributed unconsciously to this project, years before it actually started, by providing me with a rigorous German legal education. Some of the ideas in this book directly originate from it. The teaching of Prof. Steffen Detterbeck and Prof. Monika Böhm in particular helped to shape my understanding of the law. Second is the University of Kassel, which provided me with the opportunity to conduct in-depth research on German energy law. In addition, I am grateful to the city of Kassel itself, which hosted my family and me for a crucial period of time in our lives, providing us with an inspiring environment to contemplate energy law and other ideas.

This book benefited from generous financial, personal and professional support by the German Friedrich Ebert Foundation in the form of a scholarship. A number of people at the Friedrich Ebert Foundation committed themselves to this project. I am especially grateful to Dr Ursula Bitzegeio for her helping hand and dedication throughout the years.

Finally, this book would have never been completed without the loving support of my family (Ms Margaret Fleming, Ms Ida Fleming, Ms Dana Fleming, Ms Katharina Fleming, Mr Grigorij Krivzov, Ms Angela Zeuschner, Mr Ralf Zeuschner, Ms Sophie Zeuschner). In particular, my mother, Ms Angela Zeuschner, un-waveringly believed in this book and contributed in many different ways. Her unconditional support in itself is extraordinary, but even more so given the fact that the views expressed in the book are quite opposed to her personal beliefs.

The most important person has been spared for the end – it is my wife Ida. Her critical mind and unconventional way of thinking, her logical rigour, as well as her encouragement at difficult times are indispensable and invaluable. Her passion and drive, her love and empathy and her ability to withstand the strains of life enrich every single day. She and our beloved daughter Dana Ska sacrificed a lot of time and put up with, what may only be described as, challenging circumstances. Thank you both, so much. I love you!

I had the time of my life when putting this book together. I hope this transpires and it can live up to the most important dogma that exists in academics. It was formulated 100 years ago by the one writer I admire most, Kurt Tucholsky: '*Langweilig ist noch nicht ernsthaft*' (Serious is not necessarily boring).

Table of legislation

INTERNATIONAL TREATIES, UN DECISIONS AND AGREEMENTS

EUROPEAN LEGISLATION

Treaties

Directives

Regulations

Decisions

Communications

Recommendations

Protocols

NATIONAL LEGISLATION

Bulgaria

France

Germany

Italy

Netherlands

Northern Ireland

Spain

United Kingdom

JUDGEMENTS

International Arbitral Proceedings

International Court of Justice

European Court of Justice

Court of First Instance

NATIONAL COURT RULINGS

France

Constitutional Court

Germany

German Federal Constitutional Court

German Federal Administrative Court

German Federal Civil Court

Introduction

'Das Große spiegelt sich im Kleinen'
('The whole is a reflection of its parts')
(German proverb)

After triggering a 'gas glut' in the United States of America, shale gas extraction is currently arriving in Europe.[1] Several oil and gas companies already applied for exploratory licences in a number of EU Member States.[2] These applications, however, were met with public resistance.[3] Societal concerns about the environmental sustainability of shale gas extraction were fuelled by the occurrence of earth tremors in the UK and media coverage of water contamination in America.[4]

[1] Anne-Sophie Corbeau 'The Introduction of unconventional gas in Europe: Opportunities and Challenges' in: Martha Roggenkamp and Olivia Woolley (eds) '*European Energy Law Report IX*' (Intersentia Publishing Ltd, Cambridge 2012) 195/196 (hereinafter: Corbeau); Daniel Yergin '*The Quest: Energy, Security and the Remaking of the Modern World*' (Penguin Ltd., London 2011) 329 (hereinafter: Yergin).

[2] Philippe & Partners Law Firm '*Final Report on Unconventional Gas in Europe*' (2011) 5 available at: http://ec.europa.eu/energy/studies/doc/2012_unconventional_gas_in_europe.pdf [accessed 12 March 2012] (hereinafter: Philippe & Partners). Throughout this book the term 'Member State' will be used generically for EU Member States.

[3] Philippe & Partners 11–14; Dominik Greinacher and Sebastian Helmes 'Revising the Environmental Impact Assessment Thresholds: The Case of Germany' in Cecile Musialski et al. (eds) '*Shale Gas in Europe*' (Claeys & Casteels, Deventer 2013) 508; Paul Cairney, Manuel Fischer, and Karin Ingold 'Hydraulic Fracturing Policy in the United Kingdom: Coalition, Cooperation, and Opposition in the Face of Uncertainty' in: Christoph Weible, Tanya Heikkila, Karin Ingold and Manuel Fischer (eds) '*Policy debates on hydraulic fracturing: comparing coalition politics in North America and Europe*' (Palgrave Macmillan, Basingstoke 2016) 87–95 (hereinafter: Cairney/Fischer/Ingold); Sébastien Chailleux and Stéphane Moyson 'The French Ban on Hydraulic Fracturing and the Attempts to Reverse It: Social Mobilization, Professional Forums, and Coalition Strategies' in: Christoph Weible, Tanya Heikkila, Karin Ingold and Manuel Fischer (eds) '*Policy debates on hydraulic fracturing: comparing coalition politics in North America and Europe*' (Palgrave Macmillan, Basingstoke 2016) 123–36 (hereinafter: Chailleux/Moyson).

[4] Corbeau 202/203.

'Frack Off',[5] 'No Shale Gas'[6] or 'Stop Fracking Now'[7] are just some of the Non-Governmental Organizations and Initiatives that are actively opposing hydraulic fracturing and/or shale gas extraction in Europe and beyond. They are united by one demand: a prohibition of hydraulic fracturing and/or shale gas extraction.[8]

Outlawing shale gas extraction appears to be a natural choice for many, considering that the activity has been portrayed as new and dangerous. Potential repercussions for the environment and humans could be manifold and unforeseeable. By the beginning of the 2010s, the controversy about the potential threats of shale gas extraction became so eminent in societies around the globe that a Hollywood blockbuster movie on the subject was shot (called '*Promised Land*'), featuring superstar Matt Damon as lead character.[9]

Another film, a documentary called 'Gasland', which investigated the early days of the shale gas rush in the US, was even nominated for an Oscar.[10] That film entailed a sequence, which captured public imagination and turned into a symbol for potential issues of shale gas extraction: American house owners were filmed while setting their water taps on fire.[11]

The regulatory reaction to these pictures in Europe was swift and forceful:[12] a number of EU Member States put in place moratoria or out-

---

5 Frack Off Extreme Energy Action Network http://frack-off.org.uk/ [accessed 28 November 2016].

6 New Brunswick Anti-Shale Gas Alliance http://www.noshalegasnb.ca/ [accessed 28 November 2016].

7 Stop Fracking Now http://www.stopfrackingnow.com/ [accessed 28 November 2016].

8 The difference between both is discussed at Chapter 1 below.

9 Focus Features 'Promised Land Movie (2012)' available at: http://www.focusfeatures.com/promised_land [accessed 22 December 2012].

10 The homepage of the movie 'Gasland' is available at International WOW Company, http://www.gaslandthemovie.com/ [accessed 22 December 2012]. In 2013 the director launched another movie called 'Gasland 2'. For the Oscar nomination, see Mike Soraghan *New York Times* (24 February 2011), 'Groundtruthing Academy Award Nominee "Gasland"' available at: http://www.nytimes.com/gwire/2011/02/24/24greenwire-groundtruthing-academy-award-nominee-gasland-33228.html?pagewanted=all [accessed 22 December 2012].

11 It was claimed that their water supplies had been poisoned by shale gas extraction. Those claims were later thrown into question by another documentary that defended shale gas extraction, called '*Frack Nation*' available at Phelim McAleer http://fracknation.com/ [accessed 22 December 2012].

12 Occasionally, it has indeed been a direct reaction to the movies, see the debate on a shale gas moratorium for Northern Ireland in the Northern Ireland Assembly: Northern Ireland Assembly Deb 6 December 2011, Vol 69 No 6, cols. 305 and 311 (hereinafter: Northern Ireland minutes).

right bans on the activity.[13] These quick actions have been underpinned by the opinion of some legal scholars, who claim that it is very easy to prohibit shale gas extraction, whereas it is more complicated to create permissive shale gas regulation.[14]

But there is also a different view in Europe. Above all, the European Commission appears to be open to permissive shale gas regulation.[15] Shale gas extraction is viewed as a potential 'game-changer' for the security of energy supplies in Europe.[16] Supporters of this argument point towards the USA, the first country in the world to extract shale gas on an industrial scale.[17] Shale gas allowed the USA to switch its gas-importer status to

---

13 As described in much detail below in Chapter 3.

14 Tina Hunter, Emre Usenmez and John Paterson 'Future Trends in Shale Gas Law and Policy in the United Kingdom' in Tina Hunter (ed.) *'Handbook of Shale Gas Law and Policy'* (Intersentia, Cambridge 2016) 389 at footnote 40 (hereinafter: Hunter/Usenmez/Paterson). The author questions that proposition. It can be just as difficult to get prohibitive shale gas regulation right as it is to produce a coherent permissive legal framework for the activity. The most immediate example is probably Bulgarian shale gas regulation. Bulgaria put into place one of the first bans on shale gas extraction in the world. It prescribed a maximum pressure that may be used for gas extraction (20 atmospheres, according to art. 1 of the Bulgarian ban: Bulgarian National Assembly 'Decision to ban the application of hydraulic technology according to the break-in, study and/or extraction of oil and gas in the Republic of Bulgaria' published in (2012) No 7 *Official Journal of Bulgaria* 13 (*РЕШЕНИЕ за забрана върху прилагането на технология-та хидравлично разбиване при проучванеи/или добив на газ и нефт на територията на Република България published in (2012) БРОЙ 7, ДЪРЖА ВЕНВЕСТНИК 13*) (hereinafter: Bulgarian moratorium). Immediately afterwards it became clear that no gas reservoir, whether of shale gas or any other, could be reached by drills that are executed with such low pressures. As a result, not only shale gas but any form of gas extraction was impossible in Bulgaria until the ban had been revised in line with the findings of a parliamentary committee: *Временна комисия за проучване, анализ и обсъждане на добри практики и законодателни решения във връзка с регулирането на дейности по проучване и добив на подземни богатства при опазване на околната среда 'ПРОТОКОЛ № 1'* of 11 April 2012, 3 (Ad-hoc Committee to study and analyse and discuss best practices and legislative decisions concerning the regulation of activities in exploration and mining and effects on the environment 'MINUTES № 1' of 11 April 2012, page 3).

15 See below Chapter 2.

16 Maximilian Kuhn and Frank Umbach *'Strategic Perspectives of Unconventional Gas: A Game Changer with Implications for the EU'* (2011) European Centre for Energy and Resource Security (EUCERS) Strategy Paper No 1 http://www.eucers.eu/2011/05/06/eucers-strategy-paper-no1/_ [accessed 24 April 2012] (hereinafter: EUCERS).

17 Slawomir Raszewski 'Shale Gas and Energy Security' in Tina Hunter (ed.). *'Handbook of Shale Gas Law and Policy'* (Intersentia, Cambridge 2016) 124 (hereinafter: Raszewski); Corbeau 195 and 203; Yergin 329.

that of a gas-exporter.[18] It has been argued that European countries could similarly free themselves from the iron grip of their current gas-suppliers.[19]

Moreover, shale gas extraction could have the potential to lower gas prices.[20] It is not even necessary for shale gas extraction to actually commence in the EU to achieve that goal.[21] The mere prospect of domestic shale gas extraction in Europe might make current suppliers think twice about further increases in the price of their commodity.[22]

In line with the arguments from both sides, studies on shale gas extraction in Europe and its Member States identified environmental protection and energy security as the two most important aspects for the appraisal of the activity.[23] Prudent shale gas regulation should operate between these poles.[24]

It is important to note that environmental protection and energy security are not mere societal interests. They also have legal leverage and are enshrined in EU constitutional documents and in the constitutions of EU

---

18 Ibid.

19 For more on that see Chapter 1 below.

20 Affordability of energy is part of the security of supplies, see Leigh Hancher and Sally Janssen, 'Shared Competences and Multi-Faceted Concepts – European Legal Framework for Security of Supply' in Barton B et al. (eds) '*Energy Security: managing risk in a dynamic legal and regulatory environment*' (Oxford University Press, Oxford 2004) 93.

21 Commission 'Communication on the exploration and production of hydrocarbons (such as shale gas) using high volume hydraulic fracturing in the EU' (Communication) *COM (2014) 23 final/2* at 3 (hereinafter: Commission Shale Gas Communication); Ivan L G Pearson et al. '*Unconventional Gas: Potential Energy Market Impacts in the European Union*' (Joint Research Centre of the European Commission, 2012) 59 available at: http://ec.europa.eu/dgs/jrc/downloads/jrc_report_2012_09_unconventional_gas.pdf [accessed 20 May 2014] (hereinafter: Pearson et al.); EUCERS 42. More on that can be found in Chapter 1 below.

22 Ibid. Current suppliers would not want to raise the price beyond the point at which it would be cheaper for European countries to develop their domestic shale gas reserves than to buy imported gas.

23 Milieu Ltd '*Regulatory provisions governing key aspects of unconventional gas extraction in selected Member States*' (2013) available at: http://ec.europa.eu/environment/integration/energy/uff_studies_en.htm [accessed 4 September 2014] 16; EUCERS 6; Sachverständigenrat für Umweltfragen (SRU); Martin Faulstich et al. '*Fracking zur Schiefergasgewinnung Ein Beitrag zur energie- und umweltpolitischen Bewertung Stellungnahme*' available at: http://www.umweltrat.de/SharedDocs/Downloads/DE/04_Stellungnahmen/2012_2016/2013_05_AS_18_Fracking.pdf?__blob=publicationFile [accessed 2 July 2013] 39 et sqq. (hereinafter: SRU Faulstich).

24 Notwithstanding the fact that other interests might play a role as well, see Chapter 1 below.

Member States in the form of state objectives.[25] State objectives[26] are legal norms that address the state, not the individual.[27]

The legislator is asked to adhere to the objectives that are relevant to a regulated area when creating new laws. If a legal norm disregards a relevant constitutional objective it may be annulled by national constitutional courts.[28] Thus, new regulations may only be deemed legally sound if they comply with the applicable constitutional requirements.[29]

In the case of shale gas regulation, the legislator is asked to adhere to, seemingly contradictory, constitutional objectives. The strongest environmental protection can be achieved by not allowing shale gas extraction in Europe at all. The best result with a view to energy security is achieved by not applying any environmental safeguard measures to shale gas extraction. The interplay between these two interests is hence a focal point for shale gas regulation and this book.

The book is organized in three parts: after the introduction, a first part, consisting of Chapters 1, 2 and 3, deals with shale gas, EU and Member State regulation. The purpose of this part is to show that strictly prohibitive shale gas regulation, which is currently favoured in many corners of

---

25 They are enshrined in the form of (quasi-)constitutional objectives in several national constitutions and in the Treaty on the Functioning of the European Union (hereinafter: TFEU) as well as in the Treaty on the European Union (hereinafter: TEU). Although TFEU and TEU are not actually being called 'constitutions', their evolution into the current form is the result of endeavours to draw up a constitutional document for Europe, see Paul Craig and Grainne de Burca '*EU Law Texts Cases and Materials*' 5th edition (Oxford University Press, Oxford 2011) 23 and 25 (hereinafter: Craig/De Burca). TEU and TFEU contain the provisions of the failed European Constitution of 2005, with only minor modifications, and are considered to be quasi-constitutional documents, see Craig/De Burca 25 and 75. Thus, the term quasi-constitutional objectives will be used when objectives of these two Treaties are discussed in particular. In all other parts the term constitutional objectives is used generically, meaning that it shall also encompass quasi-constitutional objectives.

26 Often also referred to as constitutional objectives in countries with a written constitution; for reasons of clarity this book entertains the term constitutional objectives hereinafter.

27 See Chapter 4 below.

28 See, for instance, the rulings of the German Federal Constitutional Court in BVerfGE 14, 263 (275); 59, 57 (108); indirectly BVerfGE 102, 1 (18); Bericht der Sachverständigenkommission. 'Staatszielbestimmungen/ Gesetzgebungsaufträge' 1983 in Bundesminister des Innern/Bundesminister der Justiz (eds) *'Staatszielbestimmungen/Gesetzesbestimmungen'* (Konkordia, Bonn, 1983) paragraphs 7 et sqq.

29 Konrad Hesse '*Grundzüge des Verfassungsrechts*' 15th edition (C F Müller, Heidelberg 1985) paragraph 72 (hereinafter: Hesse).

Europe, is difficult to reconcile with constitutional law pre-settings. This reasoning applies equally to EU[30] and Member State law.

To demonstrate that issue, Chapter 1 introduces the two concepts of environmental protection and energy security. It assesses the main potential environmental issues as well as possible energy security benefits that are associated with shale gas extraction. The chapter starts by explaining what shale gas is and how it may be extracted and elaborates on terminology that the industry is using. The chapter focuses on shale gas extraction because shale gas has the biggest potential of all 'unconventionals'[31] to become commercially viable in Europe in the middle to long-term.[32]

After establishing potential environmental threats and energy security benefits of shale gas extraction, Chapter 2 investigates the legal framework that applies to shale gas extraction. The chapter focuses on the EU level and exposes a number of gaps and uncertainties in the primary and secondary law framework. This is followed by a critical assessment of the European Commission's recent efforts to close these gaps with a shale gas specific recommendation and guiding documents. The chapter concludes that these EU measures are not legally binding, so that the development of new, shale gas specific legislation is largely left to the discretion of EU Member States.

Chapter 3 provides an overview of shale gas regulation in EU Member States. It takes a closer look at three jurisdictions, France, Germany and the UK. They are representing the three different approaches to shale gas regulation that are currently dominating in Europe – strictly prohibitive, mainly prohibitive and cautious but permissive regulation. The chapter discusses advantages and disadvantages of these approaches. It shows that strictly prohibitive shale gas regulation is not the most commendable option for governing this activity. Instead, cautious, but permissive regulation would be better suited to govern shale gas extraction in a sound way.

The second part of the book, consisting of Chapters 4, 5 and 6, is looking beyond the status quo and discusses future regulation. Its purpose is to demonstrate a legally sound way to administer shale gas extraction. This part develops a particular and new methodology for energy regulation and uses shale gas as a case study. The new methodology is called

---

30 Although in this case it would be more apt to speak of a 'quasi-constitution' since the Treaty of Lisbon is the result of a failed attempt to draw-up a European Constitution in 2005, see Craig/De Burca 23, 25 and 75.

31 For the terminology see Chapter 1 below.

32 Commission Shale Gas Communication 3.

trias, as it consists of three types of legal norms: objectives,[33] principles[34] and rules.[35]

The trias methodology shows that the development of prudent energy regulation must take, rather abstract, constitutional objectives as the starting point, translate them into more clearly defined law principles, which in turn may then be translated into concrete rules. All three categories of norms – objectives, principles and rules – represent gradations of one legal thought and are interlinked with each other.

Concrete stipulations for legally sound[36] shale gas regulation can be deduced from all three steps. The resulting set of measures allows for cautious but permissive regulation of shale gas extraction, which reconciles environmental protection interests with energy security aspects. Such regulation is in line with legal prerequisites – a hurdle that strictly prohibitive shale gas regulation is not able to overcome.

To explain the three aspects of the trias in more detail, Chapter 4 zooms in on its first level, (quasi-) constitutional objectives. The chapter highlights that several constitutional objectives might conflict with each other, when regulating an energy activity like shale gas extraction. The legal nature of environmental protection and energy security as (quasi-) constitutional objectives is established by examination of EU quasi-constitutional documents and the constitutions of Germany and France.[37]

Afterwards, the chapter focuses on the competition between the two (quasi-) constitutional objectives of environmental protection and energy security in the case of shale gas extraction. Constitutional law theory provides a mechanism called practical concordance to reconcile competing constitutional objectives. The chapter concludes that this mechanism can be used in various jurisdictions. The mechanism could help the regulator in achieving legally sound energy regulation.

Chapter 5 further follows the structure of the trias and investigates its second level, legal principles. It establishes five (environmental) law principles as most relevant to shale gas extraction, namely the precautionary/prevention principle, the polluter pays principle and the principles of

---

33 (Quasi-) constitutional objectives, see Chapter 4 below.

34 Principles of law, see Chapter 5 below.

35 Rules on the regulation of an existing, comparable technology, see Chapter 6 below.

36 Legally sound in this book is to be understood as complying, to the greatest possible extent, with constitutional objectives, law principles and pre-existing rules.

37 The UK does not have a written constitution and is therefore omitted in that particular examination, see Chapter 6 below.

sustainable development, public participation and rectification at source. The chapter highlights how each of these principles can be translated into a concrete feature of cautious, but permissive shale gas regulation.

Chapter 6 addresses the third level of the trias, concrete rules. It draws analogies with the regulation of another energy technology, Carbon Capture and Storage (hereinafter: CCS). CCS has certain similarities to shale gas extraction in terms of the technology used and its potential threats.[38] The EU CCS Directive, as well as national CCS regulations in France, Germany and the UK, are scrutinized to see how particular problems have been tackled in CCS regulation. The chapter concludes that remaining gaps in the regulatory framework for shale gas extraction could be closed by *mutatis mutandis* application of some of the mechanisms used in French, German and UK CCS regulation.

The book concludes with a third part that consists of Chapter 7. It outlines two main results. The first is that cautious, but permissive shale gas regulation complies best with the relevant (quasi-) constitutional objectives. Strictly prohibitive shale gas regulation, by contrast, is hard to reconcile with constitutional law pre-requisites.

The second result is the new trias methodology. Its advantages and setbacks are discussed in these conclusions which show that the trias transcends the context of shale gas extraction. It can be used to develop regulatory regimes throughout the extractive industry and in the renewables sector. It could also facilitate the elaboration of a regulatory framework for non-producing parts of the energy sector.[39] Since energy law is a very young discipline that has to keep up with rapid technological developments,[40] a consistent methodology to develop regulations for all sorts of new and emerging energy technologies is in increasing demand.

---

[38] See Chapter 6 below.

[39] For more on that see Chapter 7 below.

[40] Adrian J Bradbrook 'Energy Law as an Academic Discipline' (1996) Vol 14 No 2 *Journal of Energy & Natural Resources Law* 193–217.

PART I

Shale gas, EU and Member State regulation

1. Shale gas extraction in a nutshell – technology, issues, benefits

1.1 INTRODUCTION

For a meaningful discussion on the regulation of shale gas extraction in Europe and its Member States some understanding of the technical processes is required. This chapter aims to provide this essential background information on shale gas extraction. However, the explanations do not progress beyond a basic level and are not meant to provide a comprehensive overview. Instead, the current chapter zooms in on those aspects of shale gas extraction that are most relevant for its legal appraisal.

It will be demonstrated that the techniques which are currently in use for the extraction of shale gas are not new in themselves. However, new threats might arise when some of those familiar technologies are applied to a different geological set-up (so called unconventional reserves),[1] as opposed to the geological circumstances in which conventional gas extraction takes place.

Oil and gas (generic term: hydrocarbons[2]) are formed from the soft parts of microscopic organisms that are preserved in certain sediments.[3] Over time these parts are gradually buried deeper and deeper in the ground, moving towards the Earth's interior, where they are gently cooked (matured) by exposure to heat.[4]

The preservation of this organic matter from early destruction requires an oxygen-free environment, the source rock; mudstone and shale are the most common and suitable source rocks for hydrocarbons.[5] Oil and gas is

---

[1] For a general introduction to petroleum exploration, see Robert Stoneley '*An Introduction to Petroleum Exploration for Non-Geologists*' (Oxford University Press, Oxford 1995) (hereinafter: Stoneley). For specifics of unconventional hydrocarbon production see particularly pages 100–102.

[2] See footnote above.

[3] Stoneley 27.

[4] Ibid.

[5] Ibid.

formed at considerable depths and then slowly squeezed out of the source rock.[6] In conventional hydrocarbon extraction this process is called 'primary migration', migration out of the, commonly tight and dense, source rock.[7]

The hydrocarbons then make their way up into layers of more permeable reservoir rocks like sandstone or limestone ('secondary migration') and gather in minute holes, gaps or pores between the grains of these rocks.[8] These naturally occurring reservoirs or 'traps'[9] are supervened by caps of impermeable rock. From these reservoirs conventional gas may be produced.[10]

Shale gas is commonly referred to as an *unconventional* gas.[11] The difference between *conventional* and *unconventional* gases is the ability of the gases to migrate *in situ*.[12] As opposed to *conventional* gas, *unconventional*

---

6 Ibid.

7 Understanding just how hydrocarbons move through the extremely fine-grained and tight source rock is one of the outstanding problems of petroleum geology. We cannot normally get a fluid or a gas to enter or move through it at all and yet it is clear that, somehow or other, the hydrocarbons just must have migrated. For more, see Stoneley 35.

8 Stoneley 27.

9 For this term, see Stoneley 35.

10 Engineers accordingly tap into these reservoirs to extract *conventional* gas, which flows with comparative ease from the reservoir rock as a result of its permeability, see Stoneley 35 and SRU Faulstich 7.

11 See, for instance, the use of terminology by the European Commission in Commission Recommendation 2014/70/EU of 22 January 2014 on minimum principles for the exploration and production of hydrocarbons (such as shale gas) using high-volume hydraulic fracturing [2014] OJ L 39/72 preamble 3 and 5. In addition, numerous scientific reports from the EU and its Member States clarified this point; for instance: Milieu Ltd., '*Regulatory provisions governing key aspects of unconventional gas extraction in selected Member States*' (2013) available at: http://ec.europa.eu/environment/integration/energy/uff_studies_en.htm [accessed 4 September 2014]; Energy and Climate Change Committee of the House of Commons 'Shale Gas' Fifth Report of Session of the House of Commons 2010–12, Vol. I and Vol. II (Crown 2011) Ev. 24 (hereinafter: UK report I and UK report II).

12 Hydrocarbons are chemical substances (gaseous or fluid) that are essentially composed of carbon and hydrogen atoms. These atoms can bind in various ways and, depending on the arrangement of atoms, either simple structures like paraffins (most commonly methane, ethane and butane) or more complex structures like naphthenes and aromatics are created. Only the four simplest forms of the paraffins, containing 1 to 4 carbon atoms, are hydrocarbons that occur as gas at atmospheric temperature and pressure; the rest are liquids (different forms of oil). For this and more, see Stoneley 28 and 113.

gas is 'trapped'[13] in the source rock and does not migrate out of it without stimulation (no 'primary migration');[14] the 'unconventional bit' is hence the impermeability of the source rock and the fact that no 'primary migration' can take place.[15]

The term shale gas as such refers to natural gas (mainly methane), which is present in very small pores of organic rich shales.[16] The gas is encapsulated in fine grains of shale rock.[17] Besides shale, several other rocks may 'host' unconventional gas, most prominently coal and sandstone.[18] However, the production methods for all types of *unconventional* gases are quite similar because the different rocks that hold unconventional gas are all low permeability structures.[19]

The structure of the current chapter reflects the title of the book and it is separated in three parts concerning shale gas, the environment and energy security. First, the technical process of shale gas extraction is explained. This part of the chapter assesses the individual stages that are required to make shale gas extraction happen. These technical explanations provide the necessary backdrop for the legal appreciation.

In the second part potential issues and benefits of shale gas extraction are demonstrated, which centre around the two concepts of environmental protection and energy security. Only the most salient potential environmental threats of shale gas extraction will be discussed.[20] In a third part,

---

13 Note that this terminology might be deceptive as it is also used in the context of conventional gas extraction, but with a slightly different meaning, see explanations in the previous text.

14 An apt explanation of the terminology has been provided by Mark Miller, CEO Cuadrilla Resources – Statement to the UK Commission, see UK report I Ev. 24: 'Unconventionals are only a term that we as an industry coined years ago to describe a type of reservoir. It is not the process. There is no such thing as an unconventional well or a conventional well; there is only an unconventional reservoir, and that only means that the gas is stored in the same place that it is generated.'

15 Stoneley 101 and 11; SRU Faulstich 7/8. Permeability is a measure of the ease with which fluids or gases can move through a rock: one may force them through and see how far they come out of the other end of a sample to determine permeability of the structure, see Stoneley 38.

16 Knut Bjorlykke *'Petroleum Geoscience – From Sedimentary Environments to Rock Physics'* (Springer Verlag, Berlin 2010) 464 (hereinafter: Bjorlykke).

17 Ibid.

18 Ibid. and SRU Faulstich 8.

19 Lars Dietrich and Till Elgeti '*Rechtliche Implikationen der Aufsuchung und Förderung von unkonventionellem Erdgas*' (2011) 127 (7–8) Erdöl Erdgas Kohle 311; Pearson et al. 56/57.

20 This is an approach that closely follows the most authoritative technical studies on shale gas extraction in Europe, see for instance SRU Faulstich 8;

these concerns are juxtaposed with the most important potential energy security benefits that shale gas extraction might bring about for Europe and its Member States.

1.2 THE TECHNOLOGICAL PROCESS OF SHALE GAS EXTRACTION[21]

A major difference between conventional and unconventional gas extraction is the number of geological prerequisites that have to be fulfilled for successful extraction. With conventional gas extraction broadly speaking five essential requirements must be met before gas extraction can start. Requirements are: a source rock, heat, a reservoir, a cap rock or seal and a 'trap'.[22] For unconventional gas, however, the existence of three components suffices: a source rock, heat and a cap rock or seal.[23] Unconventional gas should be more abundant in the world than conventional gas, inter alia because fewer geological conditions (three compared to five) have to be met. However, given that the research into the extent of recoverable unconventional gases is just beginning, verification of that hypothesis is still to surface.[24]

Shale gas may be encountered at varying depths, depending on regional geological circumstances; thus, generalizations about 'shale gas extraction' have to be made very cautiously, as each formation has its unique characteristics.[25] In Europe, shale formations that contain gas can be found at depths of 2 km or more and they may be as thick as 100 metres.[26]

Stefan Lechtenböhmer et al. 'Impacts of shale gas and shale oil extraction on the environment and on human health' (European Parliament, 2011) 11 (hereinafter: Lechtenböhmer et al.); Maximilian Kuhn and Frank Umbach *'Strategic Perspectives of Unconventional Gas: A Game Changer with Implications for the EU'* (2011) European Centre for Energy and Resource Security (EUCERS) Strategy Paper No 1 http://www.eucers.eu/2011/05/06/eucers-strategy-paper-no1/ [accessed 24 April 2012] (hereinafter: EUCERS).

21 A good overview of the process may be found at: Rick Rickman et al. '*A practical use of shale petrophysics for stimulation design optimization: All Shale plays are not clones of the Barnett Shale*' (2008) Society of Petroleum Engineers SPE 115258 available at: http://www.onepetro.org/mslib/app/Preview.do?paperNumber=SPE-115258-MS&societyCode=SPE (accessed 20 March 2012) (hereinafter: Rickman et al.).

22 Stoneley 54.

23 Ibid.

24 For more details see further below in this chapter.

25 Stoneley 54.

26 Ibid.

However, the layers of shale rock (or shale 'plays'[27]) are not fixed at certain depths as they are progressively buried under further layers of sediment or may be uplifted over the course of geological time.[28]

1.2.1 Innovative Drilling and Well-Integrity

Two different technologies are required for industry-scale shale gas extraction: first, so called innovative drilling and second, hydraulic fracturing.[29] The first aspect, innovative drilling, actually includes three features with relevance to shale gas extraction: horizontal drilling, multi-well pad drilling and improved well-casing/well-integrity.[30] Horizontal drilling constitutes an innovation in the way a drill may be brought down: until 30–40 years ago, only vertical drilling of a well, from the surface to the targeted rock formation, was technically feasible.[31] Today horizontal drilling is used for all sorts of pipe and cable-laying, including the laying of electricity cables to renewable energy facilities, like windmills.[32] Thus, horizontal drilling may not only be used for shale gas extraction but also to enable the transmission of renewable energy.

Horizontal drilling means that the drilling trajectory diverts in a

---

27 A 'play' is a group of fields or prospects in the same region that is controlled by the same set of geological features, for instance similar types of source rock or reservoirs, see Stoneley 54–59 and 106. The term shale play is used by the oil and gas industry to refer to a geographic area that has been singled out for exploration due to favourable geoseismic survey results, well logs or some other factor.

28 Stoneley 3 and 11.

29 Ivan L G Pearson et al. '*Unconventional Gas: Potential Energy Market Impacts in the European Union*' (Joint Research Centre of the European Commission, 2012) 59 available at: http://ec.europa.eu/dgs/jrc/downloads/jrc_report_2012_09_unconventional_gas.pdf [accessed 20 May 2014] (hereinafter: Pearson et al.); Harald Andruleit et al. *Bundesanstalt für Geowissenschaften und Rohstoffe (BGR) 'Abschätzung des Erdgaspotenzials aus dichten Tongesteinen (Schiefergas) in Deutschland'* (Bundesanstalt für Geowissenschaften und Rohstoffe, Hannover 2012) 35 (hereinafter: Andruleit et al. BGR Abschätzung); Bundesanstalt für Geowissenschaften und Rohstoffe (BGR) 'Schieferöl und Schiefergas in Deutschland Potenziale und Umweltaspekte' (Bundesanstalt für Geowissenschaften und Rohstoffe, Hannover 2016) 13 (hereinafter: NIKO).

30 See the text immediately below.

31 Stoneley 83; Pearson et al. 60.

32 For examples from Scotland and Germany see: Community Energy Scotland '*Horizontal directional drilling*' available at: http://www.communityenergyscotland.org.uk/news/10-mar-2015-horizontal-directional-drilling.asp [accessed 28 June 2016]; H Schmidt GmbH '*Horizontalbohrtechnik*' available at: http://www.schmidt-rohrleitungsbau.de/leistungen/horizontalbohrtechnik/ [accessed 28 June 2016].

horizontal direction at a 'kick off point'.[33] The drill commences on the surface in a vertical direction and is brought down approximately 2 km, intersecting various layers and types of rock.[34] Once the drill approaches the targeted shale layer, it is diverted horizontally to follow the route of the shale play.[35]

The horizontal drills may follow the path of shale layers over long distances: in the US, horizontal drills of up to 6 km length have been reported.[36] But this technique is also commonly used for conventional extraction, where it enables the development and production of hydrocarbons from particular fields that would otherwise be too expensive to produce.[37]

Multi-well pad drilling allows for synergy effects: several subsurface areas can be developed from one surface spot, that is to say one surface location ('pad') can be used to drill multiple wells.[38] Drilling from single surface 'pads' is not unique to shale gas extraction, but it is considered to be indispensable for European shale gas extraction.[39] As Europe is a densely populated region of the world, the advantages of the multi-well pad technology (reduction of land-use, environmental surface impacts and infrastructure costs)[40] are deemed to be key factors of successful shale gas extraction.[41]

Some understanding of the drilling process itself is required to grasp particular potential environmental threats that may arise. A typical drilling derrick stands over the well that is drilled; it lowers a string of steel drill-pipe, which carries the bit (drilling head), into the hole and draws it out again.[42]

From time to time it is necessary to protect the bit and line the well bore with steel casings. These are 10–15 metres long, strengthened steel pipes that are screwed together, lowered into the hole and cemented into position.[43] Once this has been accomplished, the well is drilled further with a smaller bit, which will pass through the casing and the next string of

---

33 Ibid.
34 Ibid.
35 Ibid.
36 Stoneley 83.
37 Ibid.
38 See Pearson et al. 62 for more details on the process.
39 UK report I, 49.
40 Pearson et al. 62.
41 UK report I, 49.
42 Stoneley 81.
43 Stoneley 82; UK report I, 39.

casing will be smaller still.[44] This process is similar for conventional and unconventional gas extraction.

The particular issue with a view to shale gas extraction is well integrity. Well integrity is a delicate topic and deserves particular attention, due to technical difficulties and non-unified procedures in Europe.[45] Well integrity includes the planning, design and execution of well completion (cementing, casing and well head placement).[46] Well integrity is fundamental to the security of shale gas wells because hot and cold fluids will be pumped through the wellbore with particularly high pressure during shale gas extraction.[47]

The aim of the casing programme is to optimize cementing operations.[48] Poorly cemented wells can offer pathways for leakage.[49] Casing and cementing programmes should be designed to provide optimal isolation of the gas-producing zones from overlying formations.[50] Multiple engineered barriers are needed to prevent communication between hydrocarbons and potable aquifers.[51]

'State of the art' is the use of a three casing system, consisting of one steel case, an inner cement case and a third additional internal steel case, to insulate the well against the geological layers it intersects.[52] However, in

---

44 Ibid.

45 The attention given to wellbore integrity was kick-started in the US but it is also becoming a focal point of the debate about shale gas extraction in Europe, see Jennifer Morrissey and Jason Schumacher 'Water quality, water use and wastewater issues related to hydraulic fracturing' in Vivek Bakshi (ed.) '*Shale Gas: A Practitioner's Guide to Shale Gas and Other Unconventional Resources*' (Globe Law Publishing, London 2012) 78; Cecile Musialski 'An External Comment on the UK/England & Wales: Towards an even more refined legal & regulatory framework by imposing A.O. additional and specific requirements for mitigating the risk of induced seismicity?' in Cecile Musialski et al. (eds) '*Shale Gas in Europe*' (Claeys & Casteels, Deventer 2013) 524 (hereinafter: Musialski UK chapter).

46 For more details on the process, see Stoneley 81–7.

47 Spencer Ferguson and Matthew T Gilbert '*Hydraulic Fracturing and Shale Gas Production: Issues, Proposals and Recommendations*' (Nova Science Publishers, New York 2013) 114 (hereinafter: Ferguson/Gilbert). Nonetheless, well integrity is, of course, also important with regard to conventional extraction. The reasons why hydraulic fracturing puts particular strain on a well will be explained in the next sub-section below.

48 Ibid.

49 Ibid.

50 Ibid.

51 Ibid.

52 International Energy Agency (IEA) '*Golden Rules for a Golden Age of Gas: World Energy Outlook Special Report on Unconventional Gas*' (International Energy Agency, Paris 2012) 23/24 (hereinafter: IEA Golden Rules); Yi Wang et

many European countries no unified standard for shale gas casing exists and operators are currently applying their own security standards.[53] There is a wide range in procedures for casing placement and cementing in shale gas drilling.[54] The lack of common leakage criteria for intervention in a well that exhibits damage and an absence of unified standards could pose risks concerning the nature of the intervention.[55]

1.2.2 Hydraulic Fracturing

After the well has been successfully established by innovative drilling technologies and lined with casings, the next step towards shale gas extraction is hydraulic fracturing (commonly referred to as fracking).[56] Hydraulic fracturing distinguishes itself from the above mentioned processes insofar as it is not a drilling method, but a stimulation treatment of an existing shale gas well.[57]

Pressure from the overlying rock and the natural movements of the Earth's crust create small extensional fissures or fractures in the shale layer, which concentrate in fracture swarms.[58] Hydraulic fracturing uses those fractures as 'highways' to extract gas[59] because the 'trapped' shale gas accumulates in these fractures.[60] Although gas could be produced from the naturally occurring fractures without stimulation, the gas would flow at a very low rate and so stimulation becomes key to the economic viability of a shale gas well.[61]

---

al. 'Study of borehole stability of Marcellus shale wells in longwall mining areas' (2014) 4 *Journal of Petroleum Exploration and Production Technology* 64.

[53] For example, Germany: Meiners et al. Bund C 70.

[54] Ferguson/Gilbert 115.

[55] Ibid.

[56] Details on hydraulic fracturing may be found at: Ralph W Veatch Jr 'Overview of Current Hydraulic Fracturing Design and Treatment Technology–Part 1' (1983) *Journal of Petroleum Technology* 677 (hereinafter: Veatch Part 1). Veatch Part 1, 677.

[57] Andruleit et al. BGR Abschätzung 35.

[58] Joseph H Frantz and Valerie Jochen '*Shale Gas – When your gas reservoir is unconventional so is our solution*' (2005) White Paper Schlumberger Ltd. at 4 http://www.pe.tamu.edu/wattenbarger/public_html/Selected_papers/--Shale%20 Gas/shale_gas-%20schlumberger.pdf (accessed 20 March 2012) (hereinafter: Schlumberger).

[59] Ralph W Veatch Jr 'Overview of Current Hydraulic Fracturing Design and Treatment Technology–Part 2' (1983) *Journal of Petroleum Technology* 853 (hereinafter: Veatch Part 2).

[60] Veatch Part 1, 681.

[61] Schlumberger 4 and 6.

Hydraulic fracturing thus aims to enlarge the naturally occurring fissures in the shale layer or to create additional ones to allow the gas to flow more readily.[62] Starting at the 'kick off point', where the vertical wellbore turns horizontal, segments of the wellbore are being isolated.[63] The casing in those isolated sections is perforated and fracturing fluid is pumped under high pressure (up to 8000 psi[64]/1000 bar)[65] into the ground.[66] The process of isolating a particular segment of the well and fracturing it will be repeated on a segment-by-segment basis. Up to 25 fracture stages can occur and the amount of fracturing fluid required for 'completion' of the whole well[67] varies considerably.[68] Careful calculation of the amount of fracturing fluid and the pressure with which it is pumped into the ground allows engineers to determine the length, extent and propagation of fractures.[69]

The fracturing fluid typically consists of 99 per cent water and proppants[70] and up to 1 per cent chemical additives.[71] However, most fracturing fluids will be mixed individually, in accordance with the geological structure and pertaining circumstances of a given well.[72] Because of this case-by-case approach that the industry is taking, generalizations about the fracturing fluids and their chemical compositions are hard to make.[73]

The fluids that re-surface after injection are referred to as 'flow back'. While a considerable amount will find its way back to the surface immediately after injection, remnants of the fracturing fluid emerge from the

---

62 Bjorlykke 464.

63 Ching H Yew '*Mechanics of hydraulic fracturing*' (Gulf Publications Co., Houston Texas 1997) 6 (hereinafter: Yew).

64 Schlumberger 4.

65 That number has been named for shale gas extraction specifically in the European context by: SRU Faulstich 8.

66 Yew 6.

67 'Completing a well' means installing equipment in the well to allow a safe and controlled gas flow from the well, see Mohd Fauzi Hamid and Wan Rosli Wan Sulaiman 'Fundamentals Of Petroleum Engineering Well Completion and Stimulation' available at: http://ocw.utm.my/file.php/12/Chapter_6-OCW.pdf [accessed 27 February 2014]; Meiners et al. Bund A 48/49.

68 Veatch Part 1, 677; SRU Faulstich 24.

69 Veatch Part 1, 681/682.

70 Proppant is the technical term for sands or ceramics; their task is to prop up the fractures and to keep them open, see Veatch Part 2 858; Rickman et al. 6.

71 UK report I, 8.

72 Veatch Part 2, 854.

73 Rickman et al. 2; Meiners et al. Bund A 42 and A 64.

ground throughout the entire lifecycle of the well.[74] However, numbers differ widely: while some reckon that 20–30 per cent returns,[75] others estimate that the number might be as high as 60 per cent.[76] The wide variation in figures has been explained by the characteristics of the individual wells and the variety in processes leading to re-surface of 'flow-back'.[77]

One single well can produce gas for 30 years or more at a relatively constant level.[78] This, however, may only be achieved by occasional re-stimulation of the well,[79] which basically means repetition of the hydraulic fracturing process.[80] One single well might be re-stimulated five times or more, but fewer than 10 per cent of the shale-gas wells drilled in the United States have, so far, been subjected to re-stimulation.[81]

Hydraulic fracturing is not only deployed for shale gas extraction: it may be used for all sorts of oil and gas extraction and even for geothermal purposes.[82] This is not a new development. Hydraulic fracturing was initially developed and introduced for such uses in America by 1949.[83] In Europe, hydraulic fracturing has been used, at least since 1961, when the first well on German soil was hydraulically fractured.[84]

---

[74] Meiners et al. A 76; Mark Miller, CEO Cuadrilla Resources – Statement to the UK Commission, see: UK report I, Ev.24.

[75] Mark Miller, CEO Cuadrilla Resources – Statement to the UK Commission, see: UK report I, Ev.24.

[76] WWF-UK – Statement to the UK Commission, see: UK report I, Ev. 104.

[77] Energy Institute of the University of Texas 'Fact-Based Regulation for Environmental Protection in Shale Gas Development' (2012) in Part 4 at page 22 '*Environmental Impacts of Shale Gas Development*' at 107, available at: http://barnettprogress.com/media/ei_shale_gas_regulation120215.pdf [accessed 14 June 2012] (hereinafter: Energy Institute Texas). Note: This study has come under intense criticism and will hence merely be used in this book to illustrate points of view and minor aspects. For the debate see: Revkin A, *The New York Times* 'Damning Review of Gas Study Prompts a Shakeup at the University of Texas' available at: http://dotearth.blogs.nytimes.com/2012/12/06/damning-review-of-gas-study-prompts-a-shakeup-at-the-university-of-texas/ [accessed 16 April 2013].

[78] Schlumberger 5.

[79] Ibid.

[80] M C Vincent 'The next opportunity to improve hydraulic-fracture stimulation' (2012) *Journal of Petroleum Technology* 119/120 (hereinafter: Vincent).

[81] IEA Golden Rules 27; Vincent 120. For the significance of the United States in shale gas extraction see the part on energy security below in this chapter.

[82] SRU Faulstich 6; UK report I, 54.

[83] Veatch Part 1, 677.

[84] Wirtschaftsverband Erdöl- und Erdgasgewinnung e.V. 'Hydraulic Fracturing-Prozess und Perspektiven in Deutschland' available at: http://www.erdoel-erdgas.de/Themen/Erdgas-aus-Deutschland/Hydraulic-Fracturing [accessed 1 July 2014] (hereinafter: Wirtschaftsverband hydraulic fracturing).

Up until 2011, at least 326 oil, gas and geothermal wells had been subjected to hydraulic fracturing in EU Member States.[85] However, hydraulic fracturing *for shale gas purposes* has been rarely deployed in Europe.[86] If hydraulic fracturing was used for shale gas purposes in EU Member States, it was only used to stimulate exploratory wells in shale plays, but not for shale gas production.[87]

This lack of experience makes it very difficult to predict how much unconventional gas can be produced from a particular well in Europe.[88] Currently there is no other way of quantifying the productive potential of a shale play than to drill and fracture it.[89] Only after application of the hydraulic fracturing technique to the targeted formation may its ability and suitability for shale gas extraction be assessed.[90]

1.3 POTENTIAL ISSUES AND BENEFITS

The outlined technological particularities result in a number of potential environmental threats and energy security benefits, which are discussed immediately below. The assessment starts by outlining the conceptual, abstract meaning of these interests and then focuses on their practical relevance for shale gas extraction. However, the discussion does not explain the legal character of environmental protection and energy security (for instance whether or not these concepts comprise constitutional objectives, principles of law or something else). These, certainly crucial, aspects are discussed in detail in Chapter 4 below, but lie beyond the scope of this introduction to the process of shale gas extraction.

---

85 According to a German study: Lower Saxony State Agency for Mining, Energy and Geology (*Landesamt für Bergbau, Energie und Geologie*) '*Hydraulische Bohrlochbehandlung*' available at: http://www.lbeg.niedersachsen.de/bergbau/genehmigungsverfahren/hydraulische_bohrlochbehandlung/hydraulische-bohrlochbehandlung-110656.html [accessed 17 July 2014]; Bundesrat '*Stenografischer Bericht 904. Sitzung*' of 14 December 2013 Plenarprotokoll 904 available at: http://dipbt.bundestag.de/dip21/brp/904.pdf [accessed 16 April 2014] 579 (hereinafter: Bundesrat Stenografischer Bericht 904. Sitzung).

86 Three times in German history, for instance see Meiners et al. Bund A 10.

87 Ibid.

88 Stoneley 67–80 lists inter alia remote sensing and preliminary studies, geological surveys, gravity surveys, magnetic surveys, seismic reflection surveys and seismic refraction surveys as part of the standard repertoire of conventional hydrocarbon extraction.

89 Submission of the Tyndall Centre Manchester to the UK Commission, see UK report I, Ev 87; IEA Golden Rules 22.

90 Ibid.

1.3.1 Potential Environmental Threats of Shale Gas Extraction

There is an interrelation between environmental protection and shale gas extraction. The environment may be defined very broadly as the natural surroundings or all *living* and non-living things occurring *naturally* on *earth*.[91] Although anthropocentric views on the environment (the environment defined from a human perspective) might compete with other views (for instance that the environment has value in its own right), all definitions agree that water, air and soil are part of the environment.[92] Environmental protection, in terms of shale gas extraction, thus, means protecting these elements.

Potential environmental threats in relation to shale gas extraction in Europe have been garnered from experiences with shale gas extraction in America, but observations made there might not be transferrable to the EU and its Member States. This is due to considerable differences in geology and regulation between the USA and Europe.[93]

Despite these differences there are studies which broadly identify four potential environmental threats of shale gas extraction as most salient in Europe as well as in America.[94] These are groundwater contamination/issues with well integrity, irresponsible disposal of 'flow-back', the repercussions of excessive land use in densely populated areas and emission of greenhouse gases/insufficient monitoring.[95]

---

[91] Donald L Johnson et al. 'Meanings of Environmental Terms' (1997) Vol 26 No 3 *Journal of Environmental Quality* 581/582.

[92] Ibid.

[93] Bundestag '*Stenografischer Bericht 178. Sitzung*' of 10 May 2012 Plenarprotokoll 17/178 available at: http://dipbt.bundestag.de/dip21/btp/17/17178.pdf [accessed 17 April 2014] 21166 (hereinafter: Bundestag Stenografischer Bericht 178. Sitzung); SRU Faulstich 27; Meiners et al. 'Fracking in unkonventionellen Erdgas-Lagerstätten in NRW Kurzfassung zum Gutachten "*Gutachten mit Risikostudie zur Exploration und Gewinnung von Erdgas aus unkonventionellen Lagerstätten in Nordrhein-Westfalen (NRW) und deren Auswirkungen auf den Naturhaushalt insbesondere die öffentliche Trinkwasserversorgung*"' of 7 September 2012 (Ministry for Climate Protection, Environment, Agriculture, Conservation and Consumer Protection of North Rhine-Westphalia 2012) 56 (hereinafter: Meiners NRW).

[94] In America, this was done by a study, commissioned by then President Barack Obama on 31 March 2011, see Spencer Ferguson and Matthew T Gilbert '*Hydraulic Fracturing and Shale Gas Production: Issues, Proposals and Recommendations*' (Nova Science Publishers, New York 2013) 94 et sqq. (hereinafter: Ferguson/Gilbert). Similar conclusions were then reached by a 2013 study, conducted by an advisory board to the German government on the specifics of German and European shale gas extraction, see SRU Faulstich, particularly 44/45.

[95] Ferguson/Gilbert 96; SRU Faulstich 44/45.

A fifth potential environmental threat has been discussed as belonging to the group of most salient: earth tremors. Although this potential threat attracted some interest in the American context,[96] in Europe earth tremors in the shale gas context are mostly associated particularly with the British geological set-up.[97] As a consequence, the described four potential threats are discussed immediately below, while the potential fifth threat of earth tremors is going to be discussed only in the context of the British regulatory regime, which is one of the focal points of Chapter 3.

1.3.1.1 Groundwater contamination/issues with well-integrity

Due to suspicions in the US about potential impacts of shale gas extraction on groundwater reserves,[98] this possible threat attracted special attention around the globe and also in Europe.[99] Several layers of rock strata have to be bored through in order to reach the shale plays.[100] These strata include groundwater-bearing aquifers in certain locations, as these are commonly situated at shallow depths, way above the shale plays.[101] Methane and noxious substances could reach those aquifers and contaminate them.[102] This could pose a direct threat to human health, since groundwater in Europe is used as a source of drinking water.[103]

---

96 IEA Golden Rules 26 and 127.

97 The Royal Society and the Royal Academy of Engineering '*Shale gas extraction in the UK: a review of hydraulic fracturing*' (London, 2012) available at: https://royalsociety.org/policy/projects/shale-gas-extraction/report/ [accessed 11 April 2014] 4–7 (hereinafter: Royal Society); Christopher A Green et al. '*Preese Hall Shale Gas Fracturing Review and Recommendations for Induced Seismic Mitigation Report on behalf of the UK Department of Energy and Climate Change*' available at: https://www.gov.uk/government/uploads/system/uploads/attachment_data/file/15745/5075-preese-hall-shale-gas-fracturing-review.pdf [accessed 11 April 2014] ii/iii and 13/14 (hereinafter: Green et al.).

98 See for instance: US Environmental Protection Agency '*Draft Plan to Study the Potential Impacts of Hydraulic Fracturing on Drinking Water Resources*' http://water.epa.gov/type/groundwater/uic/class2/hydraulicfracturing/upload/hf_study_plan_110211_final_508.pdf [accessed 25 April 2012] 30 (hereinafter: US EPA Study Plan); UK report I, 39; Osborn et al. 8175. This will be discussed below.

99 SRU Faulstich 23–30.

100 Modern Shale Gas Development ES 3/ES 4.

101 Ibid.

102 Stephen G Osborn et al. 'Methane contamination of drinking water accompanying gas-well drilling and hydraulic fracturing' (2011) Vol 108 No 20 *Proceedings of the National Academy of Science of the United States of America* (PNAS) 8175 (hereinafter: Osborn et al.); EUCERS 21.

103 SRU Faulstich 23 et sqq.

Technical experts discuss four potential exposure pathways for groundwater contamination by shale gas extraction.[104] First, groundwater contamination may stem from fracturing fluid.[105] During the process of hydraulic fracturing, some components of the fracturing fluid could migrate upwards through particularly long vertical fractures, reach aquifers[106] and contaminate them.[107] Despite considerable research, it is still difficult to accurately predict and control the location and length of fractures.[108]

Second, gases or fluids might ascend diffusely from their point of release in the deeper ground and reach groundwater-carrying rocks.[109] As opposed to the option discussed above, no clear pathway might be pinpointed here, but gases and/or fluids might, nonetheless, reach aquifers.[110] The reasons for such a diffuse ascent are poorly understood. Preliminary findings suggest that the permeability of the segregating rock between shale formations and the groundwater differs considerably from site to site.[111] Groundwater contamination through diffuse pathways might occur when fracturing fluid is physically displaced.[112] Fracturing fluid might not reach the shale plays, but accidentally be pumped into other rock strata and migrate from there into aquifers.[113] In the worst case, fracturing fluid might even be pumped directly into aquifers.[114]

Third, poor well casing and poor well integrity, more generally, has been frequently blamed for groundwater contamination.[115] A good 'casing-job', as discussed earlier, should take prudent account of the fact that various fluids will be pumped through the borehole at varying pressures.[116] If the 'well casing' is done poorly, repeated high pressure fracturing can damage the 'well casing' and contaminating substances might escape from the well and flow into aquifers.[117]

Fourth, leakage on the surface is a possible further source of contami-

---

104 Meiners et al. Bund A 13 et sqq.; Osborn et al. 8175; EUCERS 21.
105 Osborn et al. 8175.
106 US EPA Study Plan 30; UK report I, 39; Osborn et al. 8175.
107 UK report I, 39.
108 US EPA Study Plan 37.
109 Meiners et al. Bund A 15 and C 5; Osborn et al. 8175.
110 Meiners et al. Bund A 15.
111 Meiners et al. Bund C 5.
112 Osborn et al. 8175.
113 Meiners et al. Bund C 5.
114 Ibid.
115 Osborn et al. 8175; EUCERS 21.
116 IEA Golden Rules 22/23.
117 US EPA Study Plan 38; UK report I, 38; Meiners NRW 56.

nation.[118] All sorts of accidents, equipment and infrastructure failures, as well as improper handling of noxious substances during transportation, processing and storage, might lead to contamination.[119] A storage facility for fracturing fluid, for instance, could leak and release contaminating substances into the soil, from where they trickle down into the groundwater.[120]

It is important to note, however, that these are theoretically perceivable pathways, which have been contemplated by geologists, based on American models.[121] Exposure pathways are site-specific and must be assessed, depending on the unique geological composition of each location.[122] Generalizing statements about the potential threat of groundwater contamination by shale gas extraction in Europe should hence be treated with much caution.[123]

1.3.1.2 Disposal of 'flow back'

A second major issue of shale gas extraction, which is particularly prominent in Europe,[124] is the disposal of 'flow back'.[125] Environmental damage might occur during disposal of this fluid.[126] The 'flow back' may consist of water, proppants, chemicals and other substances and its composition varies considerably, in line with the different geological circumstances of each shale play and the chemical makeup of pre-existing reservoir waters.[127]

It is important to note that 'flow back' disposal differs from country to country: in America 'flow back' is regularly recycled and re-used, a

---

118 Meiners et al. Bund A 14 et sqq.; UK report I, 44; EUCERS 21.

119 Meiners et al. Bund A 14 and C 3.

120 UK report I, 44; EUCERS 21. However, this would explain a contamination with chemicals but not with methane, since methane is not an additive to the fracturing fluid, see EUCERS 21.

121 Osborn et al. 8175.

122 Meiners et al. Bund C 3.

123 Ibid.

124 Meiners et al. Bund A 76/77; C Ewen et al. 'Hydrofracking Risk Assessment-Executive Summary' available at: http://dialog-erdgasundfrac.de/sites/dialogerdgasundfrac.de/files/Ex_HydrofrackingRiskAssessment_120611.pdf [accessed 15 June 2012] 68 (hereinafter: Ewen et al.).

125 The treatment of 'flow back', however, is a general problem, see Ewen 46/47.

126 Since 2008 the US Department of Environmental Protection has reported 130 incidents of groundwater and soil contamination by 'flow back', see: EUCERS 22; UK report I, 44.

127 Energy Institute Texas part 4 at page 22.

process which can require on-site storage of the ‘flow back’.[128] Leakage from these ‘flow back’ storage basins has contaminated American soil and water on the surface and sub-surface in the past.[129] By contrast, in a major European country like Germany recycling of ‘flow back’ is rather improbable, as studies pointed out.[130] The most cost-effective disposal method for ‘flow back’ in Europe and Germany is disposal in ‘dead wells’ by deep ground injection.[131]

However, injecting ‘flow back’ into the deeper ground, as such, could trigger environmental degradation in Europe. As the long-term sealing qualities of ‘flow back’ disposal wells have never been tested before,[132] contamination might occur in a variety of possible settings: the re-injected ‘flow back’ could react with pre-existing substances in the disposal wells.[133] Furthermore, the flow from disposal wells could be noxious[134] and, due to potential geological links between the disposal well and surrounding rock strata, soil and water sources could be contaminated.[135] To sum up, the environmental repercussions of the disposal of ‘flow back’ in Europe are currently under scrutiny and scientific research has started, but no final verdict on the disposal has been reached.

1.3.1.3 Land use

The spatial demands of shale gas extraction are considerable: compared to conventional gas extraction, more shale gas wells need to be drilled to extract the same amount of gas.[136] In addition, the areas in which drill-

---

128 EUCERS 22; Daniel Yergin ‘*The Quest: Energy, Security and the Remaking of the Modern World*’ (Penguin Ltd., London 2011) 331 (hereinafter: Yergin).

129 Ibid. and UK report I, 44.

130 Ewen 47; EUCERS 22; Meiners et al. Bund C 54.

131 Ibid.

132 SRU Faulstich 45.

133 Meiners et al. Bund 53.

134 Karl-Heinz Rosenwinkel et al. ‘*Gutachten zur Abwasserentsorgung und Stoffstrombilanz* ISAH 2012’ available at: http://dialog-erdgasundfrac.de/sites/dialog-erdgasundfrac.de/files/Gutachten%20zur%20Abwasserentsorgung%20und%20Stoffstrombilanz%20ISAH%20Mai%202012.pdf [accessed 25 June 2012] 3 (hereinafter: Rosenwinkel).

135 SRU Faulstich 45.

136 For European examples see: Claudia Baranzelli et al. ‘Scenarios for shale gas development and their related land use impacts in the Baltic Basin, Northern Poland’ (2015) Vol 84 *Energy Policy* 92 (hereinafter: Baranzelli et al.); SRU Faulstich 32. For the American experiences: Andrew Blohm et al.’ The significance of regulation and land use patterns on natural gas resource estimates in the Marcellus shale’ (2012) Vol 50 *Energy Policy* 358/359; Simona L Perry ‘Development, Land Use, and Collective Trauma: The Marcellus Shale Gas

ing takes place are wider in scope.[137] While conventional gas gathers at a certain point and is extracted from there shale plays expand over long distances.[138] On top of this increased need for land, additional spatial pressures arise from the necessity to put into place transportation and processing infrastructure.[139]

This need for land to extract shale gas could aggravate pre-existing spatial pressures in many, densely populated, European countries.[140] An increased competition with other land uses like agriculture, water management, forestry, human settlements and recreation is feared.[141] The overlap of shale gas extraction with these other uses is called 'spatial resistance' (*Raumwiderstand*).[142] Areas with 'very high' and 'high' spatial resistance are generally unsuitable for shale gas extraction.[143]

Recent investigations into 'spatial resistance' for shale gas extraction in Germany and Poland yielded alarming results. In the German state of North-Rhine Westphalia[144] 49 per cent of the land mass had 'very high spatial resistance' and 17 per cent still had 'high spatial resistance' against shale gas projects.[145] The results for this state are especially relevant, since it is deemed to host considerable European shale gas reserves,[146] but it also has an important agricultural industry.[147] In northern Poland between 7 and 12 per cent of land currently dedicated to other industrial activities would be required for shale gas extraction.[148] However, a study into that subject pointed out that these numbers could increase if more intense shale gas extraction should start in Poland.[149]

Furthermore, visual and noise impacts as well as light emissions from shale gas extraction plants could add to the mounting spatial pressures

---

Boom in Rural Pennsylvania' (2012) Vol 34 Issue 1 *Culture, Agriculture, Food and Environment* 81–92.

137 UK report I, 7; Corbeau 191; Commission Shale Gas Communication 5.

138 Stoneley 101 and 104; UK report I, 7; Corbeau 191.

139 Commission Shale Gas Communication 6.

140 For examples from Poland and Germany, see Baranzelli et al. 92; SRU Faulstich 33.

141 Ibid.; Meiners NRW 3.

142 Meiners NRW 9.

143 Meiners NRW 10.

144 One of the German areas where shale gas could be produced according to Meiners NRW 1.

145 Meiners NRW 10.

146 Meiners NRW 1.

147 SRU Faulstich 33.

148 Baranzelli et al. 92.

149 Ibid.

in these densely populated European regions.[150] Potential threats and knock-on effects might endanger tourism and recreational uses as well as biodiversity.[151] Natural habitats could be diminished by the removal or alteration of existing vegetation in connection with construction activities; the drilling pad and associated transport links could intersect natural habitats and constitute potential barriers for the distribution of certain species.[152]

1.3.1.4 Climate change/insufficient monitoring

The gas that is being produced from shale is mainly methane, which makes a powerful greenhouse gas.[153] Potential environmental threats could be brought about by its emission during shale gas extraction as well as by emissions from subsequent combustion for electricity production.[154] Regarding shale gas extraction as such, scientific concerns focus on gas that could escape from the well or associated equipment (so called 'fugitive methane').[155] A study into that subject bemoaned that current monitoring arrangements would be insufficient to detect such emissions, as several pathways could lead to 'fugitive methane' emissions and consequent adverse effects on the climate.[156]

The most important source of GHG emissions (Greenhouse Gas),

---

[150] SRU Faulstich 33. For the particular issue of potential tremors see Chapter 3 below.

[151] Ibid.

[152] SRU Faulstich 34.

[153] Methane is a far more potent greenhouse gas than CO_2; for this and a good overview on the impacts of methane to the climate, particularly from shale gas extraction, see: Robert W Howarth, Renee Santoro and Anthony Ingraffea, 'Methane and the greenhouse-gas footprint of natural gas from shale formations' (2011) Vol 106 No 4 *Climatic Change* 680 and 685 (hereinafter: Howarth/Santoro/Ingraffea); furthermore, SRU Faulstich 6.

[154] Howarth/Santoro/Ingraffea 680.

[155] Ibid.

[156] Fugitive methane could occur when drills penetrate methane deposits and methane flows up the well (John Broderick and Kevin Anderson '*Has US Shale Gas Reduced CO_2 Emissions? Examining recent changes in emissions from the US power sector and traded fossil fuels*' (Tyndall Centre at the University of Manchester, Manchester 2012) available at: http://www.tyndall.manchester.ac.uk/public/Broderick_Anderson_2012_Impact_of_Shale_Gas_on_US_Energy_Emissions.pdf [accessed 20 March 2013] 66 (hereinafter: Broderick/Anderson); EUCERS 22; it could also come to the surface, accompanying 'flow back'; it could leak from the 55 to 150 connections to equipment (Howarth/Santoro/Ingraffea 680/681) and during processing, transport, storage, and distribution of natural gas (Howarth/Santoro/Ingraffea 685).

however, is combustion of extracted shale gas in gas-fired power plants.[157] Emissions from the pre-combustion stage,[158] taken together, account for roughly 10–20 per cent of total predicted shale gas emissions.[159] The combustion of shale gas for electricity generation accounts for up to 90 per cent of total predicted shale gas GHG emissions.[160]

However, studies yielded different results,[161] the validity of which is subject to an on-going scientific debate.[162] A unanimous resolution to this discussion cannot be expected any time soon, given the fact that current studies partly rely on estimates, since no shale gas production is currently taking place in Europe.[163]

Moreover, the level of GHG emissions is not the most contentious point. It is rather the question whether or not GHG emissions actually constitute a potential threat *of shale gas extraction*. Some argue that shale gas extraction could be beneficial for the climate, despite the discussed emission levels, if it could provide a substitute for coal combustion in electricity production.[164] Compared to coal, combustion of shale gas results

---

157 Daniel Forster and Jonathan Perks '*Climate Impact of potential shale gas production in the EU*' study of 30 July 2012 (European Commission 2012) available at: http://ec.europa.eu/clima/policies/eccp/docs/120815_final_report_en.pdf [accessed 29 October 2012] Figure 7 at 60 and Figure 10 at 64 (hereinafter: Forster/Perks); Broderick/Anderson 66; SRU Faulstich 35.

158 Including emissions from site-preparation, equipment, actual shale gas extraction, all sorts of transportation and processing and waste water treatment.

159 Forster/Perks Figure 7 at 60 and Figure 10 at 64.

160 Ibid.; Christopher L Weber and Christopher Clavin 'Life Cycle Carbon Footprint of Shale Gas: Review of Evidence and Implications' (2012) Vol 46 No 11 *Environmental Science & Technology* 5693.

161 A comparison of the results of different studies on the lifecycle GHG emissions of shale gas can be found at Forster/Perks 64.

162 The only study on GHG emissions of German shale gas extraction, for instance, found a higher proportion of pre-combustion GHG emissions, compared to other studies, see: Uwe R Fritsche and Jana Herling '*Energie- und Klimabilanz von Erdgas aus unkonventionellen Lagerstätten im Vergleich zu anderen Energiequellen. Endbericht zum Gutachten für Team Ewen im Rahmen des InfoDialog Fracking*' (Öko-Institut, Darmstadt 2012) 13/47 available at: http://dialog-erdgasundfrac.de/sites/dialog-erdgasundfrac.de/files/OEKO_IINAS-Fracking-Energie-Klimabilanz.pdf [accessed 17 July 2014] (hereinafter: Fritsche/Herling).

163 Forster/Perks 58.

164 Stephen Pacala and Robert Socolow '*Stabilization Wedges: Solving the Climate Problem for the Next 50 Years with Current Technologies*' available at: http://www.princeton.edu/mae/people/faculty/socolow/Science-2004-SW-1100103-PAPER-AND-SOM.pdf [accessed 26 February 2014] 17/18 and 25/26 of Supporting On-Line Material (hereinafter: Pacala/Socolow); A R Brandt et al.

in a 41–49 per cent reduction in emissions.[165] The climatic benefits of substituting coal with gas combustion is the reason why shale gas has been called a 'bridge fuel' that could power the transition to a decarbonized energy system.[166]

However, emission reduction might only materialize if the redundant coal is left *in situ*.[167] Some scientists point to the United States to show that it is more likely the coal would be extracted and exported to other, non-US countries, resulting in simultaneous use of coal and gas.[168] This would lead to an overall increase of global emissions.[169]

On top of that, the exported coal could possibly end up in combustion plants of developing countries, which lack western environmental standards, contributing even more to an overall increase in emissions.[170] Thus, the whole issue of GHG emissions is subject to an ongoing scientific debate and clear-cut numbers cannot yet be provided with any certainty.

To sum up, suspicions about environmental threats of shale gas extraction are rife. Closer examination reveals that the suspected environmental threats are subject to a lot of uncertainties and estimates. A multitude of factors influence whether or not potential environmental threats would come to pass if US shale gas extraction would be replicated in Europe.

1.3.2 The Potential Energy Security Effects of Shale Gas Extraction and the Energy 'Trilemma'

The second main interest in relation to shale gas extraction is energy security. Shale gas is a global phenomenon with potential repercussions for gas markets and the energy security of countries.[171] If all known shale gas reserves could be exploited, this form of energy alone would add 40 per cent to the world's technically recoverable gas reserves.[172] In North

---

'Methane Leaks from North American Natural Gas Systems' (2014) Vol. 343 No. 6172 *Science* 733 (hereinafter: Brandt et al.).

165 Forster/Perks 67.

166 Pacala/Socolow 17/18 and 25/26.

167 Broderick/Anderson 13 and 24.

168 Broderick/Anderson 13–15 and 21–24.

169 Ibid.

170 Broderick/Anderson 23/24.

171 Yergin 329.

172 US Energy Information Administration '*World Shale Gas Resources: An Initial Assessment of 14 Regions outside the US*' (April 2011) 4 available at: http://www.eia.gov/analysis/studies/worldshalegas [accessed 6 June 2012]. For the terminology on recoverable reserves, see Stoneley 90.

America, the 'mother country' of shale gas,[173] only 1 per cent of the overall gas demand was covered by shale gas in 2000; by 2011 that figure had rocketed to 25 per cent.[174] Current estimates show that this form of energy alone could account for 50 per cent of overall North American gas production by 2020.[175] Due to shale gas, the United States overtook Russia as the world's largest gas producer in 2009.[176]

The world's remaining unconventional gas reserves are estimated to be bigger than its remaining conventional gas reserves.[177] The estimates of shale gas reserves are likely to increase even further, due to constant improvements in geological information and test data.[178]

In EU Member States, shale gas reserves are also expected to be of considerable size. To take just two examples: Germany's recoverable shale gas reserves are currently estimated at between 700 000 000 000 m^3 and 2 300 000 000 000 m^3.[179] Polish recoverable reserves were initially estimates to stand at 187 trillion cubic feet, or roughly 5 300 000 000 000 m^3 by the US Energy Information Administration.[180] Two years later these estimates of reserves were reduced to 148 trillion cubic feet of gas or roughly 4 190 000 000 000 m^3 by the same institution.[181] Other estimates see Poland's recoverable shale gas reserves in the region between 346 000 000 000 and 768 000 000 000 m^3.[182]

The breadth of these figures shows that assessments of recoverable reserves appear to be very rough and subject to huge fluctuations. The term recoverable reserves pertains to the volume of hydrocarbons that can actually be produced to surface from an accumulation.[183] The proportion of the gas in place that can be recovered, however, depends on the

---

173 Yergin 329.

174 Ibid.

175 Bjorlykke 464.

176 EUCERS 8.

177 Ibid.; however, see Stoneley 90–92 for explanations on the shakiness of such estimates in general.

178 Ibid.

179 Meiners et al. Bund A 7 and D 1.

180 US Energy Information Administration '*Technically Recoverable Shale Oil and Shale Gas Resources: An Assessment of 137 Shale Formations in 41 Countries Outside the United States*' (US Department of Energy, Washington D.C., June 2013) 14 (hereinafter: US EIA 2013); Corey Johnson and Tim Boersma 'Energy (in)security in Poland the case of shale gas' (2013) 53 *Energy Policy* 389.

181 US EIA 2013 at 14.

182 Marynia Kruk *The Wall Street Journal* 'Poland Cuts Estimate Of Shale Gas Reserves' available at: http://www.wsj.com/articles/SB1000142405270230381 2904577295790442844470 [accessed 25 June 2016].

183 Stoneley 90.

economics – in other words how much money a company is prepared to spend to get gas out of the ground.[184] Moreover, one has to factor in that these are rough estimates and more accurate figures about the shale gas potential of EU Member States will only become available with the conclusion of further research projects.[185]

Despite these uncertainties, some experts reckon that domestic extraction of considerable amounts of gas could have a multitude of important repercussions for the security of energy supplies in EU Member States.[186] To assess these effects, however, one needs to define precisely what security of energy supplies is and what energy security means.

As a starting point, energy security has to be situated in the context of the energy 'trilemma'. Energy 'trilemma' is a term that has been coined by the World Energy Council[187] and adopted by energy law and policy scholars.[188] The 'trilemma' consists of energy security, equitable access to energy/energy cost and sustainability, or in other terms: availability, affordability and sustainability of energy supplies.[189] Energy law and policy is constantly trying to balance all three policy goals.[190] Viewed from this perspective, energy security is merely one of several goals that has to be weighed with environmental protection and equitable access to energy.[191]

---

184 Ibid.

185 This has been pointed out by German Federal Environmental Agency '*Statement: Appraisal of shale gas production in Germany*' (December 2011) at 8, (*Bundesumweltamt 'Stellungnahme: Einschätzung der Schiefergasförderung in Deutschland' (Dezember 2011)*) available at: http://www.umweltbundesamt.de/wasser-und-gewaesserschutz/publikationen/stellungnahme_fracking.pdf [accessed 26 October 2012] (hereinafter: appraisal shale gas production in Germany).

186 SRU Faulstich 5 and 19.

187 World Energy Council '*World Energy Trilemma*' available at: https://www.worldenergy.org/work-programme/strategic-insight/assessment-of-energy-climate-change-policy/ [accessed 22 September 2016] (hereinafter: World Energy Council).

188 For instance: Cristelle Maurin and Vlado Vivoda 'Shale Gas and the Energy Policy "Trilemma" in Tina Hunter (ed.) *'Handbook of Shale Gas Law and Policy'* (Intersentia, Cambridge 2016) 369–81 (hereinafter: Maurin/Vivoda); Raphael J Heffron, '*Energy Law: an Introduction*' (Springer International, Cham 2015) 3–5; Raphael J Heffron, Darren McCauley and Benjamin K Sovacool 'Resolving society's energy trilemma through the Energy Justice Metric' (2015) 87 *Energy Policy* 168 (hereinafter: Heffron/McCauley/Sovacool); Neil Gunningham 'Managing the energy trilemma: The case of Indonesia' (2013) Volume 54 *Energy Policy* 184–93.

189 For the 'trilemma' in the shale gas debate see: Maurin/Vivoda 369–77.

190 Ibid.

191 World Energy Council; Heffron/McCauley/Sovacool 169.

It is striking to see that legal treaties do not include precise definitions of the term energy security,[192] so the task of defining energy security has been largely left to scholarly debate. In 2011 Benjamin Sovacool assessed that about 45 different definitions of energy security exist.[193]

According to Haghighi, energy security is 'adequacy of energy supply at a reasonable price', which suggests that energy should be physically available and its price should be reasonable.[194] Barton et al. define energy security as 'the conditions under which a country and its citizens (. . .) and companies have access to sufficient energy resources at reasonable prices for the foreseeable future, without a serious risk of major disruption of service'.[195] Roggenkamp, while agreeing with the main aspects of these definitions, stresses that the reliability of energy supplies is a crucial component of energy security, which has, as yet, been underemphasized.[196]

Cameron provides a working definition of the concept that centres on the EU, stating that energy security means 'the ability of the energy industries (. . .) to provide their services throughout the EU to a high standard and at a reasonable cost in a competitive, fully liberalised, pan-European market.'[197] Talus, in addition, highlights that energy security as a concept could have a different meaning depending on whether it is discussed in the EC/EU or in the national context.[198]

As these different interpretations highlight, energy security can mean very different things to different individuals and different nations.[199] The

---

192 See for instance article 194 (1) TFEU.

193 Benjamin K Sovacool (ed.) *'The Routledge Handbook of Energy Security'* (Routledge Ltd., London, New York 2011) 3 (hereinafter: Sovacool).

194 Sanam S Haghighi '*Energy Security The External Legal Relations of the European Union with Major Oil- and Gas- Supplying Countries*' (Hart Publishing, Oxford and Portland 2007) 14 (hereinafter: Haghighi).

195 Barry Barton et al. (eds) '*Energy Security: managing risk in a dynamic legal and regulatory environment*' (Oxford University Press, Oxford 2004) 9.

196 Martha M Roggenkamp 'Protecting Energy Infrastructure in the EU – The Impact of External Damages on Supply Security' in Martha M Roggenkamp et al. (eds) '*Energy Networks and the Law*' (Oxford University Press, Oxford 2012) 227 et sqq.

197 Peter D Cameron '*Competition in Energy Markets – Law and Regulation in the European Union*' (Oxford University Press, Oxford 2007) 518.

198 Kim Talus 'Security of Supply – An Increasingly Political Notion' in Bram Delvaux, Michael Hunt and Kim Talus '*EU Energy Law and Policy Issues*' (Euroconfidentiel, Rixensart 2008) 129 (hereinafter: Talus 2008).

199 See for instance: Gail Luft, Anne Corin and Eshita Gupta 'Energy Security and Climate Change – A tenuous link' in Benjamin K Sovacool (ed.) '*The Routledge Handbook of Energy Security*' (Routledge Ltd., London, New York 2011) 44 (hereinafter: Luft/Corin/Gupta); Felix Ciuta 'Conceptual Notes on

concept is highly context dependent, since an energy importing country is likely to have priorities that are different from those of an energy exporting country.[200]

But there is some common ground. From the point of view of energy importing regions like Europe,[201] there are traditional and more recent definitions.[202] Traditional consumer country definitions of energy security often consist of three different aspects, the availability, reliability and affordability of energy supplies.[203]

However, some definitions of energy security also include the environmental sustainability of supplies and their socio-economic impacts as fourth and fifth dimension.[204] These modern definitions are thought to supplement the traditional consumer country definition of energy security as availability, reliability and affordability of energy supplies. Each of

---

Energy Security: Total or Banal Security?' (2010) Vol 41 No 2 *Security Dialogue* 124/125; Scott Victor Valentine 'The Fuzzy Nature of Energy Security' in Benjamin K Sovacool (ed.) '*The Routledge Handbook of Energy Security*' (Routledge Ltd., London, New York 2011) 56 (hereinafter: Valentine).

200 Luft/Corin/Gupta 44.

201 Despite a prognosed stagnasis in future consumption, Europe is and will remain heavily dependent on gas imports to meet its demand, see International Energy Agency '*World Energy Outlook 2015*' (OECD/International Energy Agency, Paris, 2015) 195/196 (hereinafter: World Energy Outlook 2015).

202 Jonathan Elkind 'Energy Security Call for a Broader Agenda' in Carlos Pascual and Jonathan Elkind (eds) '*Energy Security Economics, Politics, Strategies and Implications*' (Brookings Institution Press, Washington DC 2010) 121 (hereinafter: Elkind).

203 See, for instance, Daniel Yergin 'Ensuring Energy Security' (2006) Vol 85 No 2 *Foreign Affairs* 70/71; Elkind 121; International Energy Agency (IEA) '*What is Energy Security*' available at: http://www.iea.org/topics/energysecurity/subtopics/whatisenergysecurity/ [accessed 22 September 2015]; Michael A Toman' International Oil Security: Problems and Policies' (2002) Vol 20 No2 *The Brookings Review* 20/21; Martin Scheepers et al. '*EU Standards for Energy Security of Supply*' (Energy Research Centre of the Netherlands, The Hague 2006) 29 and 35–7; Sovacool 6/7; Ann Florini 'Global Governance and Energy' in Carlos Pascual and Jonathan Elkind (eds) '*Energy Security Economics, Politics, Strategies and Implications*' (Brookings Institution Press, Washington DC 2010) 151.

204 Bert Kruyt et al. 'Indicators for Energy Security' in Benjamin K Sovacool (ed.) '*The Routledge Handbook of Energy Security*' (Routledge Ltd., London, New York 2011) 291 et sqq. (hereinafter: Kruyt et al. 2011); Henrik Bjornebye '*Investing in EU energy security: exploring the regulatory approach to tomorrow's electricity production*' (Kluwer Law International, Alphen aan den Rijn 2010) 71 (hereinafter: Bjornebye); Sovacool 9/10; Asia Pacific Energy Research Centre '*A Quest for Energy Security in the 21st Century*' (Asia Pacific Energy Research Centre, Tokyo 2007)27 et sqq: Bert Kruyt et al. 'Indicators for energy security' (2009) 37 Energy Policy 2167 (hereinafter: Kruyt et al. 2009).

the, in total, five dimensions of the modern definition of energy security is subsequently applied to the concrete example of shale gas extraction to showcase their functioning and meaning.[205]

1.3.2.1 Availability and reliability

Availability refers to the ability of consumers to secure the energy that they need.[206] Since energy demand has increased significantly across the globe in recent decades, future oil and gas developments will involve deposits that are scarcer, farther from existing demand centres and harder to extract.[207]

Reliability refers to the extent that energy services are protected from disruption.[208] This might be achieved by the diversification of supply sources and/or by utilization of a variety of fuels and technologies.[209] For shale gas extraction the question of availability and reliability of energy supplies may, among others, be broken down into the issues of import dependency and problems with the 'Not In My Back Yard' phenomenon.[210]

Import dependency Conventional gas production has been in steep decline all over Europe for quite some time and this process is expected to accelerate in the future.[211] The downturn in domestic gas production must be contrasted with a stable European consumption of gas; the widening gap has to be bridged by gas imports.[212] To give a concrete example: the biggest European country, Germany, is reliant on gas imports for about 90 per cent of its annual gas consumption.[213] Without domestic shale gas

---

205 The effect of shale gas extraction on Europe's energy security has been concisely discussed by Rafael Leal-Arcas, Andrew Filis and Ehab S Abu Gosh '*International Energy Governance Selected Legal issues*' (Edward Elgar, Cheltenham 2014) 316/317 (hereinafter: Leal-Arcas/Filis/Abu Gosh).

206 Elkind 121.

207 Elkind 122.

208 Sovacool 9.

209 Ibid.

210 There are of course other aspects of availability and reliability, which will not be discussed here, as they have no direct relevance to shale gas extraction.

211 Eurostat '*Development of the production of primary energy (by fuel type)*, EU-28, 2003–13' available at: http://ec.europa.eu/eurostat/statistics-explained/index.php/File:Development_of_the_production_of_primary_energy_(by_fuel_type),_EU-28,_2003%E2%80%9313_(2003_%3D_100,_based_on_tonnes_of_oil_equivalent)_YB15.png [accessed 26 June 2016]; World Energy Outlook 2015, 206; Lechtenböhmer et al. 11.

212 World Energy Outlook 2015, 200.

213 German Ministry of the Economy and Energy 'Ordinance on the introduction of Environmental Impact Assessments and on mining requirements in deployment of the fracking technology and deep drills' (*Verordnung zur Einführung*

production, imports are likely to rise even further, but there is no guarantee that the required amounts of gas will always be readily available.[214]

A strong reliance on energy imports has always been viewed as problematic under energy security aspects.[215] The European Commission assessed in 2014 that the EU needs to reduce its dependency on imported fossil fuels in the long-term if it wants to guarantee energy security for its citizens.[216] Two examples from Europe immediately show the issue: around 40 per cent of German gas demand is met by deliveries from Russia.[217] This is even worse with the Baltic States Latvia, Lithuania and Estonia, which used to be entirely (100 per cent) dependent on Russian gas imports.[218] This non-diversified supply structure poses a risk, if failing infrastructure or other causes interrupt Russian gas supplies.[219]

---

von Umweltverträglichkeitsprüfungen und über bergbauliche Anforderungen beim Einsatz der Fracking-Technologie und Tiefbohrungen) 12 available at: http://www.bmwi.de/BMWi/Redaktion/PDF/V/verordnung-zur-einfuehrung-von-umweltvertraeglichkeitspruefungen,property=pdf,bereich=bmwi2012,sprache=de,rwb=true.pdf [accessed 7 April 2015]

[214] This issue has been analysed by the German parliament: Bundestag '*Beschlussempfehlung und Bericht des Ausschusses für Umwelt, Naturschutz und Reaktorsicherheit (16. Ausschuss) a) zu dem Antrag der Abgeordneten Frank Schwabe, Ingrid Arndt-Brauer, Dirk Becker, weiterer Abgeordneter und der Fraktion der SPD – Drucksache 17/7612 – b) zu dem Antrag der Abgeordneten Oliver Krischer, Hans-Josef Fell, Bärbel Höhn, weiterer Abgeordneter und der Fraktion* BÜNDNIS 90/DIE GRÜNEN – Drucksache 17/5573 –' of 26 April 2012 Bundestagsdrucksache 17/9450 available at: http://dipbt.bundestag.de/dip21/btd/17/094/1709450.pdf [accessed 16 April 2014] 5 (hereinafter: Beschlussempfehlung Fracking moratorium).

[215] Pietro S Nivola and Erin E R Carter 'Making Sense of "Energy Independence"' in Carlos Pascual and Jonathan Elkind (eds) '*Energy Security Economics, Politics, Strategies and Implications*' (Brookings Institution Press, Washington DC 2010) 105 et sqq. (hereinafter: Nivola/Carter).

[216] European Commission '*European Energy Security Strategy*' (Communication) *COM (2014) 330 final* at 2/3 (hereinafter: European Energy Security Strategy).

[217] German Ministry of Economy and Energy 'Erdgasimporte und Eigenproduktion' available at: http://www.bmwi.de/DE/Themen/Energie/Konventionelle-Energietraeger/gas,did=292324.html [accessed 18 June 2014]; EUCERS 33.

[218] Agnia Grigas 'The Gas Relationship between the Baltic States and Russia' (Oxford Institute for Energy Studies, Oxford 2012) 2 available at: https://www.oxfordenergy.org/wpcms/wp-content/uploads/2012/10/NG_67.pdf [accessed 26 June 2016] (hereinafter: Grigas).

[219] Luciani 8; Grigas 6; Sascha Müller-Kraenner '*Energiesicherheit*' (Kunstmann, München 2007) 24 (hereinafter: Müller-Kraenner 2007); Clingendael International Energy Programme 'Study on Energy Supply Security and

Previous experience highlights that this is not a purely hypothetical consideration. In the winter of 2011/2012 Member States ran close to a gas 'blackout' as Russian gas supplies fluctuated.[220] Due to declining pressure in gas supply pipes, gas fired power plants in some regions were unable to operate.[221]

Actually, this was not the first time that scientists feared Russian gas supplies to Europe could be cut off. This threat was also immanent during the Russian-Ukrainian energy crises of 2006 and 2009 and continues to play a role in the current armed conflict between Russia and the Ukraine.[222] These instances highlight the role that political influences can play in gas security.[223]

Russia's conduct during the gas disputes with Ukraine damaged the country's reputation as a reliable supplier of gas.[224] European domestic shale gas production would enable instant and uninterruptible access to gas,[225] which could potentially make Europe more energy-independent.[226]

Energy independence of a nation has been defined as meeting the

---

Geopolitics Final Report' (2004) available at: http://www.clingendael.nl/publications/2004/200401000_ciep_study.pdf [accessed 25 April 2013] 38.

220 Bundestag Stenografischer Bericht 178. Sitzung 21169; German Federal Network Agency (*Bundesnetzagentur*) 'Bericht zum Zustand der leitungsgebundenen Energieversorgung im Winter 2011/2012' available at: http://www.bundesnetzagentur.de/SharedDocs/Downloads/DE/Allgemeines/Bundesnetzagentur/Publikationen/Berichte/2012/NetzBericht_ZustandWinter11_12pdf.pdf?__blob=publicationFile&v=2 [accessed 26 June 2014] 80 et sqq.

221 Ibid.

222 For elaborations on the two Russian-Ukrainian energy conflicts, see Jonathan Stern '*The Russian-Ukraine Gas Crisis of January 2006*' (Oxford Institute for Energy Studies, 2006) 9, 14; Simon Pirani, Jonathan Stern and Katja Yafimava 'The Russo-Ukrainian gas dispute of January 2009: a comprehensive assessment' (Oxford Institute for Energy Studies, 2008) 8, 19–25, 55 (hereinafter: Pirani); Matthias Neumann, Heiko Pleines and Henning Schröder '*Russland Analysen Erdgaskonflikt mit der Ukraine*' [2009] No 176 Forschungsstelle Osteuropa an der Universität Bremen available at: http://www.laender-analysen.de/russland/pdf/Russlandanalysen176.pdf [accessed 18 July 2013] 4/5.

223 Kruyt et al. 2011 at 295.

224 Pirani 33.

225 Commission 'An Energy Policy for Europe' (Communication) *COM (2007) 1 final* at 3/4; Andruleit et al. BGR Abschätzung 48.

226 Commission 'An Energy Policy for Europe' (Communication) *COM (2007) 1 final* at 3/4; UK Energy and Climate Change Committee '*UK Energy Supply: Security or Independence? Volume I*' (Crown, London 2011) 6, 13; Bundestag Stenografischer Bericht 178. Sitzung 21169; BVerfGE 30, 292 (294); Edward N Krapels 'Oil and Security' in Gregory Treverton *'Energy and Security'* (Gower Publishing, Westmead 1980) 41–42 and 44–46; Friederike Anna Dratwa et al. '*Energiewirtschaft in Europa*' (Springer, Berlin 2010) 23.

respective country's energy needs without depending on any foreign sources.[227] However, the idea of energy independence has come under twofold criticism in recent times: one, at a conceptual level and two, in concrete relation to shale gas extraction.

Regarding the first aspect, energy independence has been described as an outdated concept. Scholars like Elkind, Nivola and Carter argue that energy security is not so much about achieving energy independence, but about realizing and, above all, managing existing dependencies.[228] But even from that point of view, shale gas could make a contribution, as it could help to counterbalance energy imports with domestic energy supplies, and thereby opening up options for the management of energy dependencies.

Second, the factual ability of shale gas extraction to reduce a given country's dependence on gas imports has been disputed.[229] Some scientists argue that the amount of shale gas that is actually recoverable in European countries is too small to have any substantial impact on the overall supply situation.[230]

Current projections are based on the American experience about how much gas might be recovered from a given well, but these figures cannot simply be transferred to Europe, as they disregard special features of certain regions, as discussed earlier.[231] The existing appraisals of European shale gas potential have been criticized for not, or not sufficiently, taking into account areas that are generally excluded from gas extraction under environmental regulations.[232] These constraints would also impact upon the economic viability of shale gas projects.[233]

In any case, estimations are rough figures as long as no exploratory drilling takes place to test the contradictory hypotheses on recoverable reserves.[234] Thus, a potential opportunity to increase Europe's domestic gas production might present itself with shale gas extraction, but the extent to which this potential might materialize cannot currently be pinned down.

---

227 Nivola/Carter 105.

228 Elkind 125; Nivola/Carter 105 et sqq.

229 This has been the case in Germany, see SRU Faulstich 10 and 15/16.

230 SRU Faulstich 15/16.

231 SRU Faulstich 10.

232 See the example of Germany: SRU Faulstich 12.

233 SRU Faulstich 13.

234 Uwe Dannwolf et al. 'Umweltauswirkungen von fracking bei der Aufsuchung und Gewinnung von Erdgas insbesondere aus Schiefergaslagerstätten Teil 2' (Umweltbundesamt, Dessau-Roßlau 2014) AP8 – 1 (hereinafter: Dannwolf et al.); SRU Faulstich 10/11.

'NIMBY' and availability of energy supplies The 'Not In My Back Yard' phenomenon has been named one of the biggest potential threats to the availability of energy supplies.[235] 'Not In My Back Yard' (hereinafter: NIMBY) means that people feel it desirable to site a potentially hazardous facility far away from where they are living.[236] Citizens who are inclined to 'NIMBY' thus tend to advocate the use of energy which has not been produced domestically.[237] 'NIMBY' would make it more difficult, in energy security terms, to guarantee the availability of energy, as a country which adopts strong 'NIMBY' policies might become over-dependent on energy imports.[238]

Regarding shale gas it has been argued that Europe is operating a 'Not In My Back Yard' policy, as it turns a blind eye to environmental degradation resulting from gas production in other parts of the world.[239] Europe is asking foreign countries to produce gas for it, despite the fact that environmental protection and technical as well as public participation standards are much worse there than they are in Europe.[240]

Russian gas production, for instance, has a terrible record of environmental devastation: Siberian soil, air and water, as well as previously pristine forests, fell victim to various contaminations.[241] Corruption, low environmental standards and an outdated infrastructure from the soviet-era make Russian gas production a noxious undertaking.[242] Gas extraction in Europe, by comparison, is rather sustainable.[243] Shale gas extraction in Europe could potentially increase the availability of, relatively sustainable, gas supplies, but this may only materialize if the 'NIMBY' opposition can be overcome.[244]

---

235 Sovacool 10; Elkind 122.

236 The term originates in political sciences, see: Frank Fischer and Michael Black *'Greening Environmental Policy: The Politics of a Sustainable Future'* (Paul Chapman Publishing, London 1995); 'NIMBY' is criticized as an irrational, emotional, self-interested or unethical response from those who are unwilling to share in the costs of industrial activities, see Dustin Mulvaney and Paul Robbins (eds) '*Green Politics*' (SAGE Publications, London 2011) 285.

237 Ibid.

238 Elkind 122; Sovacool 10.

239 Bundesrat Stenografischer Bericht 904. Sitzung 579.

240 Ibid.

241 See for instance: Müller-Kraenner 2007, 147–58; Russian Nature 'Environmental Impact of Oil and Gas Development' available at: http://www.rusnature.info/env/20.htm [accessed 22 August 2013].

242 Ibid. 148

243 Bundesrat Stenografischer Bericht 904. Sitzung 579.

244 Ibid.

1.3.2.2 Affordability

A non-diversified supply structure is putting current gas suppliers in a very strong bargaining position.[245] Ever-increasing gas import dependency might lead to rocketing gas prices.[246] According to the European Commission, this has already contributed to price increases in the past.[247]

The price of gas is a crucial component of energy security: the basic affordability of energy services is elementary to citizens and companies in Europe and all over the world.[248] A reasonable price of energy is thus a vital part of the definition of European energy security.[249]

Price increases could, potentially, be avoided or mitigated by European shale gas.[250] Current suppliers would have to make sure that the price of their gas does not exceed the point of economic viability of European shale gas projects, as gas-importing countries could otherwise be incentivized to switch supplies to these sources.[251] Although unconventional gas production in Europe has not even started, the mere prospect of increased domestic gas extraction could already influence the current gas price.[252]

The ability of European shale gas production to influence gas prices, however, is subject to controversy amongst the scientific community. Some scholars argue that the gas market is primed to become fully globalized in some years.[253] When the gas price is set at global level the, in comparison to other parts of the world, negligible[254] European

---

[245] Commission Shale Gas Communication 2.

[246] Ibid.

[247] Ibid.

[248] Ibid.; Sovacool 9/10; Kruyt et al. 2011 at 295. To give the example of a legal text in an EU Member State: § 1 of the German Energy Industry Act (*Energiewirtschaftsgesetz) (hereinafter: EnWG)* discusses national energy security and establishes that the German state shall ensure that energy supplies comply with a 'pentagon of aims', one of which is a reasonable price. More on the 'pentagon of aims' may be found at: Franz Jürgen Säcker '*Berliner Kommentar zum Energierecht Band 1*' 2nd edition (Verlag Recht und Wirtschaft, Frankfurt am Main 2010) § 1 EnWG paragraphs 1/2 (hereinafter: Säcker).

[249] Leigh Hancher and Sally Janssen, 'Shared Competences and Multi-Faceted Concepts – European Legal Framework for Security of Supply' in Barton B et al. (eds) '*Energy Security: managing risk in a dynamic legal and regulatory environment*' (Oxford University Press, Oxford 2004) 93.

[250] Commission Shale Gas Communication 2.

[251] Pearson et al. 142; Commission Shale Gas Communication 4/5.

[252] EUCERS 42.

[253] SRU Faulstich 13.

[254] No European country is in the top ten league table of countries with the most technically recoverable shale gas resources, see James R May and John C

shale gas reserves are unlikely to have a perceivable impact on gas prices.[255]

Moreover, large-scale shale gas production in Europe would be a cost-intensive undertaking. Scientists estimate that production costs would range between two and three times the cost of comparable US developments.[256] Although these production costs are likely to decrease in line with increased technological understanding, European production costs will never meet low US levels.[257] To sum up, European shale gas extraction might impact upon the affordability of energy, but that effect is not certain to materialize.

1.3.2.3 Environmental sustainability and socio-economic benefits as fourth and fifth criterion?

Two examples from Europe highlight that the concept of energy security could also include ecological and economic aspects. First, the policy of some Member States to phase out nuclear power generation (the most prominent example being the German '*Energiewende*'),[258] which incites a quest for reliable, non-intermittent[259] substitutes in electricity production. The German government stresses that the additional gas demand for the

Dernbach 'Introduction' in John C Dernbach and James R May (eds) '*Shale Gas and the Future of Energy Law and Policy for Sustainability*' (Edward Elgar Publishing, Cheltenham, 2016) 4 (hereinafter: May/Dernbach).

255 SRU Faulstich 10.

256 Quentin Philippe 'Europe's comparative disadvantage in energy intensive industries: a comparison of shale gas production costs and break-even prices in Europe and the US' in Cecile Musialski et al. (eds) '*Shale Gas in Europe*' (Claeys & Casteels, Deventer 2013) 259–61; SRU Faulstich 16.

257 Ibid.

258 See more at: Bundesregierung '*Der Weg zur Energie der Zukunft – sicher, bezahlbar und umweltfreundlich Eckpunktepapier der Bundesregierung zur Energiewende*' available at: http://www.bmwi.de/DE/Themen/Energie/energiepolitik,did=405004.html [accessed 31 May 2012] at paragraph 4 (hereinafter: Eckpunkte Energiewende).

259 Renewables are discounted here as an intermittent source of electricity production, as wind turbines are only producing when the wind blows and solar panels (mostly) when the sun shines. However, smart alternatives are currently increasing in prominence, but they lie beyond the scope of this book.

Energiewende should be covered by domestic gas production.[260] Shale gas could make a potential contribution here.[261]

Second, socio-economic development has been included by some scholars as a fifth aspect that energy security should take care of.[262] The idea that unconventional gas could help to create new jobs and act as a potential driver of overall economic growth[263] originates from the US.[264] Direct effects of shale gas extraction on the economy might include employment opportunities and increased regional investments in infrastructure.[265] Indirect effects could be the generation of additional public income via taxes and fees.[266]

However, when viewed more closely, it actually works the other way around: one key pre-condition of socio-economic development is energy security, since modern economies are based on a secure and reliable supply of energy.[267] The European Court of Justice stressed this point in drastic words decades ago in the infamous *Campus Oil* case, which illustrated

---

260 Bundesregierung '*Energiekonzept 2050*' available at: http://www.bmu.de/files/pdfs/allgemein/application/pdf/energiekonzept_bundesregierung.pdf [accessed 31 May 2012] 16 (hereinafter: Energiekonzept); Bundesregierung '*Der Weg zur Energie der Zukunft – sicher, bezahlbar und umweltfreundlich Eckpunktepapier der Bundesregierung zur Energiewende*' available at: http://www.bmu.de/energiewende/beschluesse_und_massnahmen/doc/print/47465.php [accessed 31 May 2012] 16 (hereinafter: Energiekonzept); Eckpunkte Energiewende paragraph 14.

261 Bundestag Antrag '*Leitlinien für Transparenz und Umweltverträglichkeit bei der Förderung von unkonventionellem Erdgas*' Bundestagsdrucksache 17/7612 of 8 November 2011; Bundestag Stenografischer Bericht 178. Sitzung 21169; EUCERS 6.

262 Kruyt et al. 2011 at 292; Sovacool 11.

263 Commission Shale Gas Communication 3; Northern Ireland Assembly Deb 6 December 2011, Vol 69 No 6, col. 307 (hereinafter: Northern Ireland minutes).

264 This is, however, not undisputed in America, see Patrick Artus '*US reindustrialisation poses challenge for Eurozone*' London: FTSE Global Markets, London 2012) available at: http://www.ftseglobalmarkets.com/blog/european-review/us-industrialisation-poses-challengefor-eurozone.html [accessed 17 July 2014]; Mathilde Mathieu, Thomas Spencer and Oliver Sartor 'Economic analysis of the US unconventional oil and gas revolution' VOX CEPR's Policy Portal available at: http://voxeu.org/article/limited-economic-impact-us-shale-gas-boom [accessed 26 June 2016]; SRU Faulstich 17/18.

265 Commission Shale Gas Communication 5.

266 Commission Shale Gas Communication 3 and 5.

267 Susan Y Noe and George Rock Pring 'The "Fear Factor": Why We Should Not Allow Energy Security Rhetoric to Trump Sustainable Development' in Barry Barton et al. (eds) '*Energy Security: managing risk in a dynamic legal and regulatory environment*' (Oxford University Press, Oxford 2004) 432 (hereinafter: Noe/Pring).

how society could grind to a halt without secure energy supplies.[268] Energy security is thus part of the broader field of economic development and as it is a tiny part of the definition of economic development, economic development is not part of the definition of energy security.

As this work is concerned with the interplay of environmental protection and energy security the following section focuses on the first example and the question whether environmental protection is part of the very definition of energy security. If so, does that mean the interests of environmental protection and energy security cannot contradict each other? How may such a view be reconciled with the realities and the regulation of new energy technologies like shale gas extraction?

The European Commission shares some responsibility for the emergence of the view that environmental protection is part of the definition of energy security. The Green Paper of 2000, 'Towards a European Strategy for the Security of Energy Supply'[269] includes a reference to that effect.[270] Less clear, but more authoritative,[271] is the Internal Gas Market Directive,[272] which could also be viewed as discussing environmental protection in the context of energy security.[273]

The latest EU document that has been counted in this category is the Energy Roadmap 2050.[274] Some scholars have assessed that this document is 'fusing' environmental and energy security objectives.[275] While close scrutiny of the text reveals that there are interactions, aims like achieving the Roadmap's '*decarbonisation objective while at the same time ensuring*

---

268 Case 72/83 *Campus Oil Ltd v Minister for Industry and Energy* [1984] ECR 2727 paragraphs 34/35 (hereinafter: *Campus Oil)*.

269 Commission 'Towards a European Strategy for the Security of Energy Supply (Green Paper)' (Communication) *COM (2000) 769 final* (hereinafter: Green Paper).

270 Green Paper 2.

271 As this is a Directive with a clearly defined legal effect, as opposed to a Communication which of itself, does not have any direct legal effect, see article 288 TFEU.

272 Council Directive (EC) 2009/73 of 13 July 2009 concerning common rules for the internal market in natural gas and repealing Directive 2003/55/EC [2009] OJ L 211/94 (hereinafter: Internal Gas Market Directive).

273 This, however, is ambivalent. While preamble 44 defines environmental protection and energy security as two distinct objectives that the Directive should facilitate, the body of the Directive, in particular articles 3 (1), (2) and (7) and 25 (1) Internal Gas Market Directive, mentions both interests together, without making any perceivable distinction between them.

274 Commission 'Energy Roadmap 2050' (Communication) *COM (2011) 0885 final* (hereinafter: Energy Roadmap 2050).

275 Leal-Arcas/Filis/Abu Gosh 303.

security of energy supply'[276] rather imply that two, conceptually different, objectives shall be met at the same time. The Roadmap, therefore, does not state that environmental protection is part of the energy security objective.[277]

Parts of the legal literature are pushing hard for the conceptual inclusion of environmental protection into the definition of energy security, despite the ambiguous signals from the European Commission.[278] Proponents of such an inclusion are putting forward three main arguments. First, energy infrastructure is typically long-lived: if a company decides, for instance, to construct a power plant, this will be an investment based on the presumption of long-term use of that asset.[279] In this sense, current energy decision-making is creating the environmental reality that will shape people's lives around the world for decades to come.[280]

Second, promoting energy security without including environmental aspects incentivizes the use of technologies and practices which exacerbate climate change, for instance by increasing greenhouse gas emissions.[281] Vice versa, an inclusion of environmental aspects in the definition of energy security, allegedly, leads to synergies, as the two examples of energy efficiency and renewables demonstrate.[282] Renewables are a domestic source of energy with an overall positive effect on the environment.[283] The same is true for energy efficiency: if less energy is required, the need for

---

276 Energy Roadmap 2050 at 2.

277 The EU's unclarity on this and the possibility that it might view energy security as a pre-condition to the objective of environmental protection apparently caused considerable criticism in the aftermath of the issuance of the Roadmap, see: Leal-Arcas/Filis/Abu Gosh 303.

278 Sascha Müller-Kraennar '*Energy Security: Re-measuring the World*' (Earthscan, London 2008) xi; Kruyt et al. 2009 at 2166; Marilyn A Brown and Michael Dworkin 'The Environmental Dimension of Energy Security' in Benjamin K Sovacool (ed.) '*The Routledge Handbook of Energy Security*' (Routledge Ltd., London, New York 2011) 176/177, 181 and 186 (hereinafter: Brown/Dworkin); Sovacool 9–11 and 33; Elkind 128/129; David F von Hippel et al. 'Evaluating the Energy Security Impacts of Energy Security' in Benjamin K Sovacool (ed.) '*The Routledge Handbook of Energy Security*' (Routledge Ltd., London, New York 2011) 75/76; Kruyt et al. 2011 at 291 et sqq.

279 Elkind 129.

280 Ibid.

281 Ibid.

282 Marilyn A Brown and Benjamin K Sovacool '*Climate Change and Global Energy Security*' (Massachusetts Institute for Technology, Sabon USA 2011) 84 and 122/123 (hereinafter: Brown/Sovacool); Luft/Korin/Gupta 52; Brown/Dworkin 181 and 186.

283 ECJ Case C-379/98 *Preussen Elektra* [2001] ECR I-2099 paragraphs 15, 73, 77 and 81.

energy production and imports, which are both burdening the environment, is reduced.[284]

Third, climate change will affect energy systems profoundly, for example by raising sea levels, a threat that requires adaptations in the transportation infrastructure which serves energy systems, including oil terminals, etc.[285] In addition it is feared that mass migration of climate refugees seeking asylum from ecological disaster could destabilize regions of the world, threatening energy as well as national security.[286]

All three arguments are valid in their own right but, crucially, do not explain why the concept of environmental protection must be a part of the very definition of energy security. Indeed, these arguments indicate that a strong interrelation between both interests exists. However, in order to curtail the use of, for instance, environmentally harmful energy technologies, environmental protection does not have to be a part of the concept of energy security. It may also function as an external corrective if it is viewed as a different interest, existing in its own right.

This reasoning is supported by the energy 'trilemma', mentioned above, which envisages the interplay between the two, structurally separate, interests of environmental protection and energy security. Environmental sustainability is a lone-standing pole in the energy 'trilemma'. If it would be viewed as part of the different pole of energy security, the risk arises that the, already fuzzy, concept of energy security may disintegrate altogether.

Furthermore, an inclusion of environmental protection in the definition of energy security would run into the danger of 'whitewashing' the fact that the two interests can be at odds with each other. This view has, for instance, been put forward by Talus, who describes energy security as the main opponent of a high level of environmental protection.[287] Shale Gas extraction is quite an apt example for this: if the legislator were to pursue the objective of environmental protection in a strict manner, for instance

---

[284] For that idea see: International Energy Agency '*Energy Security and Climate Policy – Assessing Interactions*' (OECD/International Energy Agency, Paris, 2007) 17, 35, 94/95 and 112; Nathalie Trudeau and Peter G Taylor 'The Energy Efficiency Dimension of Energy Security' in Benjamin K Sovacool (ed.) '*The Routledge Handbook of Energy Security*' (Routledge Ltd., London, New York 2011) entire chapter, but particularly expressed at 225 and 237; Anatole Boute 'Combating Climate Change and Securing Electricity Supply: The Role of Investment Protection Law' (2007) Vol 16 No 8 *EELR* 237.

[285] Ibid.

[286] Brown/Dworkin 177; furthermore, see Luft/Korin/Gupta 46 for a reproduction of the argument.

[287] Kim Talus '*EU Energy Law and Policy – A Critical Account*' (Oxford University Press, Oxford 2013) 186 (hereinafter: Talus 2013).

by a prohibition on shale gas extraction, this would forfeit potential benefits for Europe's energy security. If the legislator strictly adheres to the objective of energy security, this could result in legislation that might incite a rush to extract as much shale gas as possible and as quickly as possible, risking severe damage to the environment.

Moreover, the conceptual inclusion of environmental protection in the definition of energy security could open the 'floodgates' towards hundreds of new variables, which could also be included in the energy security definition.[288] This would complicate international dialogue on energy security and make policies designed to enhance energy security much more difficult to agree upon and implement.[289] It would cause sluggishness, if not total paralysis, in energy security decision-making[290] and further intellectual discord by introducing contentious debates over the weighting of multiple indicators.[291]

The European Commission, maybe because of these arguments, appears to have changed its position on the issue slightly in recent times. By 2014 the EU issued a new Communication, titled 'European Energy Security Strategy'.[292] This Communication asserts that a strong relationship between the interests of environmental protection and energy security exists, going so far to call the two 'inseparable'.[293] However, later in the Communication, it becomes clear that this wording relates to the close relationship between both interests and does not purport that they are the same: the Communication explicitly defines energy security on the basis of eight pillars.[294] None of these eight pillars refers to environmental protection or environmental sustainability.

With specific regard to the security of gas supply in Europe, the European Union established the Gas Security Regulation in 2010.[295] This

---

288 Luft/Corin/Gupta 46.

289 Ibid.

290 Luft/Corbin/Gupta 45

291 Valentine 57.

292 European Commission 'European Energy Security Strategy' (Communication) *COM (2014) 330 final* at 2/3 (hereinafter: European Energy Security Strategy).

293 European Energy Security Strategy 2.

294 European Energy Security Strategy 3.

295 Regulation (EU) 994/2010 of 20 October 2010 concerning measures to safeguard security of gas supply and repealing Council Directive 2004/67/EC [2010] OJ L 295/1 (hereinafter: Regulation 994/2010). This Regulation is about to be replaced by a new one which, however, is currently at draft stage, see European Commission 'Proposal for a Regulation of the European Parliament and of the Council concerning measures to safeguard the security of gas supply and repealing

Regulation includes, in its preamble, a comprehensive list of indicators for the security of gas supplies, but none of these indicators refers to environmental protection.[296] The document rather considers that a balance between environmental protection and energy security interests has to be struck.[297]

Overall, the conclusion that environmental protection should not be included in the conceptual definition of energy security is warranted. The examples that have been put forward by the literature do not make a sufficient case for an *inclusion* of environmental protection in the definition of energy security. These examples merely prove that the, *by nature distinct*, interests of environmental protection and energy security may be *reconciled* in individual cases.

1.4 CONCLUSION

Shale gas extraction is not a new technology. First it is not new and second it is not a technology. Instead, it is brought about by the combination of two technologies that have been deployed safely in Europe for over 30 and 50 years respectively:[298] innovative drilling and hydraulic fracturing. The new bit is the application of the combined technologies to unconventional shale gas reserves at a large scale, which was first done around the year 2000 in the Barnett shale play in Texas.[299]

Although shale gas undertakings around the globe rely on the combination of these two technologies, it would be wrong to infer that the potential environmental and/or energy security repercussions of both technologies are similar. Quite the opposite: shale plays and the overall make-up of the subsoil differ from region to region and country to country.

Shale plays are situated at different depths, their permeability and brittleness varies, the layers of rock strata that have to be bored through and the amount of gas that is contained in the shale rock is also widely different. As a consequence, the composition and amount of the fracturing fluid, the pressure with which it is pumped underground, the extent of created fractures and the ways to dispose of the 'flow back' are site-specific.

Shale gas extraction is bringing about a number of potential

Regulation (EU) No 994/2010' *COM (2016) 52 final*. Thus, this book will work with Regulation 994/2010 in its current form.

296 Preamble (7) of Regulation 994/2010.

297 Preamble (14) of Regulation 994/2010.

298 Stoneley 83; Pearson et al. 60; Wirtschaftsverband hydraulic fracturing.

299 US EIA 2013 at 13; May/Dernbach 2.

environmental issues as well as potential energy security benefits. Both, however, are still largely shrouded in scientific uncertainty. Assertions about issues and benefits in studies are often based on a mix of projections, estimations and certain American experiences, which may not be transferable to Europe.

Recent developments in EU Member States, like the *Energiewende* in Germany, are posing a conceptual question about the extent to which environmental protection objectives and energy security interests are intertwined. In this author's view, environmental protection and energy security are separate concepts, although they might interact closely with each other in certain cases. In the case of shale gas extraction both interests actually compete with each other.[300] The main question is not if environmental protection forms part of the definition of energy security, but how both separate, competing interests may be reconciled with each other.

---

[300] Mark R Robeck and Michael Bennett 'Shale Gas in the United States: An Institutional Comparison' in Cecile Musialski et al. (eds) '*Shale Gas in Europe*' (Claeys & Casteels, Deventer 2013) 36/37 (hereinafter: Robeck/Bennett). Same conclusion reached by Robeck/Bennett 36/37.

2. Shale gas and EU law

2.1 INTRODUCTION

Recent applications of various companies for shale gas licences in Europe[1] pose two questions. First, to what extent does the EU have a comprehensive environmental law framework[2] in place that could cater for the regulation of this activity? Second, how do Member States approach the regulation of shale gas extraction? While the first question will be answered in this chapter, the next chapter is dedicated to shale gas regulation at Member State level.

The main pillars of environmental and energy regulation in Member States emanated from EU law.[3] The current chapter thus not only provides an overview of the current legal regime at EU level but also forms the foundation of further discussions on the regulation of shale gas extraction in particular Member States.

The EU framework on environmental protection, mainly consisting of Directives and Regulations, will be analysed in the light of each of the, earlier identified, four main potential threats that shale gas extraction could pose, namely groundwater-, soil-, land- and air contamination.[4] If shortcomings and gaps in this framework are discovered the question of further, shale gas specific, regulatory action arises.

The discussion of EU law issues can be structured into three main parts. The first is dedicated to the identification of the relevant primary EU legislative competence for shale gas regulation, while the second explores

---

1 As at April 2017, test drills are about to take place in the UK, in North East Derbyshire according to Derbyshire County Council 'Steps taken towards shale gas test drilling application' available at: https://www.derbyshire.gov.uk/council/news_events/news-updates/2017/january/news_items/steps_taken_towards_shale_gas_test_drilling_application.asp [accessed 7 April 2017].

2 Jan H Jans and Hans H B Vedder '*European Environmental Law*' 3rd edition (Europa Law Publishing, Groningen 2008) 3–9 (hereinafter: Jans/Vedder) provide a compact overview of the development of this EU environmental law framework.

3 Germany, the biggest Member State of the EU, might serve as a generic example for this, see Michael Kloepfer '*Umweltrecht*' 3rd edition (C H Beck, Berlin 2004) § 9 paragraphs 79–81.

4 As discussed in Chapter 1 above.

the existing secondary law framework and its applicability to shale gas regulation. Finally, in a third part, the EU's recent endeavours to develop shale gas specific regulation in order to close identified gaps in the existing framework have to be discussed.

The main pieces of EU legislation with relevance to groundwater contamination/issues with well integrity, irresponsible disposal of 'flow back', the repercussions of significant land use and increased emission of greenhouse gases,[5] will be identified. The assessment of the EU legislative framework for the regulation of the four most salient potential threats of shale gas extraction is building upon the first comprehensive study on shale gas extraction in Europe by the European Commission.[6] These findings will be contrasted with results of other, more recent, studies and opinions from the legal literature, where necessary.[7] The identified pieces of secondary EU legislation are scrutinized, one by one, for the extent to which they cover one or more potential threats of shale gas extraction.

The chapter begins by discussing the primary EU law competence for introduction of shale gas specific regulation. After that, the pre-existing secondary EU law framework is assessed. This framework applies to several aspects of shale gas extraction, but it also has gaps and inadequacies, which make the introduction of shale gas specific regulation at EU level desirable. In order to close these gaps, the EU in 2014, enacted new framework legislation consisting of the 2014 Shale Gas

---

[5] SRU Faulstich 44/45.

[6] Janez Potočnik European Commissioner for Environment '*Transmission Note on the EU environmental legal framework applicable to shale gas projects*' of 22 December 2011 Annex II page 6 available at: http://ec.europa.eu/environment/integration/energy/pdf/legal_assessment.pdf_[accessed 24 April 2014] (hereinafter: Potocnik Transmission) identified seven EU Directives and one EU Regulation.

[7] Mark Broomfield '*Support to the identification of potential risks for the environment and human health arising from hydrocarbons operations involving hydraulic fracturing in Europe*' (AEA Technology, Didcot 2012) available at: http://ec.europa.eu/environment/integration/energy/uff_studies_en.htm [accessed 4 September 2014] (hereinafter: Broomfield); Lechtenböhmer et al. 61; Philippe & Partners Law Firm 'Final Report on Unconventional Gas in Europe' (2011) 101/102 available at: http://ec.europa.eu/energy/studies/doc/2012_unconventional_gas_in_europe.pdf [accessed 12 March 2012] (hereinafter: Philippe & Partners); Elen Stokes 'New EU policy on shale gas' (2014) Vol 16 Issue 1 *Environmental Law Review* 43/44 (hereinafter: Stokes).

Recommendation[8] and the 2014 Shale Gas Communication.[9] The final part of this chapter is, hence, dedicated to a critical analysis of these recent regulatory efforts.

2.2 PRIMARY EU LAW AND SHALE GAS

2.2.1 Consideration of a Possible Legal Basis

Due to the division of powers between the EU and its Member States,[10] it is necessary for the EU to have competence whenever it wishes to act on a certain issue.[11] A competence that will provide the EU with regulatory powers in a particular area[12] must have been conferred upon it[13] by the European Treaties.[14]

Shale gas extraction strikes at the core of two vital interests, environmental protection and energy security. Both interests are laid out as potential areas of EU competency in the TFEU (article 194 TFEU for energy and article 192 TFEU for the environment). Due to their factual link with shale gas extraction these two competences lend themselves naturally for consideration of a legal basis of shale gas regulation at EU level.[15]

However, prior to the enactment of the TFEU a third legal basis had been used to justify EU action in both the environmental and the energy sphere, namely article 114 TFEU.[16] Article 114 TFEU allows the EU to adopt measures that have as their objective the establishment and

---

[8] Commission Recommendation 2014/70/EU of 22 January 2014 on minimum principles for the exploration and production of hydrocarbons (such as shale gas) using high-volume hydraulic fracturing [2014] OJ L 39/72.

[9] Commission 'Communication on the exploration and production of hydrocarbons (such as shale gas) using high volume hydraulic fracturing in the EU' (Communication) *COM (2014) 23 final/2.*

[10] Paul Craig and Grainne de Burca '*EU Law Texts Cases and Materials*' 5th edition (Oxford University Press, Oxford 2011) 75–8 (hereinafter: Craig/De Burca).

[11] Craig/De Burca 74; Jans/Vedder 10.

[12] Craig/De Burca 75.

[13] Article 5 (2) Treaty on European Union (hereinafter: TEU); Jans/Vedder 10.

[14] Craig/De Burca 75/76.

[15] Stokes 43.

[16] Angus Johnston and Guy Block '*EU Energy Law*' (Oxford University Press, 2012) paragraphs 1.04 and 2.51 (hereinafter: Johnston/Block); Nicolas de Sadeleer '*Environmental Law and the Internal Market*' 350 (Oxford University Press, 2014) (hereinafter: De Sadeleer Internal Market); Jans/Vedder 94.

functioning of an internal market in Europe.[17] Environmental and energy measures that support the free movement of goods, persons, services and capital have in the past been adopted under this article and its predecessors.[18]

Establishing the legal basis of EU shale gas regulation is not merely a theoretical exercise; it has significant practical impacts. Which of the three possible articles is used as legal basis of shale gas regulation predetermines the degree to which Member States are allowed to deviate from EU regulations. Each of these articles have been interpreted as providing for different levels of harmonization.

Harmonization means that the European Union establishes standards for, inter alia, techniques, products and processes, in a certain field, which the Member States have to follow.[19] There are two kinds of harmonization, minimum on the one hand and total/maximum harmonization on the other.[20]

Article 114 TFEU has been used as a legal basis for legal measures that are harmonized to a total/maximum extent,[21] whereas articles 192 and 194 TFEU often provided the bases of measures of minimum harmonization.[22] Total/maximum harmonization occurs in cases where a definite relationship with the internal market and the free movement of goods exists, most prominently where product standards are harmonized.[23]

Measures brought under the environmental competence of article 192 TFEU often provide for minimum harmonization because of article 193 TFEU. That article establishes explicitly that Member States have

---

17 Article 114 (1) TFEU in conjunction with article 26 (1) and (2) TFEU.

18 For a concise description of the history of that particular article, see Jans/Vedder 94.

19 Craig/De Burca 148/149 and 620/621.

20 Paul Craig and Grainne de Burca '*EU Law Texts Cases and Materials*' 6th edition (Oxford University Press, Oxford 2015) 626 (hereinafter: Craig/De Burca 6th edition) talk about maximum harmonization, while Jans/Vedder 4th edition (2012) 97 and 104 et sqq. and Lorenzo Squintani '*Gold-Plating of European Environmental Law*' (University of Groningen, Proefschrift 2013) (hereinafter: Squintani) 9 et sqq. refer to total harmonization. A discussion of possible differences between these two concepts lies beyond the scope of this work.

21 Ibid.

22 Matthias Wagner '*Das Konzept der Mindestharmonisierung*' (Duncker & Humblot, Berlin 2000) (hereinafter: Wagner) 102/103; Gerd Winter 'Die Steuerung grenzüberschreitender Abfallströme' (2000) Vol 115 *DVBl* 666. Although this has also been admitted by Jans/Vedder 4th edition (2012) 119, they do not exclude the possibility of adopting total harmonization measures on the legal basis of article 192 TFEU. For that discussion and a possible solution see Squintani 24 et sqq.

23 Jans/Vedder 70.

the right to adopt *more* stringent environmental measures than those adopted under EU legislation and, thus, leaves Member States leeway for discretion.[24] However, article 193 TFEU does not confer competence on Member States to adopt *less* stringent protective measures, only *more* stringent ones.[25] Member States wishing to engage in these kind of measures will have to use a specific exit clause, if provided by the relevant EU regulation, or they are not allowed to use *less* stringent standards.[26] The reason for the strict procedure may be found in the explicit wording of article 193 TFEU, which clearly provides that Member States may only adopt *more* stringent protective measures.[27]

Energy regulation has been traditionally singled out by the legal literature as an area of minimum harmonization, where Member States can establish their own legislation within the framework provided by the EU.[28] Several decisions of the European Court of Justice (ECJ) assert that Member States indeed have the right to legislate in the field of energy if the EU has not, not yet or not comprehensively exercised its right to do so.[29]

2.2.2 The Centre of Gravity: Shale Gas Extraction and Article 114 TFEU

It is settled case law of the ECJ that the legal basis of an EU measure must be selected by the legislator on the grounds of objective factors that are accessible for judicial review, above all on the 'main objective' of the measure.[30] The European legislator may not pick a legal basis as

---

24 Jans/Vedder 69.

25 Jans/Vedder 103.

26 Ibid.

27 Ibid.

28 Talus 2013 at 221; Johnston/Block paragraph 15.03.

29 Such as Case 6/64 *Flaminio Costa v E.N.E.L.* [1964] ECR I-00585 at 593; for a discussion of this and further cases see Christian Calliess and Matthias Ruffert '*EUV/AEUV Kommentar*' 4th edition (Beck, München 2011) article 194 AEUV paragraph 23 (hereinafter: Calliess/Ruffert); Carl-Otto Lenz and Klaus Dieter Borchardt '*EU Verträge Kommentar*' 5th edition (Bundesanzeiger Verlag, Köln 2010) Vorbemerkungen zu Artikel 191–3 AEUV paragraph 3.

30 Case C-45/86 Commission of the European Communities v Council of the European Communities [1987] ECR I-1493 paragraph 11; Case C-300/89 Commission of the European Communities v Council of the European Communities [1991] ECR I-2867 paragraph 10; Case C-155/91 *Commission of the European Communities v Council of the European Communities* [1993] ECR I-939 paragraphs 10–15 and Case C-187/93 *European Parliament v Council of the European Union* [1994] ECR I-2857 paragraphs 22/23; Case C-268/94 Portuguese Republic v Council of the European Union [1996] ECR I-6177 paragraph 22;

he or she sees fit. Instead, the legislator has to look for the 'centre of gravity' of a measure, which determines its legal basis.[31] If this 'centre of gravity' is the internal market, article 114 TFEU is the appropriate legal basis.[32] If the 'centre of gravity' is protection of the environment, measures should be brought under article 192 TFEU.[33] If the measure is mainly concerned with energy, article 194 TFEU is the applicable legal basis.[34]

But this clear-cut order is penetrated by article 114 (3) TFEU, which explicitly states that the EU might legislate in the field of environmental protection to contribute to the establishment and functioning of an internal market.[35] Article 114 (3) TFEU could, thus, also provide an EU primary law basis for environmentally motivated, EU-wide shale gas regulations.

However, some pieces of EU legislation might inevitably touch upon the internal market objective, although their core content is environmental protection.[36] This is particularly the case for national environmental operating standards, since variations in these national standards amongst Member States can influence decisions by companies regarding plant location which affect the internal market.[37]

Such an issue has been showcased by the *Titanium Dioxide* case,[38] in which a Directive on Titanium Dioxide[39] set out rules prohibiting waste

---

C-176/03 Commission of the European Communities v Council of the European Union [2005] ECR I-7879 paragraph 45.

[31] Jans/Vedder 68; Mathew L Schemmel and Bas de Regt 'The European Court of Justice and the Environmental Protection Policy of the European Community' (1994) Vol 17 Issue 1 *Boston College International and Comparative Law Review* 53; Ronald van Ooik 'Cross-Pillar Litigation Before the ECJ: Demarcation of Community and Union Competences' (2008) Vol 4 Issue 3 *European Constitutional Law Review* 399.

[32] Jans/Vedder 10.

[33] Ibid.

[34] Johnston/Block paragraph 1.05.

[35] De Sadeleer Internal Market 349.

[36] De Sadeleer Internal Market 157.

[37] Albert Weale et al. '*Environmental Governance in Europe*' (Oxford University Press 2000) 35; De Sadeleer Internal Market 157.

[38] Case C-300/89 *Commission of the European Communities v Council of the European Communities* [1991] ECR I-02867 (hereinafter: Titanium Dioxide Case).

[39] Council Directive (EC) 78/176 of 20 February 1978 on waste from the titanium dioxide industry as amended by Council Directive 82/883/EEC (further amended by Council Regulation 807/2003/EC), 83/29/EEC and 91/692/EEC (further amended by Council Regulation 1882/2003/EC).

discharges into the soil and water.[40] The ECJ held that this Directive had to be adopted under ex article 100a EC (article 114 TFEU) because it intended to approximate national rules concerning production conditions with the aim of eliminating distortions of competition in that sector.[41] This judgement seemed to be pushing the whole sphere of environmental policy into the purview of the internal market.[42]

However, the judgement only applies to environmental protection measures which are inextricably linked to the completion of the internal market.[43] Acts which have a direct impact upon the internal market, and in particular those which lay down rules on production standards, must thus be adopted in accordance with article 114 TFEU.[44] By contrast, the ECJ ruled that acts which seek to achieve a high level of environmental protection, but affect the establishment of the internal market only on an ancillary basis, should be based on article 192 TFEU.[45] This is often the case with Directives that aim to protect the soil as well as underground and surface water and the air.[46]

Possible contamination of these elements by shale gas extraction is a typical potential threat that has been associated with the process. Although it might be possible that certain product standards, for instance concerning the chemical composition of the fracturing fluids, constitute a hindrance to inter-European trade, this is not the 'centre of gravity' of existing and future shale gas regulations regarding the environment. It seems likely that, when comparing article 114 TFEU with article 192 TFEU, EU shale gas regulation concerning the environment should be based upon the latter rather than on the former.

When contrasting the possible scope of application of article 114 TFEU with the energy competency of article 194 TFEU, a similar picture emerges. The literature commonly agrees that, as opposed to past practice, which relied upon article 114 TFEU, today article 194 TFEU provides the central EU competence in energy matters.[47] Shale gas, after all, is a form of energy that may be produced in individual Member States. Being a domes-

---

40 Titanium Dioxide Case paragraph 10.

41 Titanium Dioxide Case paragraph 23.

42 De Sadeleer Internal Market 158.

43 Ibid.

44 De Sadeleer Internal Market 158/159.

45 Case C-155/91 *Commission of the European Communities v Council of the European Communities* [1993] ECR I-939 paragraphs 19 and 21 (hereinafter: Case C-155/91); Case C-187/93 *European Parliament v Council of the European Union* [1994] ECR I-2857 paragraphs 24–6.

46 De Sadeleer Internal Market 158.

47 De Sadeleer Internal Market 350; Johnston/Block paragraph 1.05/1.08.

tic source of energy, other Member States might exceptionally be involved when it comes to shale gas reserves that straddle national boundaries. But other than that, the process of shale gas extraction[48] is more directly linked to the specific energy article 194 TFEU than to article 114 TFEU.[49]

To sum up, article 114 TFEU represents a rather remote primary law basis for the introduction of EU regulations on shale gas extraction. This leaves two other possibilities for further discussion, article 192 TFEU (environment) and article 194 TFEU (energy). While article 192 TFEU could be applicable, due to the described environmental threats of shale gas extraction, article 194 TFEU might apply because shale gas is a form of energy.

2.2.3 Article 192 or Article 194 TFEU as Legal Basis?

It has been argued that the sole basis of any EU regulation on shale gas has to be the environmental competence of article 192 TFEU.[50] Proponents of such a view point to the wording of article 194 (2) TFEU: '(. . .) measures shall not affect a Member State's right to determine the conditions for exploiting its energy resources, its choice between different energy sources and the general structure of its energy supply, without prejudice to Article 192(2)(c).'[51]

This paragraph has been called sovereignty clause as it says that decisions concerning the national energy mix and the issuing of licences for the exploration and exploitation of hydrocarbon resources remain the prerogative of Member States.[52] However, according to the wording of article 194 (2) TFEU, as reproduced above, this entire prerogative is subject to the condition that no prejudice to article 192 (2) (c) TFEU occurs.

Article 192 (2) (c) TFEU also includes a sovereignty clause for energy matters, which is broadly similar to the sovereignty clause of article 194 (2) TFEU. However, it differs in one important regard. While both articles prescribe that the EU shall not adopt measures significantly affecting a

---

48 It has to be emphasized that this picture might or might not change when it comes to the question what happens with the gas after extraction, further downstream. This question, however, lies outside the scope of the current work.

49 For a similar assessment see: Stokes 43/44.

50 Ibid.; Marc Pallemaerts 'Climate Change, Natural Gas and the Rebirth of EU Energy Policy' in Lillian Wylie and Pascaline Winand (eds) '*Energy and the Environmental Challenge: Lessons from the European Union and Australia*' (Peter Lang, Brussels 2011) 60 (hereinafter: Pallemaerts).

51 Article 194 (2) TFEU.

52 Stokes 44.

Member State's choice between different energy sources and the general structure of a Member State's energy supply, article 194 (2) TFEU includes a third barrier. According to this third reservation Member States have the right to determine conditions for the exploitation of domestic energy resources. Crucially, this reservation does not feature in article 192 (2) (c) TFEU. EU shale gas regulation would determine conditions under which a Member State might exploit this energy resource. As a consequence, article 194 (2) TFEU would bar the use of article 194 TFEU as legal basis for the introduction of such EU-wide measures, leaving article 192 TFEU as the only applicable legal basis.[53]

2.2.4 . . . or Both?

This, however, would be a legalistic point of view with repercussions beyond the concrete context of shale gas regulation. Following the outlined logic, the better parts of EU energy regulation would have to be adopted mainly under the environmental competence of article 192 TFEU, in spite of the existing energy competence of article 194 TFEU.[54] The EU, paradoxically, would have a competence for energy, which it might not rely upon in most cases of energy regulation.[55] As a consequence, article 194 TFEU, which is mainly concerned with the conditions for energy exploitation would be almost redundant.[56]

From a non-legalistic point of view, concentrating on the 'centre of gravity'-doctrine it seems that in a case like the regulation of shale gas extraction both competences, the energy as well as the environmental competence, should be heavily involved. After all, shale gas extraction might bring about repercussions in both areas. If an act of secondary EU law pursues more than one objective and the objectives pursued have an equal bearing on the regulated issue, the ECJ exceptionally accepts that a measure can be founded on a dual legal basis.[57]

---

[53] Leonie Reins 'In Search of the Legal Basis for Environmental and Energy Regulation at the EU Level: The Case of Unconventional Gas Extraction' (2014) 23 (1) *RECIEL* 130/131 (hereinafter: Reins legal basis); Stokes 44.

[54] A rather 'paradoxical' result, as Pallemaerts 60 rightly observes. See also Stokes 44.

[55] Ibid.

[56] This result, however, would be difficult to reconcile with the principle of the unity of the constitution, which will be discussed below in Chapter 4. Although the TFEU is not a constitution, it is a fundamental treaty of EU law and the principle of unity also applies to treaties, as will be discussed in Chapter 4.

[57] Case C-300/98 and C-392/98 *Parfums Christian Dior SA v TUK Consultancy BV und Assco Gerüste GmbH and Rob van Dijk v Wilhelm Layher GmbH & Co.*

This, however, requires that the legal procedures of both articles are not incompatible with each other.[58] While article 194 (2) TFEU makes recourse to the ordinary legislative procedure, article 192 (2) (c) TFEU establishes the use of the special legislative procedure. There are two formal differences. The first difference is whether the European Council must merely *consult* the European Parliament (special procedure according to article 289 (2) and 192 (2) TFEU)[59] or if the Parliament actually has the power to *block* a legislative proposal (ordinary procedure as provided for by articles 289 (1) and 294 TFEU).[60] The second difference is practical: while the ordinary legislative procedure of article 294 TFEU[61] requires a vote in the European Council with a qualified majority,[62] article 192 (2) TFEU states that for environmental matters, a proposal under the special legislative procedure in the Council requires unanimity if it is to be passed.

Where different decision-making procedures are being combined, their modalities must also be combined, which means the more demanding procedure must be adhered to plus additional requirements of the less demanding.[63] By adhering to the ordinary procedure (which demands that any shale gas regulation must be adopted by the European Parliament) the first of the two described differences could be reconciled.

The second difference (qualified majority v unanimity vote in the European Council) is likely to cause the bigger headache. Adhering to

---

KG and Layher BV [2000] ECR I-11307 paragraph 13; Case C-336/00 *Republik Österreich v Martin Huber* [2002] ECR I-07699 paragraph 31; Case C- 281/01 *Commission of the European Communities v Council of the European Union* [2002] ECR I-12049 paragraph 75; Case C-178/03 *Commission of the European Communities v European Parliament and Council of the European Union* [2006] ECR I-107 paragraph 42 et sqq.; De Sadeleer Internal Market 151; Jans/Vedder 69.

58 Case C-178/03 *Commission of the European Communities v European Parliament and Council of the European Union* [2006] ECR I-107 paragraph 59; Case C-155/07 *European Parliament v Council of the European Union* [208] ECR I-08103 paragraph 79; Jans/Vedder 69.

59 European Parliament 'Legislative Powers' available at: http://www.europarl.europa.eu/aboutparliament/en/0081f4b3c7/Law-making-procedures-in-detail.html [accessed 20 December 2014].

60 Article 289 (1) TFEU.

61 More specifically article 294 (8), (10) and (13) TFEU.

62 According to article 238 (2) TFEU, a qualified majority does usually consist of 72 per cent of members of the Council, representing Member States comprising at least 65 per cent of the population of the Union. This may, however, be lowered in several circumstances, see for instance article 238 (3) TFEU.

63 Jans/Vedder 69.

the higher standard would, in reality, mean that a vote of unanimity by the European Council in the special legislative procedure (see article 192 (2) TFEU) would be required to pass EU shale gas regulation. It is hard to envisage all 28 European Member States sitting around a table and unanimously agreeing upon the same approach to shale gas regulation. From a formal point of view, however, it can be presumed that no procedural issues exist that would hinder the use of a dual legal basis in shale gas cases.

From this author's perspective it is important to emphasize that article 192 TFEU (environmental) and article 194 TFEU (energy), are subject to minimum harmonization. Even if shale gas specific EU regulations are adopted, the Member States maintain the right to implement more stringent regulations alongside them, due to minimum harmonization in both, the environmental and the energy area.

From a normative perspective, a high degree of harmonization is not apt for shale gas cases because some of the main features of the procedure (required drilling depths, the chemical (or non-chemical) composition of the fracturing fluid and the disposal of waste, especially the 'flow back') alter considerably from region to region, due to different geological conditions.[64]

Since shale gas extraction has an intense bearing on both the environmental and the energy sphere, it would be most appropriate to use articles 192 and 194 TFEU as dual legal basis. However, adopting shale gas regulation on the single basis of article 192 TFEU would also be possible, but cannot be recommended.

2.3 SECONDARY EU LAW AND SHALE GAS

2.3.1 Secondary EU Law and Shale Gas Extraction

In order to regulate, control and monitor the environmental repercussions of energy activities, the European Union developed a comprehensive set of secondary law norms. Although all of these norms have not been developed with shale gas extraction in mind, they apply to certain aspects of the process.

However, the extent of this coverage of shale gas extraction by European secondary law is subject to an ongoing debate in the scientific literature.

---

[64] See above in Chapter 1.

Studies on the topic have yielded a huge variety of results, ranging from eight different applicable items of EU legislation,[65] to over 19[66] and up to approximately 40.[67] This discrepancy is mainly caused by the fact that some studies focus on the most relevant pieces of legislation, while others comprehensively address every legislative bit that might have a potential bearing on shale gas extraction.

Following on from the approach taken above in Chapter 2, this section will focus on pieces of secondary European environmental and energy legislation that cover the four main potential threats of shale gas extraction. Some pieces of European legislation apply to the whole lifecycle of a shale gas plant and may cover all four main potential issues. Thus, the analysis is structured rather by means of applicable norms than by using the individual environmental issues or stages of shale gas extraction as starting points. However, in cases where a piece of legislation does not apply to all four potential threats this is highlighted in the examination that follows.

2.3.1.1 Strategic Environmental Assessment

According to the Directive on Strategic Environmental Assessment (hereinafter: SEA Directive) Member States shall undertake environmental assessments in certain cases for big, strategic plans or programmes at policy level.[68] This applies to plans and programmes that are likely to have significant environmental impacts.[69] An SEA shall describe and appraise these likely and significant impacts on the environment. The SEA-procedure has to be contrasted with the Environmental Impact Assessment Directive (hereinafter: EIA Directive), which covers Impact Assessments for individual projects.[70]

---

[65] Potocnik Transmission Annex II page 6.

[66] Broomfield xi and 77.

[67] Lechtenböhmer et al. 48 and 54.

[68] Articles 1, 2 and 5 (1) of Council Directive (EC) 2001/42 of 27 June 2001 on the assessment of the effects of certain plans and programmes on the environment [2001] OJ L 197/30 (hereinafter: SEA Directive); Peter G Davies '*European Union Environmental Law*' (Ashgate, Hants 2004) 179/180 (hereinafter: Davies).

[69] Article 1 SEA Directive. Article 3 SEA Directive determines which plans and programmes in particular have to be subjected to an SEA.

[70] Article 1 (1) Council Directive (EU) 2011/92 of 13 December 2011 on the assessment of the effects of certain public and private projects on the environment [2012] OJ L 26/1, as amended by Council Directive (EU) 2014/52 of 16 April 2014 amending Directive 2011/92/EU on the assessment of the effects of certain public and private projects on the environment [2014] OJ L 124/1 (hereinafter: EIA-Directive). According to article 14 of Directive 2011/92/EU, this Directive repealed the pre-existing, older version of an EIA Directive, namely Council Directive (EC) 85/337/EEC of 27 June 1985 on the assessment of the effects of certain public and

A SEA should be carried out during the preparation of a plan or a programme and before its adoption or submission to the legislative procedure.[71] The SEA has been born out of the notion that impact assessments are of limited use if they only apply to specific projects,[72] since many key factors will have been determined already at an earlier stage.[73]

The SEA provides crucial information on the influence of strategic plans and programmes on a range of factors, including biodiversity, population, human health, fauna, flora, soil, water, air, climatic factors, material assets, cultural heritage including architectural and archaeological heritage, landscape and the interrelation between all of them.[74] The gathered information should be summarized in an environmental report and taken into account by the regulator, prior to the adoption of the respective plan or programme.[75]

Both the EIA and the SEA assessments may add particular value to the regulation of shale gas extraction. Project-focused assessment provides detailed information on individual impacts while planning assessment helps to put those impacts into a regional context.[76] This interplay is enshrined in preamble 10 and article 3 (2) of the SEA Directive. The latter in particular includes cross-referencing to the EIA Directive.

Article 3 SEA Directive determines which plans and programmes in particular have to be subjected to an SEA. According to article 3 (2) (a) SEA Directive this is, inter alia, the case for town and country planning programmes, which are listed in Annexes I and II of the EIA Directive and set the framework for future development consents for projects. This provision may apply to shale gas projects.[77] Annex II to the EIA Directive,

---

private projects on the environment [1985] OJ L 175/40, which had already been amended several times by Council Directive (EC) 97/11 of 3 March 1997 amending Directive 85/337/EEC on the assessment of the effects of certain public and private projects on the environment [1997] OJ L 73/5 and Council Directive (EC) 2003/35 of 26 May 2003 providing for public participation in respect of the drawing up of certain plans and programmes relating to the environment and amending with regard to public participation and access to justice Council Directives 85/337/EEC and 96/61/EC [2003] OJ L 156/17.

71 Article 4 (1) SEA Directive.

72 As the Environmental Impact Assessment (EIA), which will be discussed immediately below, does.

73 Davies 179.

74 Annex I (f) SEA Directive.

75 Article 8 SEA Directive.

76 Brian G Rahm and Susan J Riha 'Towards strategic management of shale gas development: Regional, collective impacts on water resources' (2011) 17 *Environmental Science & Policy* 20/21.

77 Broomfield 8 and 79.

in its section on the extractive industry, mentions 'd) deep drillings' and 'e) surface industrial installations for the extraction of coal, petroleum, natural gas and ores, as well as bituminous shale.' It has been argued that, because shale gas is natural gas, an SEA would be required for plans and programmes that are concerned with the town and country planning aspects of shale gas activities.[78]

However, article 3 (3) SEA Directive exempts small area plans at local level from the necessity of carrying out an SEA. But town and country planning in many Member States is a task for local or regional governments. In Germany, for instance, this responsibility is shared between the German states (*Länder*) and the local municipalities.[79] Thus, there is some likelihood that country planning, which involves shale gas extraction proposals, will be compartmentalized. Each state, and in some cases each individual municipality within that state, could draw up its own local plan.

In order to overcome that issue, article 3 (3) SEA Directive enables the Member States to demand the carrying out of an SEA, also for small areas, at nation state level if they reckon that the plan is likely to have significant environmental effects.[80] Although that is subject to the appraisal of the individual Member State, this provision allows for the request for an SEA even for small area plans which include shale gas extraction. To sum up, shale gas extraction plans can be subjected to environmental assessment under the SEA Directive, in one form or the other, depending on the views taken by the respective Member State.

2.3.1.2 Environmental Impact Assessment

According to article 1 (2) (a) EIA Directive, a project is something that requires the execution of construction works or other interventions in the natural surroundings or landscape, including activities like the extraction of mineral resources. As shale gas extraction may not be achieved without the construction of facilities and it involves interventions in both the natural surroundings and the landscape, shale gas projects fall into the general scope of the EIA Directive.

Article 2 (1) EIA Directive contains the central provision of the EIA Directive.[81] It requires Member States to ensure that an EIA is carried out

---

78 Broomfield 79.

79 Article 28 German constitution (*Grundgesetz*) in conjunction with § 8 country planning act (*Raumordnungsgesetz*) and § 2 German construction law (*Baugesetzbuch*).

80 Davies 180.

81 Jans/Vedder 312.

before consent is granted to projects which are likely to have significant effects on the environment. The EIA shall identify, describe and assess the specific effects that the respective projects have on the environment.[82] This required information is not substantially different from the information that has to be obtained under an SEA; the main difference is the scope of such information (regional or country-wide in case of a SEA; local in case of an EIA). The information, together with comments gathered in a subsequent consultation process with relevant competent authorities and the general public, is considered and taken into account during decision-making on the project.[83]

However, this does not mean that each and every project must be preceded by an EIA.[84] Article 4 EIA Directive distinguishes between two different categories of project. Projects falling under Annex I have to be accompanied by an obligatory EIA.[85] In the case of projects falling under Annex II, the Member States are asked to determine on a case-by-case basis whether or not the project in question should be subjected to an EIA.[86]

Shale gas is not explicitly listed in any of these two annexes. However, section 14 of Annex I makes the carrying out of an EIA compulsory for 'Extraction of petroleum and natural gas for commercial purposes where the amount extracted exceeds 500 tonnes/day in the case of petroleum and 500 000 cubic metres/day in the case of gas.'[87]

Since shale gas is natural gas produced for commercial purposes, this paragraph could be applicable. However, shale gas wells in Europe,[88] even in its greatest Member State,[89] cannot reasonably be expected to have a daily output of gas anywhere near 500 000 cubic metres per day.[90] Even if a project could be defined as consisting of several wells that are drilled from the same plot of land, the output would not reach 500 000 cubic metres per day.[91] Thus an EIA is not compulsory for individual shale gas projects.[92]

---

82 Article 3 first sentence EIA Directive.
83 Davies 156.
84 Jans/Vedder 314.
85 According to article 4 (1) EIA Directive.
86 Article 4 (2) EIA Directive.
87 Section 14 of Annex I Council Directive (EU) 2011/92 of 13 December 2011 on the assessment of the effects of certain public and private projects on the environment [2012] OJ L 26/1.
88 Lechtenböhmer et al. 61.
89 For the example of Germany, see Philippe & Partners 49 paragraph 145.
90 Broomfield 80.
91 Ibid.
92 Ibid.

Following this, nearly unanimous,[93] scientific observation, a fierce debate about how best to remedy this shortcoming ensued. Many scholars demanded that, given the severity of potential environmental impacts of shale gas extraction, the EIA-Directive and the pertaining EIA laws of the Member States should be amended to make EIAs obligatory for all shale gas activities, regardless of daily gas extraction rates.[94]

The idea that shale gas extraction should be included in Annex I has been taken up by the European Parliament.[95] On 9 October 2013 the parliament voted in favour of the preparation of a new EIA Directive, under which any exploration for and exploitation of unconventional hydrocarbons would be subjected to an EIA.[96] According to this proposal, hydraulic fracturing and shale gas extraction should have been inserted into Annex I of the 2011 EIA Directive as new points 14a and 14b.[97] On 16 April 2014 the European Council did indeed adopt a Directive[98] which brings about several amendments to the 2011 EIA Directive (hereinafter: 2014 EIA Directive).[99] However, the particular proposal of the European Parliament to amend Annex I with a view to shale gas extraction has not been adopted by the European Council. Hence, the detected gap in the EIA Directive has not been closed by the 2014 amendments.

---

[93] Supported by Meiners et al. Bund C 75; Broomfield 80; Lechtenböhmer et al. 61; SRU Faulstich 42. However, Broomfield suggests that instead of one well site, multiple well sites need to be considered together, as they might have 'cumulative effects' which have to be factored into the respective EIAs. See Broomfield 80.

[94] Broomfield 80; Meiners et al. Bund C 75; Lechtenböhmer et al. 61; SRU Faulstich 42.

[95] European Parliament 'Procedure 2012/0297(COD)' Amendment 31 Proposal for a directive Recital 23 a (new), available at: http://www.europarl.europa.eu/sides/getDoc.do?type=TA&language=EN&reference=P7-TA-2013-413 [accessed 12 December 2013].

[96] European Parliament 'Shale gas: new fracking projects must pass environmental test' available at: http://www.europarl.europa.eu/news/en/news-room/content/20131004IPR21541/html/Shale-gas-new-fracking-projects-must-pass-environmental-test [accessed 12 December 2013].

[97] European Parliament 'Procedure 2012/0297(COD)' Amendments 79, 112 and 126 Proposal for a directive Annex – point -1 (new) Directive 2011/92/EU Annex I, available at: http://www.europarl.europa.eu/sides/getDoc.do?type=TA&language=EN&reference=P7-TA-2013-413 [accessed 12 December 2013].

[98] Council Directive (EU) 2014/52 of 16 April 2014 amending Directive 2011/92/EU on the assessment of the effects of certain public and private projects on the environment [2014] OJ L 124/1 (hereinafter: 2014 EIA Directive).

[99] For the historical development see: European Commission '*Review of the Environmental Impact Assessment (EIA) Directive*' available at: http://ec.europa.eu/environment/eia/review.htm [accessed 12 December 2013].

Given the broad scientific and political support for a change in the EIA stipulations, this outcome seems to be surprising. One of the reasons for not amending Annex I might lie in the stipulations of article 4 (2) in conjunction with Annex II of the 2011 EIA Directive. According to this Annex II, it is for Member States to decide whether or not they wish to subject a certain project to an EIA, either through case-by-case examination or by setting thresholds or criteria (or both). Annex III of the 2011 EIA Directive establishes a list of selection criteria, with the help of which Member States shall make that choice.[100]

The criteria include certain characteristics of a project. These are the size of the project, cumulation with other projects, use of natural resources, production of waste, pollution, nuisances and risks of accidents, as well as the environmental sensitivity of the project location.[101] The 2014 EIA Directive amendments made several additions to these criteria and inserted a paragraph (g), which pertains to threats to human health (for example, by water contamination or air pollution).[102] Since shale gas projects are likely to involve potential threats of water contamination or air pollution,[103] this insertion may be seen as an indicator of the EU flagging up to Member States that shale gas projects should be made subject to an EIA.

Moreover, a 2011 guidance note of the European Commission on the application of the EIA Directive to unconventional hydrocarbons discussed further criteria of Annex II in relation to unconventional hydrocarbons.[104] It pointed out that paragraph 2 (d) Annex II of the 2011 EIA Directive refers to 'deep drillings', including examples such as geothermal drilling, drilling for the storage of nuclear waste material and drilling for water supplies.[105] It further stressed that 'the text of the EIA Directive uses the term "in particular", which implies that the enumeration of examples is indicative. Hence, unconventional hydrocarbon projects, even exploratory ones, which use deep drillings, are covered by Annex II.2.d.'[106]

---

[100] Jans/Vedder 314.

[101] Annex III 2011 EIA Directive.

[102] Annex (2) 2014 EIA Directive.

[103] See Chapter 1 above.

[104] European Commission '*Guidance note on the application of Directive 85/337/EEC to projects related to the exploration and exploitation of unconventional hydrocarbon*' of 12 December 2011 available at: http://ec.europa.eu/environment/integration/energy/pdf/guidance_note.pdf [accessed 28 April 2014] (hereinafter: 2011 Commission Guidance note).

[105] 2011 Commission Guidance Introduction and Note 1.

[106] 2011 Commission Guidance Note 3.

In addition, paragraph 2 (e) of Annex II of the 2011 EIA Directive is applicable to shale gas extraction.[107] It prescribes that surface industrial installations for, inter alia, the extraction of petroleum, natural gas and bituminous shale are projects where Member States have to decide individually on whether or not to demand an EIA.[108] Shale gas extraction requires surface industrial installations for extraction, thus, the provision applies.[109] The explicit mentioning of 'extraction' in this regard, however, means that activities during which shale gas is not actually being extracted would not be covered.[110]

The list above demonstrates that Member States might find multiple entry-points to prescribe the use of EIAs, if they want to. However, such action may be denied by the overall scope of the EIA Directive: article 1 (1) 2011 EIA Directive clarifies that EIAs may only be ordered for projects which are likely to have significant effects on the environment.

The criterion of 'significance' is not further elaborated upon in the Directive itself and the text of the directive is in fact vague.[111] Nonetheless, it has been argued that shale gas projects might meet that criterion, as the environmental repercussions of shale gas extraction are generally of unneglectable nature and the criterion of significance has not been interpreted in a strict manner in the past.[112]

This assessment is supported by the, already mentioned, 2011 Guidance Notes of the European Commission. They recommend that an unconventional hydrocarbons project should be subjected to an EIA if it cannot be excluded, on the basis of objective information, that the project will have significant environmental effects.[113] Thus, shale gas extraction might fall into the scope of the EIA Directive, but only under Annex II.[114]

It should be noted in that context that the EIA Directive is subject to minimum harmonization because it is based upon the environmental competence of article 192 TFEU.[115] Therefore, Member States may impose

---

107 Broomfield 81.
108 According to article 4 (2) 2011 EIA Directive.
109 Broomfield 81.
110 Broomfield 85.
111 Ibid.
112 Cecile Musialski 'The EU Legal and Regulatory Framework as Currently Applicable to Shale Gas Extraction and Commercialisation' in Cecile Musialski et al. (eds) '*Shale Gas in Europe – A Multidisciplinary Analysis with a Focus on European Specificities*' (Claes & Casteels, Deventer 2013) 73–5 (hereinafter: Musialski EU chapter).
113 2011 Commission Guidance Note 3/4.
114 Broomfield 81.
115 Preamble of 2011 EIA Directive, prior to point (1).

obligations that go beyond the level of what is strictly necessary under the EIA Directive. Under minimum harmonization, Member States have the undisputed right to impose *more* stringent measures than those envisaged at EU level.[116] They may thus require an EIA for every shale gas project on their territory.

If a Member State demands an EIA, several aspects of shale gas extraction would need to be considered. According to article 3 of the 2011 EIA Directive these are:

> (. . .) direct and indirect effects of a project on the following factors:
> (a) human beings, fauna and flora;
> (b) soil, water, air, climate and the landscape;
> (c) material assets and the cultural heritage;
> (d) the interaction between the factors referred to in points (a), (b) and (c).[117]

Considering the four main potential threats of shale gas extraction, the effects on groundwater by the process and by the disposal of 'flow back' would have to be investigated by an EIA, according to article 3 (b) EIA Directive, which lists soil and water. The land use issue would have to be investigated because of article 3 (a) and (b) EIA Directive, citing impacts on flora and fauna as well as on the landscape. The emission of greenhouse gases would fall under article 3 (b) EIA-Directive, which requires the scrutiny of impacts on the air and climate. Thus, all four major potential threats of shale gas extraction to the environment would be covered by the EIA Directive, if a Member State were to make EIAs obligatory for shale gas projects.

To sum up, as European law stands, EIAs are not compulsory for shale gas extraction projects and it is up to the Member States to decide on a case-by-case basis whether to demand the carrying out of an EIA.[118] Despite multiple complaints about this shortcoming, the EU legislator decided against making EIAs compulsory for shale gas projects. Instead, the EU legislator broadened the possibilities for Member States to demand an EIA for shale gas projects. The EU decision not to make EIAs obligatory for all shale gas projects is difficult to comprehend, given that

---

116 See discussion above.

117 Article 3 Council Directive (EU) 2011/92 of 13 December 2011 on the assessment of the effects of certain public and private projects on the environment [2012] OJ L 26.

118 Same view taken by Davies 159; Philippe & Partners 49 paragraph 145; Lechtenböhmer et al. 61.

EIAs are compulsory for other projects with environmentally detrimental potential, like Carbon Capture and Storage or crude oil refinement.[119]

2.3.1.3 Integrated Pollution Prevention and Control Directive and Industrial Emissions Directive

The Directive on Integrated Pollution Prevention and Control (hereinafter: IPPC Directive)[120] was repealed with effect from 7 January 2014 and its provisions have been incorporated into the new Directive 2010/75/EC (hereinafter: Industrial Emissions Directive or IED).[121] However, transitional provisions of the new IED ordered that certain parts of the IPPC Directive remained operative until June, and some even until December 2015.[122] Thus, shale gas wells that were established prior to that date were still subject to the relevant provisions of the IPPC Directive. Both the IPPC Directive and the IED will be scrutinized together for their relevance to shale gas extraction in due course.

The IPPC Directive and IED establish a pollution control regime that seeks to prevent and minimize emissions from large industrial operators.[123] According to article 1 of the IPPC Directive, its purpose is to achieve integrated prevention and control of pollution arising from, inter alia, installations of the energy industry and the mining industry and installations which handle hazardous waste or pursue industrial activities.[124] The Directive lays out measures designed to prevent or, where that is not practicable, to reduce emissions into the air, water and land from the abovementioned activities.[125] It aims to achieve a high level of protection of the environment as a whole.[126] This objective is supported by several requirements, including an IPPC/IED permit, the setting of emission standards and the prescription of the use of best available technology

---

119 See Annex I EIA Directive. For the analogy between these activities and shale gas extraction see Chapter 7 below.

120 Council Directive (EC) 2008/1 of 15 January 2008 concerning integrated pollution prevention and control [2008] OJ L 24/8 (hereinafter: IPPC Directive).

121 Article 81 Council Directive (EC) 2010/75 of 24 November 2010 on industrial emissions (integrated pollution prevention and control) (Recast) [2010] OJ L 334/17. In addition to that, the new IED recasts a number of other Directives, see preamble 1 IED.

122 See articles 82 (2) (5) (7–9), 32 (2), 33 (1) (c) IED.

123 See, with regard to the old IPPC Directive, Bettina Lange '*Implementing EU Pollution Control*' (Cambridge University Press, Cambridge 2008) 3 (hereinafter: Lange).

124 Article 1 and Chapters II–IV in case of the IED.

125 Article 1 IPPC Directive and Article 1 IED.

126 Ibid.

and its monitoring and documentation.[127] Control of emissions is thus not achieved by recourse to one single measure, but the control relies on a closely integrated system.[128]

Annex I IPPC Directive in conjunction with article 1 IPPC Directive (Annex I in conjunction with article 10 in case of the IED) specifies the nature of activities that fall into the scope of the Directive. There are multiple entry points for shale gas extraction. First, if the shale gas installations were to be classified as combustion installations with a thermal input of more than 50 Megawatts, they would fall under paragraph 1.1 of Annex I IPPC Directive/IED. The average thermal input of a shale gas installation, however, is around 8 Megawatts,[129] well below this threshold. But article 1 of Annex I IPPC/IED allows for cumulating the thermal input from several activities on one site to one overall input value. In the case of very large shale gas extraction sites, the threshold of 50 Megawatts could thus be reached. However, given the compartmentalized structure of current unconventional gas sites in Europe, this seems to be highly unlikely, at least for the foreseeable future.[130] Thus, paragraph 1.1 of Annex I IPPC Directive/IED does not currently apply to shale gas extraction.

Second, however, the fracturing fluid that re-surfaces from the treated well ('flow back') and the fluid that remains in the soil could both be classified as hazardous waste.[131] Installations for the disposal or recovery of hazardous waste are principally subjected to the IPPC Directive and the IED under paragraph 5.1 of Annex I IPPC/IED, if their capacity exceeds 10 tonnes per day. Paragraph 5.1 Annex I IPPC Directive/ IED does not define hazardous waste directly, but it points towards article 3 (2) of Directive 2008/98/EC.[132,133] Article 3 (2) Waste Framework Directive

---

127 Articles 4–15 and 17 IPPC Directive/articles 11–20 IED and Lange 3.

128 Lange 3.

129 Broomfield 86.

130 Ibid.

131 Ibid.

132 Council Directive (EC) 2008/98 of 19 November 2008 on waste and repealing certain Directives [2008] OJ L 312/3.

133 Paragraph 5.1 Annex I IPPC refers to a list of Article 1(4) of Directive 91/689/EEC for the definition of hazardous waste. However, this list in Directive 91/689/EEC has now been amended and replaced by Directive 2008/98/EC (hereinafter: Waste Framework Directive), according to Article 41 (b) of Council Directive (EC) 2008/98 of 19 November 2008 on waste and repealing certain Directives [2008] OJ L 312/3. Paragraph 5.1 of Annex I IED does not include this reference to the Waste Framework Directive but the definition of hazardous waste remains the same as preamble 38 of IED now directly refers to article 3 (2) of the Waste Framework Directive (2008/98/EC) for definitional purposes.

defines hazardous waste as waste which displays one or more of the hazardous properties listed in Annex III to the Waste Framework Directive. Paragraphs H6 and H7 of this Annex III name 'carcinogenic' and 'toxic' substances that shall be disposed of in that regard.

An influential 2012 study, commissioned by the European Commission, argues that scientific investigations into the composition of fracking fluids in the USA found that some of their components may be characterized as 'carcinogenic' and 'toxic'.[134] The fracking fluid, hence, would fall within the definition of hazardous waste and the scope of the IPPC Directive. But the argument is not entirely convincing.

As discussed above in Chapter 2 of this book, fracturing fluids are mixed individually for the treatment of each well because of the differences in geology at different sites. The ingredients of US fracking fluids are purposely compounded to apply to the geological makeup of a certain region in America. In Europe, a different geology exists and the fracking fluid might contain other chemical ingredients or a considerably different amount of chemicals than those fracking fluids that are being used in the USA.[135] Thus, no general assertion that the IPPC Directive applies to shale gas extraction can be made; instead, an assessment of the concrete constituents of the fracturing fluid on a case-by-case basis is required.[136]

It may be deduced that the IPPC Directive and the IED do not generally apply to shale gas extraction, but are operable under certain circumstances. This appraisal is bolstered by an assessment of applicable EU Directives to German shale gas extraction, which has been conducted at the behest of the German Environmental Ministry.[137] That treatise assessed that waste (as discussed above, the fracking fluid could be considered as waste) from shale gas extraction plants is not regulated by the IPPC Directive, but instead governed by the more specific provisions of the Mining Waste Directive.[138]

2.3.1.4 Mining Waste Directive

Directive 2006/21/EC (hereinafter: Mining Waste Directive)[139] provides for measures, procedures and guidance to prevent or reduce as far as pos-

---

134 Broomfield 86.
135 See Introduction above.
136 This has even been conceded by the study, which advanced this argument in the first place, see: Broomfield 87.
137 Meiners et al. Bund B 108.
138 Ibid.
139 Directive (EC) 2006/21/EC of 15 March 2006 on the management of waste from extractive industries [2006] OJ L 102/15.

sible any adverse effects on the environment, in particular to water, air, soil, fauna and flora and landscape from mining waste.[140] According to article 1 Mining Waste Directive, this also applies to any threats to human health that are brought about as a result of the management of waste from the extractive industries.

In spite of the definition of the scope of the Directive, the term waste is not defined as such in the Mining Waste Directive; however, its article 3 (1) refers to the definition of waste as given in the Waste Directive.[141] According to article 3 (1) of the Waste Directive, waste is any substance or object which the holder discards or intends or is required to discard. Waste is also defined in a Commission Decision pertaining to the Waste Directive, Council Decision 2000/532/EC,[142] which orders that drilling muds and other drilling wastes can be considered as waste.[143]

It could be discussed whether or not wastewater, such as the 'flow back'[144] and the fracturing fluid that is left *in situ*, actually constitute waste, since they are liquids and not necessarily of a muddy nature. However, given the fact that article 3 (1) Waste Framework Directive considers *every* substance which the holder wishes to discard to be waste and Paragraph 01 05 of the Annex to the Commission Decision pertaining to the Waste Directive also refers to drilling waste in addition to drilling muds, the definition of a substance as waste is independent from its liquid or non-liquid status.[145] The key mining waste from shale gas extraction is, therefore, the fracturing fluid and the 'flow back'.[146]

The Mining Waste Directive imposes a duty of care on operators of mining plants.[147] Under this duty a permit is required to operate waste facilities[148] and hence for shale gas extraction sites, since they handle waste. Article 7 (2) Mining Waste Directive sets out a list of documents

---

140 Article 1 Mining Waste Directive.

141 Waste Framework Directive.

142 Council Decision 2000/532/EC of 3 May 2000, replacing Decision 94/3/EC establishing a list of wastes pursuant to Article 1 (a) of Council Directive 75/442/EEC on waste and Council Decision 94/904/EC establishing a list of hazardous waste pursuant to Article 1(4) of Council Directive 91/689/EEC on hazardous waste (notified under document number C (2000) 1147) [2000] OJ L 226/3 (hereinafter: Commission Decision pertaining to the Waste Directive).

143 Paragraph 01 05 Annex to the Commission Decision pertaining to the Waste Directive.

144 See Chapter 1 for a definition.

145 Jans Vedder 491.

146 For a similar assessment, see Broomfield 120; Meiners et al. Bund B 108.

147 Jans Vedder 491.

148 Article 7 (1) Mining Waste Directive.

that are needed to obtain such a permit. According to this list, inter alia, a waste management plan under article 5 Mining Waste Directive needs to be submitted to the relevant authorities.[149] It is the objective of such a management plan to prevent or reduce waste production and its harmfulness, inter alia, by

> placing extractive waste back into the excavation void after extraction of the mineral, as far as is technically and economically feasible and environmentally sound in accordance with existing environmental standards at Community level and with the requirements of this Directive where relevant[150]

As described above in Chapter 2, other than in the US, the deep ground injection (*Verpressen)* of 'flow back' is currently considered to be the only feasible disposal method for shale gas extraction remnants in major EU Member States like Germany.[151] According to article 5 (2) (a) (iii) Mining Waste Directive this method is encouraged, if it can be deployed in an environmentally sound manner.

However, if this method is chosen, the waste management plan must include appropriate monitoring measures for waste water that has been disposed of by deep ground injection, according to article 5 (2) (e) and (f) Mining Waste Directive. This also requires the monitoring of groundwater in the direct vicinity of the well, in order to detect possible leakages.[152]

The key component of a waste management plan is the proposed classification of the waste facility,[153] here the shale gas extraction plant. Article 5 (3) (a) Mining Waste Directive establishes two possibilities in that regard. One is that the waste facility is a category A facility, with the consequence that a major-accident prevention policy, a safety management system and an internal emergency plan are required.[154] The second possibility is that the waste facility is a non-category A plant, in which case a mere identification of possible accident hazards is required.[155] The criteria for this delimitation are provided by Annex III of the Mining Waste Directive.

According to Annex III, one of the key criteria for determining the appropriate category is whether or not the waste, coming from the well or remaining underground, is hazardous under the criteria of the Directive

---

149 See also ibid. on the waste management plan.
150 Article 5 (2) (a) (iii) Mining Waste Directive.
151 See above Chapter 1.
152 Broomfield 96.
153 Jans/Vedder 491.
154 Article 5 (3) (a) Mining Waste Directive.
155 Ibid.

on hazardous waste.[156] As discussed above in the context of the IPPC Directive and the IED, this assessment may be made neither in general for the fracturing fluid nor for the 'flow back' nor for the residual components remaining in the subsoil. It rather requires case-by-case assessments, as the chemical components of each individual fracturing fluid need to be known to make that appraisal.

Ultimately, Member States are responsible for ensuring that the operator is taking all measures necessary to prevent or reduce as far as possible any adverse effects on the environment and human health, brought about as a result of the management of extractive waste.[157] Thus, the operator must obtain a permit for his facility from the Member States, according to article 7 Mining Waste Directive. Member States must guarantee that the operator himself has suitable monitoring and inspection plans in place, that he reports monitoring results to the competent authorities and that he is able to handle emergencies appropriately.[158] According to article 17 Mining Waste Directive, Member State authorities must inspect waste facilities regularly.

The conditions under which a permit is being issued must be reviewed periodically by Member States to ensure that the operator uses the best available techniques.[159] As the EU wishes to ensure that the practices demanded by Member States to ensure environmentally safe operations are adequate, it often summarizes the best techniques available in Best Available Technique References Documents (so called EC-BREFs).[160] BREFs are widespread in many technological sectors,[161] they are being developed together with stakeholders, such as the industry and non-governmental organizations[162] and their main function is to guide Member States when they draw up their own regulations.[163] Such a BREF does not yet exist for shale gas operations.[164] This absence of a BREF on shale gas extraction could affect the adequacy of measures to manage impacts

---

156 Council Directive (EEC) 91/689 of 12 December 1991 on hazardous waste [1991] OJ L 377/20.

157 Article 4 (2) Mining Waste Directive.

158 Articles 11 (2) and 11 (3) and 12 Mining Waste Directive.

159 Articles 7 (4) and 21 (3) Mining Waste Directive.

160 European Commission '*IED and BREF Revision*' page 13 and 14 available at: http://www.cepi.org/system/files/public/epw-presentations/2012/BREFseminar/Paper%20week.pdf [accessed 9 March 2013] (hereafter: Commission BREF); Lechtenböhmer et al. 62.

161 Commission BREF 13/14.

162 Ibid.

163 Lechtenböhmer et al. 62.

164 Broomfield 120 and 92.

related to mining waste, since Member States may request the use of varying techniques from operators.[165]

The European Commission, therefore, pledged in 2012 to establish a BREF for shale gas extraction and mining waste.[166] The European Parliament adopted a similar resolution in the same year, calling for a BREF on shale gas extraction.[167] The European Commission decided to opt for the development of a so-called Hydrocarbons BREF, which was to address Best Available Techniques (BAT) in the context of hydrocarbon extraction; but this BREF should also include BATs specific to shale gas extraction.[168] This process started in August 2015, but has not yet yielded any tangible results. For the time being a BREF that includes particular regulations on shale gas extraction is absent and this might be considered as a potential shortcoming of the EU's regulatory framework on the environment with regard to shale gas extraction.

To sum up, the Mining Waste Directive applies to waste water stemming from shale gas extraction, as this water is waste in the sense of the Directive and shale gas plants should be considered as waste facilities, accordingly.[169] Thus, shale gas plants must obtain a permit under article 7 Mining Waste Directive. Whether shale gas extraction facilities constitute category A or non-category A plants in the sense of the Directive will mainly depend on whether or not the waste that is being produced there may be categorized as hazardous. Such an assessment can only be made on a case-by-case basis. Shale gas operators must use best available techniques under the Mining Waste Directive. However, a level playing field for best available techniques in the general process of shale gas extraction does not yet exist among Member States and an EU-wide BREF for shale gas extraction has not yet been produced. This failure of the EU may be

---

165 Similar opinion given at Broomfield 120 and 92.

166 European Union Technical Working Group on Environmental aspects of unconventional fossil fuels, in particular shale gas '*First Meeting on 27 January 2012, Brussels Final Summary Report*' at page 5 available at: http://ec.europa.eu/transparency/regexpert/index.cfm?do=groupDetail.groupDetailDoc&id=8487&no=1. [accessed 10 December 2014].

167 European Parliament '*European Parliament resolution of 21 November 2012 on the environmental impacts of shale gas and shale oil extraction activities* (2011/2308(INI))' paragraph 12 available at: http://www.europarl.europa.eu/sides/getDoc.do?pubRef=-//EP//NONSGML+TA+P7-TA-2012-0443+0+DOC+PDF+V0//EN [accessed 23 April 2014].

168 European Commission '*Energy and Environment Hydrocarbons BREF*' available at: http://ec.europa.eu/environment/integration/energy/hc_bref_en.htm [accessed 6 June 2016] (hereinafter: Draft Hydrocarbons BREF).

169 Potocnik Transmission 7.

considered as a shortcoming with regard to the regulation of shale gas extraction.

2.3.1.5 Seveso I–III Directives

The three Seveso Directives[170] issued to date provide for measures to protect the public and the environment from the consequences of major industrial accidents, which involve dangerous substances.[171] According to articles 2 and 3 (1) of the latest Seveso III Directive, this protection extends to facilities in which dangerous substances are present in one or more installations. Dangerous substance means a substance or mixture covered by part 1 or listed in part 2 of Annex I of the Directive.[172] Both of these parts, essentially, set out a list of substances and a threshold amount from which onwards, the Seveso III Directive shall apply. In part 2 number 18 of Annex I Seveso III, natural gas is listed as a relevant substance.

However, part 2 number 18 of Annex I Seveso III Directive specifies that more than 50 tonnes of natural gas must be present at an installation for the Directive to apply. Shale gas, however, is not usually stored on site but immediately transported to the main gas infrastructure.[173] Moreover, the presence of shale gas underground cannot be counted towards that threshold, as it is not present at the installation as such.[174] Thus, the threshold is not likely to be met at a shale gas extraction site and the Seveso Directives are hence unlikely to apply to shale gas extraction.

2.3.1.6 Water Framework Directive and Groundwater Directive

Two of the four potential major threats of shale gas extraction are related to groundwater.[175] Potential threats could on the one hand result from the contamination of groundwater by chemicals or hydrocarbons during shale gas extraction. On the other hand, aquifers and groundwater reservoirs

---

[170] Council Directive (EC) 82/501 of 24 June 1982 on the major-accident hazards of certain industrial activities [1982] OJ L 230/1 (hereinafter: Seveso I); Council Directive (EC) 96/82/EC of 9 December 1996 on the control of major-accident hazards involving dangerous substances [1997] OJ L 13 (hereinafter: Seveso II); Council Directive (EU) 2012/18 of 4 July 2012 on the control of major-accident hazards involving dangerous substances, amending and subsequently repealing Council Directive 96/82/EC OJ L 197/1 (hereinafter: Seveso III).

[171] According to article 1 Seveso III.

[172] According to article 3 (10) Seveso III.

[173] Broomfield 93.

[174] See, for that line of reasoning, Broomfield 93.

[175] See Chapter 1 above.

could become polluted after the actual extraction process ceased, namely during the disposal of 'flow back' in deep ground injection wells.[176]

There are principally three main Directives that could be applicable to the potential threat of groundwater contamination. These three Directives might be grouped into two categories: the first dealing with the quality of the groundwater as such and the second dealing with the nature of chemicals that might be injected. With regard to the state of the groundwater in a shale gas extraction area, Directive 2000/60/EC (hereinafter: Water Framework Directive)[177] and Directive 2006/118/EC (hereinafter: Groundwater Directive)[178] might apply. With regard to the chemicals that are artificially injected into the soil and which could migrate to the groundwater, Regulation 1907/2006[179] (hereinafter: REACH) might be applicable. In line with this distinction, the current section deals with the quality of the groundwater as such, while the next section will discuss the REACH Regulations applicable to the nature of artificially injected chemicals.

The Water Framework Directive and the Groundwater Directive deal with the quality of groundwater and try to achieve a good quality of this resource from two different angles. The Groundwater Directive recommends to the Member States specific measures to prevent and control groundwater pollution. According to its article 1, the Water Framework Directive establishes the obligation for Member States to elaborate a comprehensive framework for the protection and conservation of groundwater.[180]

Groundwater is defined as all water which is below the surface of the ground in the saturation zone and in direct contact with the ground.[181] Articles 3 and 4 of the Groundwater Directive put forward criteria and a procedure for assessing the chemical status of groundwater, inclusive

---

[176] Chapter 1 above and Potocnik Transmission 7.

[177] Council Directive (EC) 2000/60 of 23 October 2000 establishing a framework for Community action in the field of water policy [2000] OJ L 327.

[178] Council Directive (EC) 2006/118 of 12 December 2006 on the protection of groundwater against pollution and deterioration [2006] OJ L 372/19.

[179] Regulation (EC) 1907/2006 of 18 December 2006 concerning the registration, evaluation, authorization and restriction of chemicals (REACH) [2006] OJ L 396/1.

[180] Marleen Van Rijswick 'The Water Framework Directive' in Marleen Van Rijswick (ed.) '*The Water Framework Directive: Implementation into German and Dutch Law*' (Centrum voor Omgevingsrecht en Beleid, Utrecht 2003) 2 (hereinafter: Van Rijswick).

[181] Article 2 (2) Water Framework Directive.

of a monitoring scheme.[182] Article 6 Groundwater Directive supplements these criteria and monitoring procedures with concrete provisions, which prevent or limit the input of pollutants into the groundwater. Member States are obliged to take all necessary measures to prevent pollutants from entering the groundwater.[183] This provision gives Member States the right to shut down production plants in extreme cases, a measure that could also be deployed at shale gas extraction sites.[184] However, as the Groundwater Directive does not directly regulate polluting installations, but aims to ensure the overall quality of the groundwater as such, it might only have an indirect impact on shale gas extraction.

Similar to the Groundwater directive, the Water Framework Directive also addresses groundwater as such: article 1 (d) of the Water Framework Directive, for instance, establishes that the groundwater shall be protected by the progressive reduction and prevention of pollution. This shall be achieved by the establishment of a framework for groundwater protection.[185] Measures that are deemed necessary to prevent or limit the input of pollutants into the groundwater and to prevent the deterioration of the status of all bodies of groundwater shall be implemented by the Member States.[186] The necessary measures are further described in articles 6, 7, 8 and 11(3) (j) Water Framework Directive.

Member States must, inter alia, establish a register of all areas which have been designated as requiring special groundwater protection.[187] This register had to be completed four years after the coming into force of the Directive.[188] Moreover, Member States must identify, protect and monitor areas in which water is being abstracted for drinking water purposes.[189] Finally, Member States are asked, by article 8 (1) Water Framework Directive, to establish monitoring programmes to check the chemical and quantitative status of the groundwater. All of these measures apply if Member States wish to allow the introduction of chemicals into the soil during shale gas extraction.

Concerning the disposal of 'flow back' by deep ground injection, article 11 (3) (j) Water Framework Directive prohibits the direct discharge of

---

182 Van Rijswick 6–8.
183 Article 6 (a) Groundwater Directive.
184 Broomfield 108/109.
185 Van Rijswick 4 and 7/8.
186 Article 4 (1) (b) (i) Water Framework Directive.
187 Article 6 (1) Water Framework Directive.
188 Ibid.
189 Article 7 (1)–(3) Water Framework Directive.

pollutants into groundwater, subject to specific provisions.[190] But there is an exception: according to article 11 (3) (j) Water Framework Directive Member States may allow

> injection of water containing substances resulting from the operations for exploration and extraction of hydrocarbons or mining activities, and injection of water for technical reasons, into geological formations from which hydrocarbons or other substances have been extracted or into geological formations which for natural reasons are permanently unsuitable for other purposes. Such injections shall not contain substances other than those resulting from the above operations[191]

The then European Commissioner for the environment, Janez Potočnik, concluded that this exemption, which may only be granted by Member States, does not apply to the deep ground disposal of 'flow back' from shale gas operations.[192] His report provides two arguments for this position: first, one of the objectives of the Water Framework Directive is to reach a good status of water resources, which would be endangered by shale gas operations.[193] Second, the exception of article 11 (3) (j) Water Framework Directive has been specifically designed for conventional hydrocarbons operations, as the negotiation history of the Directive highlights.[194]

This argumentation has been endorsed by a later study on shale gas extraction, prepared for the European Commission.[195] This study added a third argument that builds upon the wording of article 11 (3) (j) Water Framework Directive. According to this argument, only the injection of water containing substances *resulting* from the operations for exploration and extraction of hydrocarbons is subject to the exemption.[196] According to this position, the last sentence of article 11 (3) (j) Water Framework Directive ('Such injections shall not contain substances other than those resulting from the above operations') would preclude the deep ground injection of 'flow back'; 'flow back' does not only contain substances resulting from hydrocarbons extraction, but also the, artificially prepared, fracturing fluid plus substances coming from the geological formations.[197]

---

190 Broomfield 110.
191 Article 11(3) (j) Water Framework Directive.
192 Potocnik Transmission 7.
193 Ibid.
194 Ibid.
195 Broomfield 111.
196 Ibid.
197 Ibid.

The fracturing fluid, the argument goes, serves a special purpose (maximizing the flow of hydrocarbons) and its existence is not a *consequence* of the operations.[198]

All three arguments have to be dismissed as invalid. The first argument, that shale gas extraction would endanger the good status of groundwater is very broad-brush. In cases where shale gas extraction is not achieved by the use of chemical substances but by other means, like pneumatic fracturing or gel-based fracturing fluids, the existence of a potential threat might be excluded.[199] These methods do not use water-endangering chemicals.[200] Moreover, as discussed above in Chapter 2, where chemicals are being used, fracturing fluids are mixed for each well individually and therefore no general assumptions about their risk-potential can be made.

The second argument, that the exemption of article 11 (3) (j) Water Framework Directive was originally elaborated with conventional hydrocarbon extraction in mind and does not apply to unconventionals, is also not convincing. Legal interpretation requires, first and foremost, consideration of the plain meaning of the provision in question.[201] The wording of the Directive does not mention any restriction to conventional hydrocarbons.

Furthermore, law evolves in the sense that it may eventually be applied to facts and circumstances that were not foreseen when the regulation was originally promulgated.[202] After all, this whole chapter is concerned with the question of how to apply European environmental legislation to a technological development that was not thought of at the time when EU environmental laws were adopted.

The third argument has been summed up in the following terms:

> in neither case does the flow back water only contain substances resulting from the extraction process itself – that is, only substances that were originally present in the geological formation and which have been removed from the formation by the respective practice.[203]

---

198 Ibid.

199 Lucca Gandossi '*An overview of hydraulic fracturing and other formation stimulation technologies for shale gas production*' (Publications Office of the European Union, Luxembourg 2013) 10–50 (hereinafter: Gandossi).

200 Gandossi 49–53.

201 See Chapter 3 below on Savigny's 'canons' of interpretation.

202 Christine Parker and John Braithwaite 'Regulation' in Peter Cane and Mark Tushnet (eds) *The Oxford Handbook of Legal Studies* (Oxford University Press, 2003) 122 et sqq.

203 Broomfield 111.

This sentence captures the essence of the argument: the exemption of article 11 (3) (j) Water Framework Directive does not apply to 'flow back', because the latter is not only composed of substances that were originally present in the shale play.

Closer examination of the article, however, reveals that this definition is incorrect. The wording of article 11 (3) (j) Water Framework Directive has the purpose of precluding the disposal of substances which are not in any way associated with the extraction process.[204] This might be waste water from other industries or general waste that has nothing to do with the extraction of hydrocarbons.[205] It is, however, not the aim of the article to exclude the disposal of any substance that does, indeed, result from the extraction process. The natural meaning of the word 'resulting' in article 11 (3) (j) Water Framework Directive is 'as a result' and it therefore applies to the entire end-product of the process. 'Flow back' is a natural result of hydraulic fracturing.[206]

However, it has been argued that the 'flow back' contains remnants of the fracturing fluid and substances coming from the geological formations and may therefore not be considered as a substance resulting from the operations.[207] This suggestion appears to be rather odd, given that the liberation of substances in a geological formation is clearly the result of the hydrocarbons extraction process. The substances in the geological formations were liberated as a result of the treatment. To say that the existence of remnants from the shale gas extraction process is not a result of the shale gas extraction process defies logic.

Finally, a possible gap in the Water Framework Directive pertains to its monitoring requirements.[208] Given the potential threat of contamination that shale gas extraction poses to a precious resource like groundwater, the monitoring of aquifers in the vicinity of shale gas wells should always be required. Under paragraph 2 (4) (3) of Annex V Water Framework Directive, the monitoring should be carried out by Member States at least once a year for groundwater bodies which have been identified as being at risk of failing to meet the objective of good water status.[209] However, the identification of groundwater bodies which are at such a risk is carried out periodically every six years, according to articles 4 (3) (b) and 13 (7) Water Framework Directive. Hence, no monitoring is required for areas which

204 Van Rijswick 18/19 and 25/26; Meiners et al. Bund B 98.
205 Ibid.
206 See above Chapter 1.
207 Broomfield 111 at (ii).
208 A similar opinion has been put forward by Dannwolf et al. AP8-1.
209 As defined by article 4 Water Framework Directive.

have not been identified as being at risk from shale gas extraction within the last six years, a period of time during which the prospects of European shale gas extraction only started to evolve.[210] Potential threats of shale gas extraction might hence not be sufficiently included in current monitoring cycles.

To sum up, the Groundwater Directive does not directly apply to shale gas installations and may only have an indirect effect on the technology, as it is more concerned with monitoring the general quality of the groundwater. Although this is also true for the Water Framework Directive, this law creates a comprehensive framework, which compels Member States to actively manage their groundwater resources. Despite the overall sufficient legal protection that is provided by the Water Framework Directive against potential threats of shale gas extraction, a gap in the monitoring requirements of this Directive exists: the possibility of large-scale shale gas extraction in Europe only emerged 4–5 years ago,[211] but the identification of groundwater sources which might be at risk from pollution must only be carried out every six years.

Deep ground disposal of 'flow back', a crucial feature of shale gas extraction in some Member States,[212] has been explicitly endorsed by article 11 (3) (j) Water Framework Directive. Although three arguments have been advanced as to why this article does not cover shale gas extraction, none of them is convincing or coherent.[213]

2.3.1.7 REACH

Chemical products that are being used in shale gas operations must be regularly disclosed by companies and certified as safe for the specific use by the European Chemicals and Health Agency under Regulations (1907/2006) (hereinafter: REACH).[214] The explicit coverage of all chemical substances has been a key demand of the European Commission during the development of REACH.[215] In terms of the four main

---

210 Broomfield 96 and 109.

211 See Introduction.

212 For instance Germany, see Chapter 1 above.

213 This position is supported by an assessment of the German Environmental Ministry on German shale gas extraction, which arrived at the same conclusion, see Meiners et al. Bund B 98-B 100.

214 Regulation (EC) 1907/2006 of 18 December 2006 concerning the registration, evaluation, authorization and restriction of chemicals (REACH) [2006] OJ L 396/1; Potocnik 8.

215 Bjorn Hansen 'Background and Structure of REACH' in Lucas Bergkamp (ed.) *'The European Union REACH Regulation for Chemicals Law and Practice'* (Oxford University Press, 2013) paragraph 1.11 (hereinafter: Hansen).

potential threats of shale gas extraction, REACH could help to counter the risk of groundwater contamination. In order to achieve this goal, REACH holds two different groups of persons accountable for ensuring the safe use of chemicals: first, the producer or importer of a substance[216] and second, the downstream user of the substance,[217] the operator of a shale gas facility.

Manufacturers or importers must register any such substance which they produce in quantities of one tonne or more per year with the European Chemicals and Health Agency.[218] There must be an intention to release the substance into the environment under normal or reasonably foreseeable conditions of use.[219] The registration shall entail two main components: first, a technical dossier providing information on the producer/importer, the substance, its intended use, guidance on safe use and study summaries.[220] Second, a chemical safety report for the substance in question,[221] if the substance is produced or imported in quantities of 10 tonnes or more per year per registrant.[222] Below this threshold, a chemical safety report is only required in cases where thresholds of other Directives are exceeded, according to article 14 (2) REACH.

Although downstream users of chemicals do not have to register directly under REACH, they must make sure that the chemicals are fully registered by manufacturers and importers for their intended use (exposure scenarios). Otherwise downstream users are simply not allowed to utilize them.[223] Hence, REACH obliges the operator of a shale gas extraction facility (being the downstream user in this case) to assist the supplier in preparing a registration and to check compliance with exposure scenarios.[224] Their primary source of information for making this assessment is an extended safety data sheet that includes exposure scenarios, which has

---

216 Articles 3 (32) and 7 REACH; Lucas Bergkamp and DaeYoung Park 'The Organizational and Administrative Structures' in Lucas Bergkamp (ed.) '*The European Union REACH Regulation for Chemicals Law and Practice*' (Oxford University Press, 2013) paragraph 2.13 (hereinafter: Bergkamp).

217 Article 3 (34) and Title V REACH.

218 Articles 6 (1) and 7 (1) REACH.

219 Ibid.

220 Article 10 (a) REACH.

221 Article 10 (b) REACH.

222 Article 14 (1) REACH.

223 Dieter Drohmann and Matthew Townsend '*REACH Best Practice Guide to Regulation No 1907/2006*' (Beck, Muenchen 2013) 366 (hereinafter: Drohmann/Townsend).

224 Drohmann/Townsend 366/367.

to be provided by the suppliers of the substance.[225] However, downstream users are responsible on their own for taking the risk management measures, which are appropriate for their intended usage.[226] A downstream user shall prepare a chemical safety report on his own for any use outside of the exposure scenarios or for any use his supplier advises against, if the use of the substance would exceed one tonne per year.[227] The downstream user has to communicate this chemical safety report to the European Chemicals and Health Agency.[228]

Alternatively, the downstream user may notify his supplier about the intended use outside of the exposure scenarios and the supplier must then include this use in his exposure scenarios.[229] In any case, the substances may not be used in shale gas operations until these conditions are met.[230] The enforcement of these obligations is up to the Member States.[231] This means that a Member State can and must act if chemical substances are used outside of the scope of intended use or without registration,[232] for instance for unregistered shale gas extraction purposes.

Finally, article 118 REACH provides a stipulation for the disclosure of information. Information may not be made publicly available if this would undermine the protection of commercial interests of the person concerned.[233] However, this non-disclosure only pertains to certain information, like the full composition of a preparation or the precise tonnage of substances.[234] It is also important to note that this merely applies to disclosures to the public. The European Chemicals and Health Agency must be informed in any case.[235] In urgent situations, where the environment or human health is at risk, the Agency themselves may disclose sensitive, but relevant information to the public.[236]

---

225 Articles 31 (1) and 37 (4) and (5) REACH; Nicolas Herbatschek and Lucas Bergkamp and Meglena Mihova 'The REACH Programmes and Procedures' in Lucas Bergkamp (ed.) '*The European Union REACH Regulation for Chemicals Law and Practice*' (Oxford University Press, 2013) paragraph 4.97 (hereinafter: Herbatschek/Bergkamp/Mihova); Drohmann/Townsend 367.

226 Article 37 (5) REACH; Drohmann/Townsend 367/368.

227 Article 37 (4) REACH; Herbatschek/Bergkamp/Mihova paragraph 4.116.

228 Article 38 (1) REACH.

229 Drohmann/Townsend 367/368.

230 Herbatschek/Bergkamp/Mihova paragraph 4.117.

231 Articles 117 and 127 REACH.

232 Herbatschek/Bergkamp/Mihova paragraph 4.117/4.118.

233 Article 118 (2) REACH.

234 Ibid.

235 See article 118 (1) in relation with article 118 (2) REACH.

236 Article 118 (2) REACH.

To sum up, the provisions of REACH provide a comprehensive net of norms, which could help in countering the potential threat of groundwater contamination brought about by shale gas extraction. The Regulations provide an adequate control-mechanism for the use of chemicals and biocides that are associated with shale gas extraction.[237] The approach of REACH to target the manufacturer of a substance as well as, indirectly, its user, effectively covers all parties involved in shale gas extraction. Non-disclosure of chemicals to the regulator is not possible in Europe and REACH demonstrates that the interests of the public are valued higher than proprietary interests of corporations.

However, an in-depth study about the applicability of REACH on shale gas extraction concluded that three slight amendments to the current reporting praxis of REACH could be made.[238] This would increase the availability of information on use, exposure and risk management for shale gas extraction.[239] First, the industry could create a more specific, unequivocal term, which should be applied consistently to the use of substances for shale gas extraction. A review of current data sheets and exposure scenarios highlighted that these documents do not specifically address hydraulic fracturing or shale gas extraction.[240] Second, the current system of use descriptions under REACH may be complemented by an additional model description, covering the case of a substance that is intentionally introduced into the environment to carry out its technical function.[241] Third, the exposure scenarios may benefit from the development of a model that covers the direct introduction of substances into the underground and their possible upwards migration.[242]

While the first and the third issue have not yet been addressed, the European Chemicals Agency (ECHA), responsible for administering REACH, responded to the second issue in 2015 and introduced a standard label for the use name 'hydraulic fracturing for oil and gas exploration and extraction' and a corresponding product category.[243] This

---

[237] Stefania Gottardo et al. '*Assessment of the use of substances in hydraulic fracturing of shale gas reservoirs under REACH*' (Publications Office of the European Union, Luxembourg 2013) 8 (hereinafter: Gottardo et al.).

[238] Gottardo et al. 8.

[239] Ibid.

[240] Ibid.

[241] Ibid.

[242] Ibid.

[243] ECHA 'ECHA clarifies how to report substances used in hydraulic fracturing' available at: http://echa.europa.eu/documents/10162/21779840/annex+to+a+news+item+20150318.pdf [accessed: 02 September 2015) (hereinafter: ECHA update).

label covers substances that are typically used for hydraulic fracturing purposes.[244] ECHA encourages registrants to update their dossiers with this new use description if they intend to use substances for hydraulic fracturing.[245]

2.3.1.8 Biocidal Products Directive and Regulation 528/2012

In addition to the REACH Regulations, biocidal products have, since 1998, been governed by a specific Directive on biocidal products (hereinafter: Biocidal Products Directive).[246] This Directive, just like REACH, has also had a very strict authorisation process in place, applied to some 15 000 products that were grouped into 23 categories.[247] However, the Biocidal Products Directive was replaced on 1 September 2013 with Regulation Number 528/2012 (hereinafter: Regulation 528/2012).[248] Despite this recent replacement, the authorization procedure of the Biocidal Products Directive remains unaltered under the new Regulations.[249]

In the case of shale gas extraction, the Biocidal Products Directive, and now Regulation 528/2012, govern the use of biocidal products in fracturing fluids.[250] Such products are being used where biota needs to be killed off to ensure efficient gas extraction.[251] Biocidal products are mainly used in shale gas extraction processes to combat slurry that might build at the drill head.[252] Only if the biocidal products have been authorized in a special procedure, may they be placed on the market and used for the purpose of shale gas extraction, according to articles 4 and 17 (1) Regulation 528/2012.

Even if a substance, like the fracturing fluid, only contains a minor amount of a biocidal product,[253] it needs to be approved of by the relevant authorities; authorizations may be obtained for the period of ten years.[254] According to article 19 (1) (b) (iv) of Regulation 528/2012, a biocidal

---

244 ECHA update.

245 Ibid.

246 Council Directive (EC) 98/08 of 16 February 1998 concerning the placing of biocidal products on the market [1998] OJ L 123.

247 Ludwig Krämer *'EU Environmental Law'* 7th edition (Sweet & Maxwell, London 2011) paragraphs 6–41 (hereinafter: Krämer EU Environmental Law).

248 Regulation (EU) No 528/2012 of 22 May 2012 concerning the making available on the market and use of biocidal products [2012] OJ L 167/1.

249 Broomfield 113.

250 Ibid.

251 Meiners et al. Bund A 62.

252 Meiners et al. Bund B 38/39.

253 See above Chapter 1 for the composition of the fracturing fluid.

254 Article 4 (1) Regulation 528/2012.

substance shall only be approved of if an evaluation establishes that the product has no unacceptable effects on the environment.

Once authorized, the Member States are obliged to monitor the use of the biocidal product and undertake controls.[255] From 1 September 2015 onwards, Member States are obliged to issue to the European Commission a comprehensive report on the information obtained during these investigations.[256] Article 66 (1) Regulation 528/2012 deals with confidentiality and establishes that information obtained by these monitoring measures will be made publicly available on request. However, the information may be withheld by the European Commission in cases where disclosure would undermine the protection of the commercial interests or the privacy or safety of the persons concerned.[257] Similarly to article 118 (2) REACH, article 66 (3) Regulation 528/2012 orders that the regulator may, nonetheless, disclose withheld information to the public, in cases of urgency.[258]

Given that the provisions of Regulation 528/2012 are generally in line with those of the REACH Regulation, no other overall assessment is justified here than the one that has been made above with regard to REACH. Thus, the coverage of biocidal products in shale gas extraction operations by the pertaining EU Regulation is considered comprehensive and effective.

2.3.1.9 Environmental Liability Directive

Adopted in 2004, the main aim of Directive 2004/35/EC (hereinafter: Environmental Liability Directive)[259] is to make Member States responsible for ensuring that damage to water, land and biodiversity is either

---

255 Article 65 (1) and (2) Regulation 528/2012.

256 Article 65 (3) Regulation 528/2012.

257 Article 66 (2) Regulation 528/2012.

258 Article 66 (3) Regulation 528/2012.

259 Council Directive (EC) 2004/35 of 21 April 2004 on environmental liability with regard to the prevention and remedying of environmental damage [2004] OJ L 143/56 (hereinafter: Environmental Liability Directive). The Directive has been amended three times since by Council Directive (EC) 2006/21 of 15 March 2006 on the management of waste from extractive industries and amending Directive 2004/35/EC [2006] OJ L 102/15; Council Directive (EC) 2009/31 of 23 April 2009 on the geological storage of carbon dioxide and amending Council Directive 85/337/EEC, European Parliament and Council Directives 2000/60/EC, 2001/80/EC, 2004/35/EC, 2006/12/EC, 2008/1/EC and Regulation (EC) No 1013/2006 [2009] OJ L 140/63; Council Directive (EU) 2013/30 of 12 June 2013 on the safety of offshore oil and gas operations and amending Directive 2004/35/EC [2013] OJ L 178/66.

prevented and/or remedied.[260] The Environmental Liability Directive thus has a dual function. *Prior* to a possible incident, preventive measures against an imminent threat of environmental damage have to be implemented. *After* an incident where environmental damage occurred it demands the taking of necessary restorative measures.[261]

Article 3 (1) Environmental Liability Directive distinguishes between environmental damage and imminent threats which are caused by an activity that is listed in Annex III Environmental Liability Directive and those that are not. In the latter case, the Directive shall only apply if the damage pertains to endangered species and habitats and if the operator has been at fault or negligent.[262] In the former case the Directive applies regardless of fault or negligence.[263]

Strict liability applies to environmental damage caused by the operation of any of the activities listed in Annex III to the Directive.[264] Those activities that are referred to in Annex III Environmental Liability Directive and which are most salient for shale gas extraction are those covered by the IPPC Directive (paragraph 1 Annex III), the Water Framework Directive (paragraph 5 Annex III) and the Mining Waste Directive.[265] The aspects of shale gas extraction coming under one of these three Directives would thus be covered by the strict liability regime of the Environmental Liability Directive.[266] This is most likely the case for groundwater issues and issues with the 'flow back', falling into the scope of the Mining Waste Directive and the Water Framework Directive respectively.

Fault liability occurs when damages to protected species and natural habitats are caused by the operation of any occupational activity, other than those listed in Annex III and/or any imminent threat of such damage, whenever the operator was at fault or negligent.[267] This is of particular importance for shale gas extraction. The Environmental Liability Directive only covers environmental damage of diffuse character, if a

---

260 Article 1 Environmental Liability Directive; Barbara J Goldsmith and Lucas Bergkamp '*The EU Environmental Liability Directive: A Commentary*' (Oxford University Press, 2013) paragraph 00.03 (hereinafter: Goldsmith/Bergkamp).

261 Goldsmith/Bergkamp paragraph 00.04.

262 Goldsmith/Bergkamp paragraph 03.01.

263 Ibid.

264 Article 3 (1) (a) Environmental Liability Directive; Goldsmith/Bergkamp paragraph 03.02.

265 See the new paragraph 13 Annex III Environmental Liability Directive that has been inserted by article 15 Mining Waste Directive.

266 For a similar opinion, see Broomfield 94.

267 Article 3 (1) (b) Environmental Liability Directive; Goldsmith/Bergkamp paragraph 03.03.

causal link between the damage and the activities of individual operators may be established.[268] This will, most likely, be an issue in shale gas cases, given that air emissions and overbearing land use might have several sources and may hence be classified as diffuse environmental damage.[269]

To sum up, the Environmental Liability Directive applies to all of the four main potential environmental threats of shale gas extraction. However, it only orders a strict liability regime for groundwater issues and the deep-ground injection of 'flow back', as these issues are covered by the Mining Waste and the Water Framework Directive. The other two potential threats, greenhouse gas emissions and overbearing use of land, are also covered by the Directive, but are not subject to a strict liability regime. This might be perceived of as a gap in the legislation, especially with regard to air pollution, which is of diffuse character and proof of fault or negligence could be hard to establish.

2.3.1.10 Air Quality Directive

Air pollution by greenhouse gas emission is a potential threat of shale gas extraction to which Directive 2008/50/EC (hereinafter: Air Quality Directive)[270] might be applicable. But there are also other Directives covering emissions into the air, for instance the EIA Directive and the IPPC/IED.[271] However, as discussed above, the applicability of the IPPC/IED is unlikely and the EIA Directive may only be applied in singular cases. Therefore, coverage of air emissions by other Directives is a worthwhile consideration.

The Air Quality Directive establishes common methods, criteria and objectives for assessing ambient air quality in Member States.[272] The Directive sets limit values of air polluting substances in ambient air.[273] Article 13 Air Quality Directive obliges Member States to ensure that the amount of certain pollutants does not exceed an 'alert threshold', which is set in Annex XI Air Quality Directive. Where emissions exceed these limits, Member States are obliged to act by article 19 Air Quality Directive in conjunction with Annex XII Air Quality Directive.[274] However, according to article 19 Air Quality Directive, the required actions merely amount

---

[268] Article 4 (5) Environmental Liability Directive.

[269] Broomfield 94.

[270] Council Directive (EC) 2008/50 of 21 May 2008 on ambient air quality and cleaner air for Europe [2008] OJ L 152/1.

[271] Krämer EU Environmental Law paragraph 8-02-8-05.

[272] Article 1 Air Quality Directive.

[273] Articles 13 and 14 Air Quality Directive.

[274] Krämer EU Environmental Law paragraph 8-07.

to informing the public and the European Commission about the level of emissions.

Actual remedies for the observed bad air quality are described in chapter IV of the Air Quality Directive and include air quality plans[275] and short term plans,[276] which have to be established and implemented by Member States. However, there are no unified measures in the Directive that have to be taken under these air quality and short term plans. Thus, the measures have to be decided upon individually by the Member States.[277]

Since the Air Quality Directive deals with ambient air emissions and not with emissions from specific sites[278] it is not directly applicable to shale gas extraction, but could have an indirect effect. This indirect effect, however, depends on the sort of measures that Member States are willing to take. This high level of discretion might or might not be desirable, but it does not constitute a gap in the European environmental law framework as such.[279]

2.3.1.11 Emissions Trading System

In 2003, the European Emissions Trading System Directive (hereinafter: ETS Directive)[280] established a system for the trading of Greenhouse Gas Emissions (hereinafter: GHG emissions) allowances, which became operational in 2005 and has been amended several times since.[281] Article 2 ETS Directive requires that any stationary installation, which carries out one of the activities listed in Annex I of the Directive, must hold a valid GHG emission permit. Such a permit holder must surrender allowances each year equivalent to the number of tonnes of carbon dioxide that the installation emits.[282]

Annex I of the original ETS Directive was replaced in 2009 with a new, more detailed version.[283] However, neither this, nor the older version lists an activity that would directly relate to shale gas extraction. Only indirect

---

275 Article 23 Air Quality Directive.

276 Article 24 Air Quality Directive.

277 Krämer EU Environmental Law paragraph 8-07.

278 Ibid.

279 For a similar result, see Broomfield 100.

280 Council Directive (EC) 2003/87 of 13 October 2003 establishing a scheme for greenhouse gas emission allowance trading within the Community and amending Council Directive 96/61/EC [2003] OJ L 275/32.

281 Stefano Clò '*European Emissions Trading in Practice*' (Edward Elgar Publishing, Cheltenham 2011) 2/3 and 59 (hereinafter: Clò).

282 Article 12 ETS Directive and Clò 63.

283 Article 1 No 30 of Council Directive (EC) 2009/29 of 23 April 2009 amending Directive 2003/87/EC so as to improve and extend the greenhouse gas emission

application would pertain to Annex I ETS Directive, if a shale gas plant combust fuels and this fuel combustion exceeds a total rated thermal input of 20 MW.[284] This is theoretically possible, because No 3 of Annex I establishes that the rated thermal input is calculated by adding together the thermal output of all technical units of an installation in which fuels are combusted.[285] The Annex enumerates several units of power facilities in this regard.[286] For shale gas extraction plants, combustion by engines and flares could be of particular relevance.

However, whether or not the combined combustion capacity of these technical units of a shale gas plant is likely to exceed the level of 20 MW in practice is not known. Therefore, two studies, commissioned by the European Commission, concluded independently from each other in 2012[287] and 2014[288] that it is unlikely that shale gas extraction would be covered by the ETS Directive.[289]

However, this current state of the law is rather unsound as the ETS Directive has already been amended once to include a new, recently evolving technology in the system, Carbon Capture and Storage.[290] Given the similarities between shale gas extraction and Carbon Capture and Storage, which will be discussed comprehensively below in Chapter 8, it would be reasonable to take similar action and include the former in the ETS-Directive.[291] The current non-inclusion might be considered as a shortcoming of the existing EU regulatory framework that governs shale gas extraction.

2.3.1.12 Habitats Directive

European wildlife is protected under Directive 92/43/EEC (hereinafter: Habitats Directive)[292] and Directive 2009/147/EC (hereinafter: Birds

---

allowance trading scheme of the Community [2009] OJ L 140/63 (hereinafter: ETS Amendment 2009 Directive).

284 Annex I ETS Amendment 2009 Directive.

285 No 3 Annex I ETS Amendment 2009 Directive.

286 Ibid.

287 Forster/Perks 77.

288 ICF International Ltd. '*Mitigation of climate impacts of possible future shale gas extraction in the EU: available technologies, best practices and options for policy makers*' (2014) available at: http://ec.europa.eu/environment/integration/energy/uff_studies_en.htm [accessed 4 September 2014] (hereinafter: ICF).

289 Forster/Perks 77; ICF 44.

290 Annex I ETS Amendment 2009.

291 A similar conclusion has been reached by ICF 44.

292 Council Directive (EC) 92/43 of 21 May 1992 on the Conservation of natural habitats and of wild fauna and flora [1992] OJ L 206/7.

Directive).[293] Both Directives are the 'backbone' of Europe's nature conservation law.[294] These legal instruments aim to protect conservation areas on the one hand[295] and particular species on the other hand.[296] They are of interest in relation to shale gas extraction and the potential issue of its impact on the use of land.

The main provisions of the Habitats Directive[297] are formulated in a strict way and leave Member States without any real discretion.[298] Among the key measures to protect certain areas is the designation of special areas of conservation, with the aim to create the Natura 2000 network of specifically protected sites which extends all over Europe.[299] Activities in Natura 2000 sites shall only be carried out if a prior determination of impacts on the site asserts that the site's integrity will not be affected by the project.[300] Article 6 (4) Habitats Directive, however, provides that, in cases of overriding public interests, Member States might exceptionally endorse the activity in question if the loss of nature is being compensated for in other areas.

In addition, according to articles 12 and 16 of the Habitats Directive, the Directive might also apply outside of Natura 2000 sites, if the breeding sites or resting places of protected species might be affected by the project. However, just as with Natura 2000 sites, exceptions apply and intrusions into, for example, breeding sites are possible under specific circumstances;[301] these exemptions are listed in article 16 Habitats Directive.

Whether or not Natura 2000 sites or other breeding sites of protected species are affected by shale gas projects cannot be determined in general. This depends on the actual location of individual shale gas projects. If none of the described sites is affected by a shale gas project, the Natura 2000 legislation does not apply to shale gas extraction.

---

293 Council Directive (EC) 2009/147 of 30 November 2009 on the conservation of wild birds [2009] OJ L 20/7.

294 Saskia van Holten and Marleen van Rijswick 'The governance approach in European Union environmental directives and its consequences for flexibility, effectiveness and legitimacy' in Marjan Peeters and Rosa Uylenburg (eds) '*EU Environmental Legislation*' (Edward Elgar, Cheltenham 2014) 24 (hereinafter: Van Holten/Van Rijswick).

295 Articles 3–11 Habitats Directive; Van Holten/Van Rijswick 24.

296 Articles 12–16 Habitats Directive; Jans/Vedder 459.

297 Articles 12, 13 and 15, according to Van Holten/Van Rijkswick 24.

298 Van Holten/Van Rijkswick 24.

299 Jans/Vedder 459.

300 Article 6 (3) Habitats Directive.

301 Van Holten/Van Rijkswick 25.

2.3.1.13 Hydrocarbons Licensing Directive

Directive 94/22/EC (hereinafter: Hydrocarbons Licensing Directive)[302] is a key measure for the regulation of gas extraction[303] which has to be discussed in the shale gas context. However, in contrast to the pieces of legislation considered so far, the Hydrocarbons Licensing Directive does not have environmental protection as its main aim.[304] Instead, it focuses on the proper functioning of the internal market.[305] The Directive wants to combat discriminatory practices by Member States which restricted access by foreign undertakings to prospecting, exploration and production of hydrocarbons in the past.[306] However, care had to be taken not to impinge upon Member States' rights to determine their own energy policies.[307] Thus, the resulting Directive is very much of a framework nature, in the sense that it sets broad principles rather than concrete rules concerning access to hydrocarbons.[308]

The focus on the energy-sovereignty of Member States is reflected by article 2 (1) of the Directive, which asserts that Member States are free to decide which areas of their territory should be opened up for hydrocarbon extraction. Articles 4 and 5 (1) Hydrocarbons Licensing Directive lay down technical and financial capability of an entity as general requirements for the granting of licences.[309] It could be useful to discuss whether the environmental capability of an entity should be inserted alongside these two criteria, as a third criterion for the issuance of licences. This third criterion would then be applicable to the potential environmental threats of shale gas extraction.

Against such an insertion it could be argued that the ability to prevent environmental devastation could be subsumed under the term 'technical feasibility' and certain environmental protection aspects might already be entailed in the current procedures. The bottom line, however, is that the Hydrocarbons Licensing Directive cannot be blamed for a failure to address potential threats of shale gas extraction, as its main task is to ensure fair competition in the internal market.[310]

---

302 Council Directive (EC) 94/22 of 30 May 1994 on the conditions for granting and using authorizations for the prospection, exploration and production of hydrocarbons OJ L 164/3.

303 Johnston/Block paragraph 15.02/03.

304 Ibid.

305 Ibid.

306 Johnston/Block paragraph 15.03.

307 Ibid.

308 Ibid.

309 Johnston/Block paragraph 15.08/15.09.

310 Broomfield 85.

However, article 6 (2) Hydrocarbons Licensing Directive allows Member States to attach conditions to the exercise of authorizations if this is justified, inter alia, by environmental protection needs. This provision makes it possible for Member States to draft authorization conditions that are aimed at preventing or mitigating environmental impacts.[311] Environmental considerations are hence not absent from the Hydrocarbons Licensing Directive, but may only come into play indirectly as additional criteria.

Thus, in summary it appears that the Hydrocarbons Licensing Directive is not an apt legal tool to regulate any of the potential environmental threats of shale gas extraction, because it focuses on ensuring fair competition in the internal market.[312] This is not a shortcoming of the EU's regulatory framework, but rather a logical consequence of the ways in which it is built. Although it might be argued that environmental aspects should play a bigger role in the issuing of licences, it lies beyond the scope of this book to investigate that point.

2.3.1.14 Internal Gas Market Directive

In 2009 the European Union adopted the Internal Gas Market Directive 2009/73/EC.[313] This Directive forms one important component of the so called 'Third Energy Package' of the EU and its essence is the introduction of freedom of choice to engage in the business of and trade in energy.[314] This includes the choice to build power plants and pipelines, the choice to export and import energy and the choice to select and negotiate with suppliers, shippers and customers.[315] Talus describes this as market opening, which has as its *conditio sine qua non* the elimination of exclusive rights.[316] That opening and improved integration of the European gas market is counterbalanced by several provisions that shall ensure the security of energy supplies in this open market.[317]

Arguably, the Internal Gas Market Directive has not so much of an impact on shale gas as shale gas has on the Internal Gas Market Directive. This inversion of the common correlation has been aptly described by Talus.[318] He explains that gas production, unlike electricity generation,

---

311 Ibid.

312 Same result reached by Musialski EU Chapter 69.

313 Already discussed in a different context above in Chapter 1.

314 Talus 2013 at 66.

315 Ibid.

316 Ibid.

317 For instance merchant exemptions for new infrastructure, monitoring mechanisms and Public Service Obligations, see Talus 2013 at 106/107.

318 Talus 2013 at 67.

is not common within the EU, as the EU is rather dependent on a very small number of external suppliers; this results in the need to attract alternative sources of natural gas. Shale gas is an opportunity for new suppliers to enter the market, which could facilitate endeavours to increase competition.[319]

However, article 1 (1) of the Internal Gas Market Directive specifies the scope of the Directive and clarifies that the Directive 'establishes common rules for the transmission, distribution, supply and storage of natural gas.'[320] The Internal Gas Market Directive is thus relevant to shale gas (or vice versa), mainly in terms of marketing of the gas. However, the Directive does not have a direct impact on the regulation of shale gas *extraction* as such and does not address its four main potential environmental threats. As these form the focus of the current work, the Internal Gas Market Directive, being an instrument for downstream regulation,[321] will not be discussed further in the current context of shale gas extraction. The same is true for the Gas Security Regulation.[322]

2.3.2 The Evolution of Specific European Shale Gas Regulation

Several EU bodies, including the European Parliament and the Committee of the Regions,[323] since 2012, put pressure on the European Commission to introduce firm legal action on shale gas.[324] Their demands were

---

[319] Ibid.

[320] Article 1 (1) Internal Gas Market Directive.

[321] See above and Talus 2013 at 108.

[322] See Chapter 1 above.

[323] An advisory body representing local and regional authorities in the European Union, see European Union 'Committee of the Regions' available at: http://europa.eu/about-eu/institutions-bodies/cor/index_en.htm [accessed 5 September 2014].

[324] European Parliament '*Resolution of 21 November 2012 on the environmental impacts of shale gas and shale oil extraction activities* (2011/2308(INI))' paragraph 4 available at: http://www.europarl.europa.eu/sides/getDoc.do?pubRef=-//EP//NONSGML+TA+P7-TA-2012-0443+0+DOC+PDF+V0//EN [accessed 23 April 2014]; European Parliament '*Resolution of 21 November 2012 on industrial, energy and other aspects of shale gas and oil* (2011/2309(INI))' paragraph 2 available at: http://www.europarl.europa.eu/sides/getDoc.do?type=TA&reference=P7-TA-2012-0444&language=EN [accessed 4 September 2014]; EU's Assembly of Regional and Local Representatives '*Draft Opinion of the Committee of the Regions local and regional authorities perspective on shale/tight gas and oil (unconventional hydrocarbons)*' paragraphs 8–11 available at: http://cor.europa.eu/en/news/Pages/fracking-environmental-impact.aspx [accessed 4 September 2014].

underpinned by a couple of scientific reports on shale gas, in favour of their demand.[325] By 2014 the Commission responded to that request and put into place a new framework, which is specifically designed to regulate the extraction of hydrocarbons.[326]

The framework consists of two components, a Communication and a Recommendation, which should be read together.[327] The main act is the Recommendation on exploration and production of shale gas (hereinafter: 2014 Shale Gas Recommendation).[328] This Recommendation has been supplemented by a Communication on the exploration and production of hydrocarbons (such as shale gas) using high-volume hydraulic fracturing in the EU (hereinafter: 2014 Shale Gas Communication).[329] The Recommendation not only targets shale gas extraction, but also every kind of unconventional gas extraction that uses 'high-volume hydraulic fracturing'.[330] The term means the injection of 1 000 m^3 or more of water per fracturing stage or 10 000 m^3 or more of water during the entire fracturing process into a well.[331]

The overall aim of the Commission's unconventional extraction framework is threefold. It wishes

---

325 AMEC Ltd. '*Technical Support for Assessing the Need for a Risk Management Framework for Unconventional Gas Extraction*' (2014) available at: http://ec.europa.eu/environment/integration/energy/uff_studies_en.htm [accessed 4 September 2014] viii, xiii, xiv, 100/101, 104/105 and 107; ICF International Ltd. '*Mitigation of climate impacts of possible future shale gas extraction in the EU: available technologies, best practices and options for policy makers*' (2014) available at: http://ec.europa.eu/environment/integration/energy/uff_studies_en.htm [accessed 4 September 2014] 2/3 and 88/89; Broomfield xv; Lechtenböhmer et al. 9/10 and 78/79; Gottardo 8 and 42.

326 However, its main aim is shale gas extraction, which is already made clear by the title of the relevant documents. In the titles of both, the Recommendation and the Communication, shale gas is the only form of energy that has been explicitly mentioned, which highlights its importance.

327 2014 Commission Shale Gas Communication page 4.

328 Commission Recommendation 2014/70/EU of 22 January 2014 on minimum principles for the exploration and production of hydrocarbons (such as shale gas) using high-volume hydraulic fracturing [2014] OJ L 39/72.

329 Commission 'Communication on the exploration and production of hydrocarbons (such as shale gas) using high volume hydraulic fracturing in the EU' (Communication) *COM (2014) 23 final/2* (hereinafter: 2014 Commission Shale Gas Communication).

330 See title of both documents.

331 Article 2 (a) Commission Shale Gas Recommendation. For reasons of coherence and readability, the author will use the generic term 'shale gas extraction' when referring to 'high-volume hydraulic fracturing'.

> – to ensure that opportunities to diversify energy supplies and improve competitiveness can be safely and effectively taken up in those Member States that choose to do so,
> – to provide clarity and predictability for both market operators and citizens, including for exploration projects,
> – to fully consider greenhouse gas emissions and management of climate and environmental risks, including to health, in line with public expectations.[332]

Because of these aims, the Recommendation and the Communication can be seen as an attempt by the Commission to strike a balance between energy security interests and environmental protection needs. The indirect acknowledgement that existing European legislation has not yet succeeded in striking that balance constitutes a remarkable U-turn by the European Commission. Two years earlier, in 2012, the then Environmental Commissioner, Janez Potočnik, stated that existing EU legislation covered shale gas projects comprehensively 'from planning until cessation'.[333] Only two years later, the Commission acknowledged gaps in the existing legislation, by stating in the 2014 Shale Gas Communication that

> Since the EU environmental legislation was developed at a time when high volume hydraulic fracturing was not used in Europe, certain environmental aspects associated with the exploration and production of fossil fuels involving this practice are not comprehensively addressed in current EU legislation.[334]

The Commission aims to close existing gaps. The framework, however, takes care not to impose a particular approach to shale gas regulation on Member States. Instead, it wants to create a level playing field for those states that decide to take a permissive approach to shale gas extraction.[335] The framework encourages these Member States to apply a set of protective measures as common minimum regulatory standards in addition to existing EU legislation.[336]

These protective measures address a number of the gaps in EU legislation that have been identified earlier in this chapter. The individual measures will now be scrutinized with a view to their effectiveness in closing such gaps.

First of all, the Recommendation urges Member States to carry out a

---

332 2014 Commission Shale Gas Communication page 3.
333 Potocnik Transmission 6.
334 2014 Commission Shale Gas Communication page 8.
335 Preamble 9 Commission Shale Gas Recommendation; 2014 Commission Shale Gas Communication page 4.
336 Article 1 and Preamble 5 and 9 Commission Shale Gas Recommendation.

Strategic Environmental Assessment (SEA).[337] This SEA should always be carried out prior to issuing a licence that *may* lead to the use of 'high volume hydraulic fracturing'. By taking such a comprehensive approach, the Commission clarifies that it deems the carrying out of an SEA by Member States necessary, a point that was not clear under hitherto existing SEA legislation.

Particular attention has to be payed to the fact that the term 'high volume hydraulic fracturing' is consistently used throughout the 2014 Recommendation and Communication, but is not defined in these documents. The term 'High Volume Hydraulic Fracturing (HVHF)' has been coined in the USA:

> As compared to (. . .) smaller volume hydraulic fracturing activities, there are additional concerns associated with HVHF well completions. For example, well site activity for HVHF well completions is normally more intense. There is overall more equipment, personnel, and truck traffic during the initial few months of development. The hydraulic fracturing process is usually a one-time event for that particular well, so once the initial well completion has occurred, the site activity (equipment, personnel, and truck traffic) is reduced and any further activity does not significantly differ from typical oil and gas development. Additional concerns with HVHF include:
>
> - Larger overall water usage for well completions.
> - Increased use of chemical additives.
> - Larger initial volumes and handling of flowback fluids.[338]

The Commission addresses the issue of Environmental Impact Assessments (EIAs) for individual shale gas projects in a way that is similar to how it addresses SEAs and calls upon Member States to take the necessary measures to ensure that an EIA is carried out for each project.[339] However, by simply passing the ball to Member States, the EU did not adequately address the main gap in EU EIA legislation, that EIAs are not *obligatory* for shale gas projects under the EIA Directive. Instead the Commission is calling upon Member States to resolve the issue. They shall exercise their *discretionary* powers to order an EIA for every individual shale gas project. If the Commission, as is apparent from the Recommendation, takes the view that an EIA should be carried out for every shale gas project, it should have inserted shale gas extraction into Annex I of the

---

337 Article 3.1 Commission Shale Gas Recommendation.

338 Department of Environmental Quality Michigan 'Hydraulic Fracturing in Michigan' available at: http://www.michigan.gov/deq/0,4561,7-135-3311_4231-262172--,00.html [accessed 30 June 2016].

339 Article 3.3 Commission Shale Gas Recommendation.

EIA Directive. Thus, this gap in the EIA Directive has not been adequately addressed by the framework.

With regard to the IPPC Directive and the IED, the preamble of the Recommendation merely mentions that they apply to shale gas extraction insofar as specific parts of the shale gas extraction process fall under Annex I IED.[340] However, that assessment does not clarify the main question which specific part of Annex I could be applicable to shale gas extraction. Thus, the issue of the overall application of the IPPC and the IED has not been addressed adequately.

The 2014 Shale Gas Communication includes a pledge by the EU to look into the issue of a specific Best Available Technique Reference Document (BREF) for shale gas extraction.[341] It generally urges Member States to ensure that shale gas operators use the Best Available Techniques (BAT) for shale gas operations.[342] Concrete action by the EU in this regard shall involve the review of the existing BREF on extractive waste under the Mining Waste Directive.[343] This action is designed to ensure that waste is appropriately handled and treated and the risk of water, air and soil pollution is minimized.[344] Since the non-existence of a BREF on shale gas extraction has been identified in the analysis above as a gap in the existing Mining Waste Directive, this action could be adequate to address that shortcoming. The same is true for a proposed new Hydrocarbons BREF, as discussed above. However, it has to be emphasized that the framework itself does not introduce such BREFs, but merely includes a vow to look into the matter in the future.

The issue that the Seveso Directives do not directly apply to shale gas extraction is addressed by the framework in the same way in which the application of the IPPC Directive and IED have been treated. The 2014 Shale Gas Recommendation merely mentions that the Seveso Directives apply if the thresholds determined in Annex I of the Seveso III-Directive are exceeded.[345] However, this does not clarify if and under which exact circumstances shale gas projects meet these thresholds and would, accordingly, be covered by the Seveso Directives. Thus, the issue of the overall application of the Seveso Directives to shale gas extraction has not been settled.

The 2014 framework, however, addresses the issue of water management under the Water Framework Directive and the Groundwater Directive. It

340 Preamble 7 Commission Shale Gas Recommendation.
341 2014 Commission Shale Gas Communication page 9.
342 Article 9.1 Commission Shale Gas Recommendation.
343 2014 Commission Shale Gas Communication page 9.
344 Ibid.
345 Preamble 7 Commission Shale Gas Recommendation.

reinforces the, currently prevailing, view that direct discharges of pollutants into the groundwater are prohibited[346] and calls upon Member States to urge operators to use as little water as possible in shale gas extraction processes.[347] Furthermore, it asks Member States to establish regulation for minimum distances between a shale gas site and a water protection area and determine minimum depth limitations between the fracturing area and groundwater to avoid pollution and contamination.[348] Crucially, 'baseline studies' of shale gas sites with regard, inter alia, to water, soil, and air quality shall be carried out[349] and the influence of shale gas extraction on these elements should be monitored consistently.[350] Since the latter, in particular, has been identified above as a gap in both Directives, the framework can be said to tackle these gaps.

The use of chemicals during shale gas extraction is of considerable concern to the Commission. The framework asks Member States to ensure that as few chemicals as possible are used in shale gas operations.[351] If chemicals are used, information about their nature must be disseminated amongst the public.[352] The framework addresses the current shortcoming in the REACH regulations that no consistent, unified term for the use of substances for shale gas extraction exists.[353]

Moreover, the Commission vows to propose to the European Chemicals Agency certain changes in the existing database of registered chemicals under REACH to facilitate the search of information on substances used for hydraulic fracturing purposes.[354] Overall, the framework addresses the identified gaps in REACH appropriately.

Furthermore, the framework dedicates one article to environmental liability issues.[355] It calls upon Member States to apply the provisions on environmental liability to all activities that are taking place at a shale gas extraction site.[356] This request explicitly includes those activities that do not currently fall into the scope of the Environmental Liability Directive.[357] This clarification could remedy the situation that

346 Article 5.4 Commission Shale Gas Recommendation.
347 Articles 10.2, 9.2 and 9.3 Commission Shale Gas Recommendation.
348 Article 3.2 Commission Shale Gas Recommendation.
349 Article 6 Commission Shale Gas Recommendation.
350 Articles 11.2 and 11.3 Commission Shale Gas Recommendation.
351 Articles 10.1 (b) and 10.2 Commission Shale Gas Recommendation.
352 Article 15 Commission Shale Gas Recommendation.
353 Article 10.1 (a) Commission Shale Gas Recommendation.
354 2014 Commission Shale Gas Communication page 9.
355 Article 12 Commission Shale Gas Recommendation.
356 Article 12.1 Commission Shale Gas Recommendation.
357 Ibid.

the Environmental Liability Directive does not include strict liability for greenhouse gas emissions and overbearing use of land.

As discussed above, the regulation of greenhouse gas emissions from shale gas extraction was one of the three main purposes of the 2014 framework.[358] Measured against this aspiration, the framework does not deliver sufficiently. Although greenhouse gas emissions are now included in the strict liability regime, the issue that shale gas extraction as an activity is not covered by the emissions trading system has not been resolved and is not addressed by the framework.

Finally, the framework includes a number of stipulations that relate to the issuing of licences and pertain to the Hydrocarbons Licensing Directive. The framework urges Member States to ensure that their national regulatory bodies are efficiently co-ordinated[359] and have adequate resources at their disposal[360] to carry out their tasks of appraising applications for shale gas licences.

2.3.3 Critical Assessment of the 2014 Shale Gas Framework

The crucial aspect of the Commission's 2014 framework on shale gas extraction is not so much its actual content, but its legal guise as soft law.[361] The adoption of a Recommendation and a Communication on shale gas extraction is in line with article 288 TFEU, which outlines the acts of secondary EU law. Possible actions available to the EU legislator include the issuance of Regulations, Directives, Decisions, Recommendations and Opinions.[362] While Regulations, Directives and Decisions are legally binding to varying degrees, according to article 288 TFEU Recommendations and Opinions shall have no binding force. A Recommendation thus has no immediate binding effect upon Member States.[363] It constitutes a form of soft law[364] and the EU may not enforce Recommendations directly.[365]

However, this does not mean that Recommendations have no legal effect: the ECJ decided that national courts must take them into account, especially where the Recommendation helps to clarify national provi-

358 2014 Commission Shale Gas Communication page 7.
359 Articles 4 and 5 Commission Shale Gas Recommendation.
360 Article 13 Commission Shale Gas Recommendation.
361 Reins Minimum Principles 26/27; Stokes 47/48.
362 Article 288 TFEU.
363 Craig/De Burca 107.
364 Ibid.
365 Calliess/Ruffert article 288 AEUV paragraph 95.

sions.[366] Thus, a Recommendation is a specific tool for the interpretation of national law, directed to national courts and administrative authorities that are involved in formal legal processes related to shale gas extraction.[367] Member States are not obliged to comply with a Recommendation.[368] They maintain the right to pick their own approach to shale gas regulation. The EU merely *recommends* a permissive, but protective approach.

This EU approach to choose a legally non-binding tool for shale gas regulation attracted intense criticism from environmental pressure groups.[369] The European Commission itself initially put forward four different policy options for EU shale gas regulation, ranging from the now adopted Recommendation to a firm, legally binding Directive.[370] The Commission prepared an impact assessment to outline repercussions of all four options, which concluded that a Recommendation 'with its non-binding character, is the least effective of the policy options analysed. Clarification of the legal requirements for shale gas operations would provide a more secure environment for investment and therefore enable shale gas developments.'[371]

The fear is that individual states could simply ignore the recommended measures if they do not fit in with their respective agenda on shale gas extraction.[372] After all, no sanctions for non-compliance with these measures have been outlined in the Recommendation or Communication. The adoption of a legally non-binding Recommendation will hardly lead to stringent shale gas regulation among European countries. Instead of creating a level playing field for all Member States, as claimed by the 2014 Shale

---

366 Case C-322/88 *Salvatore Grimaldi v Fonds des maladies professionnelles* [1989] ECR I-4407paragraph 18.

367 Michael Tarka '*The legal consequences of European Commission recommendations on minimum principles for shale gas in Poland*' available at: http://www.shale-gas-information-platform.org/categories/legislation/expert-articles/tarka-article-poland.html [accessed 18 December 2014] (hereinafter: Tarka).

368 Calliess/Ruffert article 288 AEUV paragraph 95.

369 See for instance Friends of the Earth '*Europe opens doors to dangerous fracking*' available at: http://www.foeeurope.org/shale_gas_framework_220114 [accessed: 18 December 2014].

370 European Commission 'Executive Summary of the Impact Assessment on Communication *COM (2014) 23 final*' of 22 January 2014 available at: http://eur-lex.europa.eu/legal-content/EN/TXT/PDF/?uri=CELEX:52014SC0022&from=EN [accessed 2 April 2014] (hereinafter: Commission Impact Assessment).

371 Commission Impact Assessment 4.

372 Reins Minimum Principles 26/27; Stokes 47/48.

Gas Recommendation[373] and Communication,[374] these documents could facilitate piecemeal-regulation on shale gas in Europe.

With the 2014 framework (Recommendation and Communication) the EU leaves the essence of shale gas regulation to the Member States and that could instigate competition amongst them, resulting in a 'race to the bottom' of environmental standards.[375] Member States are keen to attract investors to their territory and this is no different in the case of shale gas. In order to prevail in this competition for foreign money, some Member States might choose a strategy of lowering their environmental standards to bring down costs for investors.[376] If other Member States start adapting to that strategy this could result in a 'race to the bottom'.[377] One of the most effective ways to suspend this race is the establishment of a legally binding European minimum standard;[378] the 2014 framework fails to do just that.

However, the argument turns a blind eye to the fact that the 2014 Shale Gas Recommendation and Communication are not the only measures regulating shale gas extraction at EU level; they are merely the ones that apply most directly. Even if a Member State would choose not to implement the 2014 framework, this does not result in a situation where no EU environmental standards apply to shale gas extraction. At least eight different pieces of secondary EU law are applicable to shale gas extraction, all having environmental protection as their main goal.[379] This net of Directives and Regulations effectively provides a minimum EU-wide standard that constrains the risk of a 'race to the bottom' in shale gas regulation.

Moreover, the fear that Member States might simply ignore the Recommendation because it is not legally binding appears to be ill-founded. The 2014 Shale Gas Recommendation has provisions which indicate that the current voluntary approach is about to change rather rapidly if the Member States do not comply with it. Member States only

---

[373] 2014 Shale Gas Recommendation Preamble 9.

[374] 2014 Shale Gas Communication 4.

[375] More on the 'race to the bottom' issue may be found at Maria Lee *'EU Environmental Law: Challenges, Change and Decision-Making'* (Hart Publishing, Oxford 2005) 11 (hereinafter: Lee); Joanne Scott 'Flexibility in the Implementation of EC Environmental Law' (2000) 1 *YEEL* 56 (hereinafter: Scott Flexibility); Jans/Vedder 13.

[376] Scott Flexibility 56.

[377] Lee 11; Scott Flexibility 56.

[378] Jans/Vedder 13; Lee 11; Scott Flexibility 56.

[379] As discussed above in this chapter.

have six months to implement the recommended measures[380] and, thereafter, shall submit crucial data on the implementation of measures every twelve months.[381] The Commission will publish this data in the form of a 'scoreboard'.[382] This mechanism is designed to highlight to the public the implementation of and compliance with the framework by Member States.[383] Crucially, the Commission vowed to review the submitted data 18 months after its publication[384] and, should it not be satisfied with progress, reserved the right to establish legally-binding provisions.[385]

This threat is not hollow as Recommendations often functioned as forerunners to legally binding Directives in the past.[386] The current 2014 Recommendation might hence represent a mere first step on an EU ladder leading towards more robust shale gas regulation. Thus, Member States are subjected to an implicit threat by the EU to introduce binding legislation, if they do not comply with the 2014 Shale Gas Recommendation.

Furthermore, the EU's current soft law approach to shale gas regulation circumvents some legal pitfalls associated with overbearing EU regulations. These are consistency with article 194 (2) TFEU, and the principles of subsidiarity and proportionality. Although article 194 (2) TFEU might preclude *legally binding* EU-wide shale gas regulations, as discussed earlier, this article cannot stop the EU from giving its view on an important issue like shale gas extraction in a *legally non-binding form.* Opting for a non-binding form of EU shale gas regulation is an adequate way for the EU to transpose the factual concurrency of energy and environmental issues into a law with an adequate dual legal basis in article 192 TFEU as well as in article 194 TFEU.

Moreover, Members of the European Parliament (hereinafter: MEPs) raised the concern that legally binding, EU-wide shale gas regulations could conflict with the principle of subsidiarity.[387] During a European

---

380 2014 Shale Gas Recommendation preamble 11.

381 2014 Shale Gas Recommendation article 16.1.

382 2014 Shale Gas Recommendation article 16.2.

383 Ibid.

384 2014 Shale Gas Recommendation article 16.3.

385 2014 Shale Gas Recommendation article 16.4. The EU has collected all of the reports and is currently in the process of revising them, see European Commission '*Roadmap Report on the effectiveness of Commission Recommendation 2014/70/EU*' available at: http://ec.europa.eu/smart-regulation/roadmaps/docs/2015_env_021_shale_gas_fracking_en.pdf [accessed 7 June 2016].

386 Stokes 48; Tarka.

387 European Parliament debate of 20 November 2012 on '*Environmental Impacts of Shale Gas and Shale Oil Extraction Activities – Industrial, Energy and*

Parliament debate, MEP Jim Higgins argued with respect to shale gas extraction that, in line with the principle of subsidiarity, Member States should be free to decide on their own energy policy.[388]

The principle of subsidiarity, included in article 5 (3) TFEU, prescribes that the EU shall act in areas of shared competence, only if and insofar as objectives of a regulatory action cannot be sufficiently achieved by Member States.[389] This principle of subsidiarity has been explicitly designed to curb and confine EU activities.[390] Matters should be dealt with at the level closest to those affected by them and EU action should be the exception, reserved for cases where the Union is better placed to act.[391]

There is no indication that prudent and effective shale gas regulation may not be achieved by Member States. Conversely, however, it could be argued that effective shale gas regulation can be achieved better at Member State than at EU level. As highlighted above in Chapter 2, shale gas extraction is not a process that lends itself to complete standardization. Crucial geological features, such as the proximity of shale plays to aquifers, the depths at which the shale plays are buried and how brittle they are will alter, sometimes substantially, from region to region. That is why the fracturing fluid is prepared for each well individually. And that is why disposal and treatment methods for the 'flow back' differ widely. Thus, it seems to be reasonable for the EU to constrain itself to recommending processes and offering help to Member States.

In addition, the principle of subsidiarity is closely related to the principle of proportionality,[392] a relationship epitomized by the fact that subsidiarity and proportionality have been defined together in one additional protocol to the TFEU on subsidiarity and proportionality.[393] Thus, subsidiarity must be viewed in the context of proportionality, meaning that actions

---

Other Aspects of Shale Gas and Oil' CRE 20/11/2012 – 11 available at: http://www.europarl.europa.eu/sides/getDoc.do?type=CRE&reference=20121120&secondRef=ITEM-011&language=EN [accessed 2 September 2014] (hereinafter: European Parliament 20 November 2012) Speeches of Jan Březina, Peter Liese, Françoise Grossetête and Jim Higgins.

388 European Parliament 20 November 2012 Speech of Jim Higgins.

389 Article 5 (3) TFEU; Craig/De Burca 94/95.

390 Craig/De Burca 94.

391 This may be deduced from the systematic of article 5 (3) TFEU; see also Craig/De Burca 94.

392 Craig/De Burca 94.

393 Protocol (No 2) on the application of the principles of subsidiarity and proportionality [2012] OJ C 326/206 (hereinafter: Protocol on subsidiarity and proportionality).

must not go beyond what is necessary to achieve a regulatory objective.[394] The European Council established that the principle of proportionality requires the use of non-binding instruments, such as Recommendations, wherever possible.[395] This statement was made with particular regard to EU environmental legislation[396] and received widespread support in the legal literature.[397]

2.4 CONCLUSION

Recent technological advancements make shale gas extraction a real possibility in Europe.[398] If this development is to be regulated, the extent to which current EU law applies to shale gas extraction has to be assessed. Such an assessment provides a crucial starting point for discussions on the regulation of shale gas extraction in Member States, since the main pillars of national environmental and energy regulation emanate from EU law.[399] For this work it is particularly important to understand how EU regulations predetermine the interplay of environmental protection and energy security in the case of shale gas extraction.

With regard to primary EU law, the main issue is to identify a suitable legal basis for the regulation of shale gas extraction among the multitude of EU competences. There are three different EU competences that lend themselves to becoming the legal bases of shale gas regulation: the internal market competence of article 114 TFEU, the environmental competence of article 192 TFEU and the energy competence of article 194 TFEU.

The regulation of shale gas extraction mainly serves the aim of protecting the groundwater, the air, the soil and the environment as a whole and not the aim to promote the internal market. Therefore, it seems rather unlikely that the bulk of EU regulations with a bearing on shale gas extraction is based upon article 114 TFEU

---

394 Article 5 (4) TFEU; for its interpretation see House of Lords '*European Union – Fourteenth Report*' paragraphs 74/75 available at: http://www.publications.parliament.uk/pa/ld200405/ldselect/ldeucom/101/10105.htm [accessed 8 September 2014].

395 Council Resolution on the drafting, implementation and enforcement of Community environmental law [1997] OJ C 321/1; Jans/Vedder 14.

396 Jans/Vedder 14.

397 Gerd Winter 'Subsidiarität und Deregulierung im Gemeinschaftsrecht' (1996) 3 *Zeitschrift für Europarecht* 261/262.

398 2014 Commission Shale Gas Communication 2/3.

399 For the example of Germany, see Kloepfer § 9 paragraphs 79–81.

As opposed to its predecessor, the TFEU includes a long-standing competence for energy regulation.[400] The pre-existing environmental competence has also received a substantial revamp over the last 30 years.[401] This raises the question of whether shale gas regulation should be based on the environmental or the energy competence or both of them at the same time.

Given that shale gas extraction may have substantial repercussions for both spheres its regulation could, exceptionally, be based on both articles 192 and 194 TFEU, although adopting shale gas regulation on the single basis of article 192 TFEU would also be possible. Articles 192 and 194 TFEU differ slightly in the procedures that need to be followed to adopt secondary law but the differences are not insurmountable and the two procedures can be reconciled.

Having established the primary law competence to regulate shale gas extraction at EU level, the second step was to investigate to which extent the existing secondary law framework actually covers the potential issues of shale gas extraction. The framework mainly consists of legally binding measures, like EU Regulations, Directives and Decisions, which regulate a certain, concrete environmental issue. This framework is not shale gas specific.[402]

Overall, a number of secondary environmental and energy law norms cover certain aspects of shale gas extraction. However, a number of gaps or possible inadequacies in the EU environmental legislative framework exist which need to be addressed by shale gas specific supplements to the respective laws.[403]

In 2014 the Commission took action in that regard. They adopted the 2014 framework, consisting of the 2014 Shale Gas Recommendation and the 2014 Shale Gas Communication. Both documents recommend to Member States a set of measures and operating standards in order to create a level playing field among European states.

However, the 2014 framework on shale gas extraction has a big sting in the tail. It does not actually implement the measures, but merely recommends to Member States that they take these measures into account when designing their national frameworks. The framework has been moulded

---

400 Stokes 44.

401 Ludwig Krämer '*Focus on European Environmental Law*' (Sweet & Maxwell, London 1992) paragraph 3-09 (hereinafter: Krämer Focus); Streinz article 191 paragraph 36.

402 Jans/Vedder 10.

403 For a similar assessment, see Broomfield xi–xv.

into a Recommendation and a Communication, secondary EU law measures with no direct binding force.

From this author's point of view, the non-binding character of the 2014 Shale Gas Recommendation is in some respects an advantage. Opting for a non-binding form allows the EU to adopt shale gas regulations on the dual legal basis of article 192 TFEU and article 194 TFEU and to respect the principles of subsidiarity and proportionality.

As a consequence, the national regulatory frameworks take centre stage for shale gas regulation. Member States have the right to take different approaches to shale gas extraction. As will be seen in further chapters, permissive and prohibitive regulatory views on shale gas extraction vary from country to country. As EU law leaves the direction of decisions to Member States, undecided Member States should carefully assess the prevailing approaches of other Member States for their effectiveness. These Member States should also look at their own constitutional system to develop shale gas regulation along the lines of the two main interests at stake, environmental protection and energy security.

3. Shale gas and national law

3.1 INTRODUCTION

Current opinion of EU Member States on shale gas extraction is notably diverse: while some take a permissive approach to the process, others try to stop shale gas extraction on their respective territories.[1] France,[2] Germany,[3] Bulgaria,[4] the Netherlands,[5] the UK,[6] the Czech

[1] Philippe & Partners Law Firm '*Final Report on Unconventional Gas in Europe*' (2011) available at: http://ec.europa.eu/energy/studies/doc/2012_unconventional_gas_in_europe.pdf [accessed 12 March 2013] 11–14 (hereinafter: Philippe & Partners); Anne-Sophie Corbeau 'The Introduction of unconventional gas in Europe: Opportunities and Challenges' in: Martha Roggenkamp and Olivia Woolley (eds) '*European Energy Law Report IX*' (Intersentia Publishing Ltd., Cambridge 2012) 198.

[2] Law No. 2011-835 of 13 July 2011 to ban the exploration and mining of hydrocarbon liquids and gases by hydraulic fracturing and repealing the exclusive licenses to include projects using this technique (1) (*LOI n° 2011-835 du 13 juillet 2011 visant à interdire l'exploration et l'exploitation des mines d'hydrocarbures liquides ou gazeux par fracturation hydraulique et à abroger les permis exclusifs de recherches comportant des projets ayant recours à cette technique (1)*) (hereinafter: French moratorium).

[3] Bundestag 'Drucksache 18/4713 Bill of the German Federal Government on the alteration of water and environmental protection norms with the aim of prohibiting and minimizing risks associated with the procedures of the fracking technology' (*Gesetzentwurf der Bundesregierung 'Entwurf eines Gesetzes zur Änderung wasser- und naturschutzrechtlicher Vorschriften zur Untersagung und zur Risikominimierung bei den Verfahren der Fracking-Technologie*) available at: http://dipbt.bundestag.de/doc/btd/18/047/1804713.pdf [accessed 7 April 2015] (hereinafter: Drucksache 18/4713).

[4] Bulgarian National Assembly 'Decision to ban the application of hydraulic technology according to the break-in, study and/or extraction of oil and gas in the Republic of Bulgaria' published in (2012) No 7 Official Journal of Bulgaria 13 (*РЕШЕНИЕ за забрана върху прилагането на технология-та хидравлично разбиване при проучванеи/или добив на газ и нефт на територията на Република България published in (2012) БРОЙ 7, ДЪРЖА ВЕНВЕСТНИК 13*) (hereinafter: Bulgarian moratorium); Tomasz Daborowski and Jakub Groszkowski 'Shale Gas in Bulgaria, the Czech Republic and Romania political context – legal status – outlook' (Centre for Eastern Studies, Warsaw 2012) 9/10 available at: http://www.osw.waw.pl/sites/default/files/shale_gas_in_bulgaria_the_czech_republic_

Republic,[7] Romania,[8] Denmark,[9] Italy,[10] and areas like Northern Ireland,[11] Scotland,[12] the German states of North Rhine-Westphalia,[13] Lower Saxony[14] and Hesse,[15] the Dutch region of Boxtel,[16] Flanders (a Belgian region)[17] and Cantabria, La Rioja and Navarre (Spanish regions)[18] have all introduced bans or moratoria on shale gas extraction.[19]

---

and_romania_net_0.pdf [accessed 23 May 2014] (hereinafter: Daborowski/Groszkowski).

5 Minister of Economic Affairs 'Letter to the House of Minister Kamp (Economic Affairs) concerning follow-up action on shale gas extraction' (*Ministerie van Economische Zaken Brief aan de Tweede Kamer van minister Kamp (Economische Zaken) over vervolgstappen winning schaliegas*) 2/3 available at: http://www.rijksoverheid.nl/documenten-en-publicaties/kamerstukken/2013/09/18/brief-aan-de-tweede-kamer-vervolgstap-schaliegas.html [accessed 14 April 2014] (hereinafter: Netherlands Statement Minister of Economic Affairs); Martha M Roggenkamp 'Energy Law in the Netherlands' in Martha M Roggenkamp et al. (eds) '*Energy Law in Europe*' 3rd edition (Oxford University Press, Oxford 2016) paragraph 10.98 (hereinafter: Roggenkamp 2016); Tim Boersma and Corey Johnson 'Twenty Years of US Experience – Lessons Learnt For Europe' in Cecile Musialski et al. (eds) '*Shale Gas in Europe*' (Claeys & Casteels, Deventer 2013) 27; Yvonne Schavemaker 'Shale Gas in the Netherlands' available at: http://www.shale-gas-information-platform.org/areas/the-debate/shale-gas-in-the-netherlands.html [accessed 14 April 2014] (hereinafter: Schavemaker).

6 Edward Davey 'Written Ministerial Statement by Edward Davey: Exploration for shale gas 13 December 2012' available at: http://www.decc.gov.uk/en/content/cms/news/wms_shale/wms_shale.aspx [accessed 17 December 2012] (hereinafter: Written Statement Davey); *BBC* 'Blackpool Shale Gas drilling suspended after quake' available at: http://www.bbc.co.uk/news/uk-england-lancashire-13599161 [accessed 7 April 2014] (hereinafter: *BBC* Blackpool); *The Telegraph* 31 May 2011 'Shale gas drilling suspended after earthquake near Blackpool' available at: http://www.telegraph.co.uk/earth/energy/8549340/Shale-gas-drilling-suspended-after-earthquake-near-Blackpool.html [accessed 7 April 2014] (hereinafter: Telegraph 31 May 2011). This moratorium, however, was lifted again by the end of 2012 for England and Wales; for more, see the discussion below in this chapter.

7 Zpravy E 15 'Chalupa wants to stop shale gas exploration until 2014' (*Chalupa chce stop průzkumu plynu z břidlic do roku 2014)* available at: http://zpravy.e15.cz/byznys/prumysl-a-energetika/chalupa-chce-stop-pruzkumu-plynu-z-bridlic-do-roku-2014-910421 [accessed 31 March 2014] (hereinafter: Zpravy E 15); Daborowski/Groszkowski 17.

8 Daborowski/Groszkowski 25; Irina Savu and Andra Timu Bloomberg Business Week 'Romania Shale-Gas, Mine Projects on Hold in 2012, Ponta Says' available at: http://www.businessweek.com/news/2012-06-07/romanian-shale-gas-mining-projects-on-hold-in-2012-ponta-says [accessed 3 April 2014] (hereinafter: Savu/Timu); Truth Live 'Prime Minister Victor Ponta: "I would be more popular if I tried to suspend Basescu but the price for Romania is too high"' (*ADEVĂRUL LIVE Premierul Victor Ponta: "Eu aş fi mai popular dacă m-aş apuca să-l suspend pe Băsescu dar preţul pentru România ar fi prea mare")*

This chapter focuses on countries and regions with prohibitive shale gas regulation and measures in place. Part II of the book, by contrast, explains in detail what permissive shale gas regulation could be looking like. The reason to focus on prohibitive regulation in the current chapter

---

http://adevarul.ro/news/politica/live-video-premierul-victor-ponta-direct-studioul-adevarul-ora-1315-1_51483b3d00f5182b85288f32/index.html [accessed 9 June 2014] (hereinafter: Truth Live Ponta); Silviu Molnar Natural Gas Europe 'Shale Gas Exploitation in Romania can be Postponed at Least Another Two Years' available at: http://www.naturalgaseurope.com/shale-gas-exploitation-in-romania-postponed [accessed 3 April 2014].

[9] The responsible Minister announced in 2012 that no licences beyond those already issued will be approved until results of test drillings become available, see Anita Ronne 'Energy Law in Denmark' in Martha M Roggenkamp et al. (eds) '*Energy Law in Europe*' 3rd edition (Oxford University Press, Oxford 2016) paragraph 6.152.

[10] Italian Parliament 'Article 38 No 11-quarter of LAW 11 November 2014, n. 164 Conversion, with amendments, of Decree-Law 11 September 2014, n. 133, Urgent measures for the opening of the sites, the construction of public works, the digitization of the country, bureaucratic simplification, the emergence of the hydrogeological instability and for the resumption of productive captivity' (*LEGGE 11 novembre 2014, n. 164 Conversione, con modificazioni, del decreto-legge 11 settembre 2014, n. 133, Misure urgenti per l'apertura dei cantieri, la realizzazione delle opere pubbliche, la digitalizzazione del Paese, la semplificazione burocratica, l'emergenza del dissesto idrogeologico e per la ripresa delle cattività produttive*)' available at: http://www.bosettiegatti.eu/info/norme/statali/2014_0164.htm [accessed: 14 July 2016] (hereinafter: Italian ban); Giuseppe Franco Ferrari 'Energy Law in Italy' in Martha M Roggenkamp et al. (eds) '*Energy Law in Europe*' 3rd edition (Oxford University Press, Oxford 2016) paragraph 9.48 (hereinafter: Ferrari).

[11] Northern Ireland Assembly Deb 6 December 2011, Vol 69 No 6, cols 300 and 336 (hereinafter: Northern Ireland minutes).

[12] Greg Gordon, Aileen McHarg and John Paterson 'Energy Law in the United Kingdom' in Martha M Roggenkamp et al. (eds) '*Energy Law in Europe*' 3rd edition (Oxford University Press, Oxford 2016) paragraph 14.30 (hereinafter: Gordon/McHarg/Paterson).

[13] Ministerium für Klimaschutz, Umwelt, Landwirtschaft, Natur- und Verbraucherschutz des Landes Nordrhein-Westfalen '*Pressemitteilung Umweltministerium und Wirtschaftsministerium legen Risikogutachten zu Fracking vor*' 7 September 2012 available at: http://www.umwelt.nrw.de/ministerium/service_kontakt/archiv/presse2012/presse120907_a.php [accessed 15 April 2014].

[14] Niedersächsisches Ministerium für Umwelt, Energie und Klimaschutz '*Zulassung von Vorhaben zur Aufsuchung und Gewinnung von Erdgas aus konventionellen Lagerstätten mittels hydraulischer Bohrlochbehandlung zur Risserzeugung in einem Verfahren mit Umweltverträglichkeitsprüfung*' available at: http://www.umwelt.niedersachsen.de/aktuelles/pressemitteilungen/strenge-auflagen-bei-der-ergasfoerderung-122495.html [accessed 14 April 2014]; Hannoversche Allgemeine 'Die Fracking-Pause dauert bis 2016' available at: http://www.haz.de/Nachrichten/

is three-fold: first, the focus lies with the precautionary principle and that principle has been commonly invoked in the context of prohibitive shale gas regulation.[20] Second, the rare legal works on prohibitive shale gas extraction that do exist include some factual inadequacies that need to be corrected.[21] Third, a number of EU Member States have not yet decided on their approaches towards shale gas extraction. Whether or

---

Politik/Niedersachsen/Die-Fracking-Pause-dauert-bis-2016-in-Niedersachsen [accessed 9 January 2015].

[15] Hessisches Ministerium für Umwelt, Klimaschutz, Landwirtschaft und Verbraucherschutz 20.08.2014 'Fracking-Klage gegen das Land Hessen zurückgenommen' available at: https://www.hessen.de/presse/pressemitteilung/fracking-klage-gegen-das-land-hessen-zurueckgenommen-0 [accessed 9 January 2015].

[16] Boxtel City 'Motion by the Boxtel Board in public meeting on 5 March 2012 '(*Motie De raad van de gemeente Boxtel in openbare vergadering bijeen op 5 maart 2012*) available at: http://www.breda.nl/system/files/20120305motieschaliegasvrijegemeenteboxtel.pdf [accessed 15 June 2012] (hereinafter: Boxtel Moratorium).

[17] HLN.be 'Vlaamse regering legt tijdelijk verbod op fracking op' (*Flemish government imposes temporary ban on fracking*) available at: http://www.hln.be/hln/nl/2764/milieu/article/detail/1936420/2014/07/04/Vlaamse-regering-legt-tijdelijk-verbod-op-fracking-op.dhtml [accessed 19 September 2015] (hereinafter: Flemish moratorium); Fracturing Law Blog 'The Minimum Principles Applied in Practice: Temporary Ban on Fracking in the Flemish Region in Belgium' available at: http://fraclawblog.com/2014/07/05/the-minimum-principles-applyed-in-practice-temporary-ban-on-fracking-in-the-flemish-region-in-belgium/ [accessed 25 September 2015].

[18] Parliament of Cantabria 'Act of independent region Cantabria to regulate a ban on the hydraulic fracturing technique and unconventional gas exploration and extraction on the territory of Cantabria No 8L/1000-0011' ('*Ley de Cantabria por la que se regula la prohibición en el territorio de la Comunidad Autónoma de Cantabria de la técnica de fractura hidráulica como técnica de investigación y extracción de gas no convencional, número 8L/1000-0011*') available at: http://www.parlamento-cantabria.es/sites/default/files/bop/1000-0011-7.pdf [accessed 28 April 2014] (hereinafter: Cantabrian moratorium). The respective acts of Cantabria, La Rioja and Navarre are Acts No 1/2013, 7/2013 and 30/2013; however, the Constitutional Court of Spain declared all three acts irreconcilable with the Spanish constitution, due to a lack of competence of all three regions (Decisions No 106, 134 and 208). For details, see Inigo del Guayo 'Energy Law in Spain' in Martha M Roggenkamp et al. (eds) '*Energy Law in Europe*' 3rd edition (Oxford University Press, Oxford 2016) paragraph 13.79 (hereinafter: del Guayo).

[19] However, Romania and the UK have already lifted their respective moratoria and the moratoria in Boxtel and Cantabria only apply to a certain areas.

[20] Suggestions that the precautionary principle applies to shale gas extraction have been plentiful in recent scientific literature, see for instance: Johnson/Boersma 391; Madelon L. Finkel and Adam Law 'The Rush to Drill for Natural Gas: A Public Health Cautionary Tale' (2011) Vol 101 No 5 *American Journal of Public Health* 785.

[21] Described in the discussion on Germany, below.

not existing moratoria and/or bans on shale gas extraction could be used as blueprints for regulatory action in these states, is vigorously discussed, but depends on the reconcilability of the respective prohibitive regulation with constitutional pre-settings. Only if prohibitive regulation on shale gas extraction does not conflict with fundamental legal principles may prudent legislators in other EU Member States act in similar manner.

As a preliminary remark to this chapter, the actual meaning of the terms 'moratorium' and 'ban' require clarification. Unfortunately, some scholars have adopted a very lax, imprecise and sometimes plainly wrong use of the terminology with regard to shale gas extraction.[22] A 'moratorium' refers to the temporary suspension of a specific activity,[23] whereas a 'ban' addresses a general prohibition by legal means.[24] A moratorium is thus a temporary measure with the aim to *suspend* an activity, whereas a ban is of indefinite duration with the goal of *suppression and prohibition* of an activity.

Existing prohibitive shale gas regulation in EU Member States may be grouped into three categories. First, outright bans, second moratoria of binding character, which have been adopted by national parliaments, third, more or less loose governmental agreements to operate a policy of non-consent to shale gas extraction applications. While the first two categories constitute binding laws, the last may be revoked at any time; the latter form of prohibitive regulation is subject to the discretion of national governments. Bans and moratoria that have been approved by the legislator will be called 'bans/moratoria by law', while the third category will be referred to as 'political moratoria'.[25]

For each of the three categories ('ban by law', 'moratorium by law', 'political moratorium') one country has been picked as an example: the French ban has been passed by the French National Assembly and the Senate ('ban by law') and the German moratorium has been approved by both chambers of parliament ('moratorium by law').[26] The UK moratorium, by contrast, was grounded in a government agreement and had been

---

22 See for example Maurin/Vivoda 377, who refer to 'temporary or permanent moratoria'. For the impossibility of the latter see immediately below.

23 Bryan A Garner (ed.) '*Black's Law Dictionary*' 8th edition (Thomson/West, St. Paul 2004) 1031 (hereinafter: Black's Law Dictionary).

24 Black's Law Dictionary 154.

25 This terminology was originally proposed by Dąborowski/Groszkowski 23.

26 See immediately below.

announced to the public in a written statement by the responsible minister ('political moratorium').[27]

In an attempt to improve readability and coherence of this chapter, only these three prohibitive shale gas regulations will be comprehensively analysed. The prohibitive measures of other Member States fall into one of the three categories and the observations made may be transferred to them. However, where measures of other Member States deviate from the examples of the French, the German and the British, this will be highlighted.

The jurisdictions have also been picked due to their respective leading roles in the Civil Law System and the Common Law System.[28] This choice allows the reader to see beyond the approach of three particular countries to the regulation of shale gas extraction, and understand how different legal systems are dealing with shale gas regulation.

This chapter will undertake legal comparison of regulatory measures on the basis of the 'functional method',[29] an overarching approach to comparative law.[30] According to the 'founding fathers' of modern comparative law, Konrad Zweigert and Hein Kötz,[31] legal instruments must not be analysed in isolation, but need to be compared in the context of their function as resolutions to societal issues.[32] Incomparables cannot usefully be compared and the only things in law that are comparable are those fulfilling the same function.[33] The legal comparatist must find the rules in a foreign system which are functionally equivalent to those which interest him in his base-jurisdiction.[34] This 'functional method'[35] works in law when different legal systems confront the same problem.[36] This is the case with shale gas extraction in Europe, where Member States discuss their individual handling of similar technological components.[37]

---

27 Ibid.

28 For explanations on both systems and their categorization, see Konrad Zweigert, Hein Kötz and Tony Weir '*Introduction to comparative law*' 3rd edition (Clarendon, Oxford 1998) 74 et sqq. and 178 et sqq. (hereinafter: Zweigert/Kötz/Weir).

29 For explanations, see James Gordley 'The functional method' in Pier Giuseppe Monateri (ed.) '*Methods of comparative law*' (Edward Elgar, Cheltenham 2012) 107 et sqq. (hereinafter: Gordley).

30 Heiko Sauer 'Juristische Methodenlehre' in Julian Krüper (edt.) '*Grundlagen des Rechts*' 2nd edition (Nomos, Baden-Baden 2013) § 9 paragraphs 17–21 (hereinafter: Krüper).

31 According to Gordley 107.

32 Zweigert/Kötz/Weir 34/35.

33 Zweigert/Köhn/Weir 34.

34 Zweigert/Köhn/Weir 36.

35 See: Gordley 107 et sqq.

36 Gordley 119.

37 See Chapter 1 above 'Shale Gas in a Nutshell'.

A genuine understanding of how state regulations address the questions posed by shale gas extraction may not be achieved by superficial reading and comparison of paragraphs. The law texts rather need to be interpreted in the light of their respective contexts in order to distil the genuine meaning of the norms.[38] Friedrich Carl von Savigny developed a rigorous methodology of legal interpretation two hundred years ago, which is based on four 'canons', or methods.[39] These methods have been widely accepted in the comparative law literature[40] and are still in use today all over the world.[41]

The four methods consist of grammatical interpretation (the interpretation of the wording of the law) and systematic interpretation (the logical interaction of different pieces of the law among themselves and with the overall legal system).[42] They also comprise historical interpretation (considering the legal situation and pertaining circumstances at the point in time when the law was enacted) and teleological interpretation (interpretation in view of the underlying aims and rationale of the law).[43] The bans/moratoria that exist in many EU Member States will be analysed with the help of Savigny's canons.

3.2 'BAN BY LAW' IN FRANCE

When the shale gas debate reached France, the government decided not to approve applications for shale gas extraction permits, putting in place a 'political moratorium'.[44] Despite this early action, protests against shale

---

38 Zweigert/Kötz/Weir 35/36.

39 Friedrich Carl von Savigny '*System des heutigen römischen Rechts Band I*' (Veit, Berlin 1840) 212–14 (hereinafter: Savigny).

40 Krüper § 9 para. 17; Larenz 231.

41 Günter Hager '*Rechtsmethoden in Europa*' (Mohr Siebeck, Tübingen 2009) 26/27 and 37 (hereinafter: Hager); Krüper § 9 para. 5; Jean-Louis Bergel '*Méthodologie juridique*' (Presses Universitaires de France, Paris 2001) 231 et sqq.; Jean-Louis Bergel '*Théorie générale du droit*' 4th edition (Dalloz, Paris 2003) No 231 et sqq.; the German Federal Constitutional Court ruled that Savigny's means of legal interpretation must be adhered to by all users of the law and that a conclusion from interpreting a norm which is not reached by one of these methods is invalid, see: BVerfGE 93, 37 (81); 113, 88 (104).

42 Savigny 212–14; further elaborations on the individual methods will be provided at apt spots throughout the chapter.

43 Ibid.

44 French National Assembly 'Information report on the implementation of the conclusions of the information report (No. 2719) 8 July 2010 on the evaluation of the implementation of Article 5 of the Charter of the environment on the appli-

gas extraction gathered momentum and on 16 April 2011, inhabitants of many French *arrondissements* (local districts) took to the streets.[45] They demanded the formal outlawing of shale gas extraction.[46] After intense discussions, law no. 2011-835,[47] which orders a ban on unconventional oil and gas extraction by hydraulic fracturing,[48] was passed on 21 June 2011 by the French National Assembly[49] and, on 30 June 2011, by the Senate.[50]

Alongside France, Bulgaria[51] and Italy[52] put into place outright bans on shale gas extraction, which were all formally adopted by the respective national parliaments. The French ban exemplifies this group and the

---

cation the precautionary principle' (*17/11/2011 RAPPORT D'INFORMATION déposé en application de l'article 146-3, alinéa 8, du Règlement par le comité d'évaluation et de contrôle des politiques publiques sur la mise en œuvre des conclusions du rapport d'information (n° 2719) du 8 juillet 2010 sur l'évaluation de la mise en œuvre de l'article 5 de la Charte de l'environnement relatif à l'application du principe de précaution*) paragraph 1) a) available at: http://www.assemblee-nationale.fr/13/pdf/rap-info/i3970.pdf [accessed 12 June 2012] (hereinafter: France Report on PP); French National Assembly 'PROHIBITION OF HYDRAULIC FRACTURE Solemn Vote Official Journal of the French Republic Regular Session 2010-2011 175th meeting Minutes integral 1st meeting of Wednesday, 11 May 2011' (*INTERDICTION DE LA FRACTURATION HYDRAULIQUE Vote solennel JOURNAL OFFICIEL DE LA RÉPUBLIQUE FRANÇAISE SESSION ORDINAIRE DE 2010–2011 175e séance Compte rendu integral 1re séance du mercredi 11 mai 2011*) 2974 available at: http://www.assemblee-nationale.fr/13/pdf/cri/2010-2011/20110175.pdf [accessed 7 May 2014] (hereinafter: French National Assembly Minutes 11 May 2011).

45 According to: French National Assembly 'Thirteenth legislature Ordinary Session 2010–2011 Report First meeting of Tuesday, 21 June 2011 ELECTRONIC SUMMARY' (*XIIIe législature Session ordinaire de 2010-2011 Compte rendu intégral Première séance du mardi 21 juin 2011 SOMMAIRE ÉLECTRONIQUE*) 4381 available at: http://www.assemblee-nationale.fr/13/pdf/cri/2010-2011/20110216.pdf [accessed 2 May 2014] (hereinafter: French National Assembly Minutes 21 June 2011).

46 French National Assembly Minutes 21 June 2011, 4381.

47 French moratorium.

48 Article 1 French moratorium.

49 French National Assembly Minutes 21 June 2011, 4392.

50 French Senate 'Meeting of 30 June 2011 (Verbatim Report of Debates)' (*Séance du 30 juin 2011 (compte rendu intégral des débats)*) available at: http://www.senat.fr/seances/s201106/s20110630/s20110630.pdf [accessed 2 May 2014] (hereinafter: French Senate 30 June 2011).

51 Bulgarian National Assembly 'Stenographic report of 304th Plenary Sittings Sofia, Wednesday, 18 January 2012' (*Стенограми от пленарни заседания ТРИСТА И ЧЕТВЪРТО ЗАСЕДАНИЕ София, сряда, 18 януари 2012 г.*) available at: http://www.parliament.bg/bg/plenaryst/ns/7/ID/2688 [accessed 29 April 2014] (hereinafter: Stenographic Report Bulgarian Assembly).

52 Italian ban; Ferrari paragraph 9.48.

subsequent analysis of its main points can be largely applied to the other European bans.

The French moratorium refers environmental concerns explicitly and directly in its text.[53] This is typical for the scrutinized 'bans by law', which all have references to environmental concerns featuring prominently in the text.[54] These concerns have also been discussed in almost every official document associated with the respective ban.[55] Energy security interests, by contrast, do not explicitly feature in any of the scrutinized 'bans by law'.

A first superficial reading of the respective bans thus suggests that these documents disregard energy security interests altogether. This assessment contrasts with the fact that the importance of implementing regulation that takes equal account of energy security aspects and environmental concerns has been widely debated in countries with a 'ban by law'.[56] However, as closer scrutiny of the French ban highlights, matters are not straightforward. The law texts might indeed address potential energy security benefits, but not in a way that immediately meets the eye of the reader.

---

53 Although the actual reference is rather brief, see article 2 French moratorium.

54 Northern Ireland Minutes columns 337/338; paragraphs 2–4 Boxtel Moratorium; preamble Cantabrian moratorium. The only exception being the Bulgarian moratorium, which does not include an environmental reference in the actual law text.

55 See for instance: Stenographic Report Bulgarian Assembly Speech of Volen Siderov; Bulgaria Explanatory Statement; Research and Information Service of the Northern Ireland Assembly Briefing Paper 'Key Concerns Surrounding "Fracking" 1/2' available at: http://www.niassembly.gov.uk/Documents/RaISe/Publications/2011/Enterprise-Trade-Investment/17111.pdf [accessed 31 March 2014]; Northern Ireland minutes columns 301–4, 308 and 311; Cantabrian Parliament Minutes 3886, 3889/3890 and 3892–3894..

56 French National Assembly 'PROHIBITION OF HYDRAULIC FRACTURE Discussion of a bill Official Journal of the French Republic Regular Session 2010-2011 173rd meeting Minutes integral 1st meeting of Tuesday, 10 May 2011' (*INTERDICTION DE LA FRACTURATION HYDRAULIQUE Discussion d'une proposition de loi JOURNAL OFFICIEL DE LA RÉPUBLIQUE FRANÇAISE SESSION ORDINAIRE DE 2010-2011 173e séance Compte rendu integral 1re séance du mardi 10 mai 2011*) available at: http://www.assemblee-nationale.fr/13/pdf/cri/2010-2011/20110173.pdf [accessed 7 May 2014] 2897 (hereinafter: French National Assembly Minutes 10 May 2011 part 1); French National Assembly Minutes 21 June 2011 4378; Stenographic Report Bulgarian Assembly Speeches of Dian Chervenkondev and Mihail Mihailov; Northern Ireland Minutes columns 302, 309/310 and 333; Cantabrian Parliament Minutes 3885/3886 and 3894.

3.2.1 Article 1 – Prohibiting Shale Gas Extraction or Hydraulic Fracturing?

There are essentially three possible gateways for energy security considerations to come into play under the French ban indirectly, namely, article 1, article 3 and article 4 of law no. 2011-835. Article 1 prohibits the exploration and exploitation of oil and gas mines by hydraulically fracturing the rock.[57] The French stipulation thus outlaws the technique of hydraulic fracturing and makes no reference to shale gas.[58] Thus, it appears as if article 1 of the French ban does not target shale gas extraction but merely one of the two technical components required to bring about shale gas extraction: hydraulic fracturing.[59]

This is relevant because hydraulic fracturing can be replaced with other means, like pneumatic fracturing, as French MPs pointed out.[60] Crucially, shale gas extraction in France would remain licit, as long as it does not involve hydraulic fracturing. Shale gas could be extracted by the use of alternative technologies, despite the existing ban.[61] The produced gas could be marketed and the potentially positive effects of shale gas extraction for France's energy security could materialize.

The problem arises because the term hydraulic fracturing has not been defined in the text of law no. 2011-835. The ban would hence succeed in providing for environmental protection against possible perils of hydraulic fracturing, while enabling shale gas to contribute to France's energy security.[62]

---

[57] This part of article 1 French moratorium reads: '*l'exploration et l'exploitation des mines d'hydrocarbures liquides ou gazeux par des forages suivis de fracturation hydraulique de la roche*' in the original text.

[58] As discussed above, all other 'moratoria by law' include hybrid provisions, outlawing hydraulic fracturing and shale gas extraction likewise.

[59] See above, Chapter 1, 'Shale Gas in a Nutshell'.

[60] Some French MPs stressed that alternative techniques of shale gas extraction, such as pneumatic fracturing (tested in the USA) or fracturing by arcing, are currently being developed; they reckon that this technology could also be deployed in France, see French National Assembly 'Thirteenth Legislature Ordinary Session 2011–2012 Report Thursday, 6 October 2011 ELECTRONIC SUMMARY' (*XIIIe Législature SESSION ORDINAIRE DE 2011–2012 Séances du jeudi 6 octobre 2011 Compte rendu integral)* 5992/5993 available at: http://www.assemblee-nationale.fr/13/pdf/cri/2011-2012/20120007.pdf [accessed 7 August 2014] (hereinafter: French National Assembly Minutes 6 October 2011).

[61] French MPs stated that the required licences would not fall under the scope of law no 2011-835 and therefore would be approvable from their point of view, see: French National Assembly Minutes 6 October 2011, 5992/5993.

[62] For explanations of the potential environmental perils, see Chapter 1 above. For the potential of shale gas to contribute to France's energy security,

Thierry Lauriol called this 'surprising' in his legal treatise on energy law in France and concluded that 'this lack of definition, which leaves a wide margin for interpretation, could lead to litigation'.[63]

The interpretation of the words 'hydraulic fracturing' in article 1 of the French ban showcases an old problem of the grammatical method of legal interpretation: shall the wording of a norm be understood in a common parlance sense or in a technical sense?[64] Amongst the general public the term hydraulic fracturing is frequently used as a synonym for shale gas extraction, or even for unconventional gas in general.[65]

However, the question of which term shall be used in the French law gave rise to considerable controversy during the preparation of the law in the sustainable development committee of the French parliament.[66] Nevertheless, an actual definition of hydraulic fracturing does not feature

---

see for instance French National Assembly 'PROHIBITION OF HYDRAULIC FRACTURE Discussion of a bill Official Journal of the French Republic Regular Session 2010-2011 174th meeting Minutes integral 2nd meeting of Tuesday, 10 May 2011' (*INTERDICTION DE LA FRACTURATION HYDRAULIQUE Discussion d'une proposition de loi JOURNAL OFFICIEL DE LA RÉPUBLIQUE FRANÇAISE SESSION ORDINAIRE DE 2010–2011 174e séance Compte rendu integral 2e séance du mardi 10 mai 2011)* available at: http://www.assemblee-nationale.fr/13/pdf/cri/2010-2011/20110174.pdf [accessed 7 May 2014] 2928/2929 (hereinafter: French National Assembly Minutes 10 May 2011 part 2).

63 Thierry Lauriol 'Energy Law in France' in Martha M Roggenkamp et al. (eds) '*Energy Law in Europe*' 3rd edition (Oxford University Press, Oxford 2016) paragraph 7.86 (hereinafter: Lauriol).

64 For the general conflict between both possible interpretations of a norm, see Kramer 84.

65 See for instance the multitude of 'anti-fracking' initiatives and citizen groups, which not only oppose shale gas extraction by hydraulic fracturing, but often unconventional gas extraction altogether, see for instance Frack Off http://frack-off.org.uk/fracking-hell/frontline/ [accessed 4 August 2014]; Frack Free Wales www.frackfreewales.org [accessed: 4 August 2014]; Kein Fracking www.kein-fracking.de [accessed 4 August 2014]; Fracking freies Hessen www.fracking-freieshessen.de [accessed 2 August 2014].

66 French National Assembly Commission for Sustainable Development and Spatial Planning 'Report on the proposed law to ban the exploration and exploitation of oil and gas by hydraulic fracturing mines and to repeal the exclusive licenses including projects using this technique 4 May 2011 (*RAPPORT FAIT AU NOM DE LA COMMISSION DU DÉVELOPPEMENT DURABLE ET DE L'AMÉNAGEMENT DU TERRITOIRE SUR LA PROPOSITION DE LOI, visant à interdire l'exploration et l'exploitation des mines d'hydrocarbures liquides ou gazeux par fracturation hydraulique et à abroger les permis exclusifs de recherches comportant des projets ayant recours à cette technique 4 mai 2011*) available at: http://www.assemblee-nationale.fr/13/pdf/rapports/r3392.pdf [accessed 12 June 2012] pages 2 and 15 (hereinafter: French Sustainable Development Commission).

in the preparatory documents, a fact that must, however, be viewed in the specific context of oil and gas/mining laws.

It is common practice in this area of law to incorporate industry terms into the actual law text without defining them, to ensure that the addressees of the norm know what the norm means.[67] Hydraulic fracturing is merely a technical denomination for one specific way to extract shale gas and/or other unconventional hydrocarbons.[68] Hydraulic fracturing is not a term that originates from everyday life but from the oil and gas industry.[69]

If the legislator decides to put specific professional terminology into a law, it must be assumed that the term in question is also to be understood in the professional sense of the word.[70] Laws referring to techniques and technical activities must hence be interpreted against the background of the specific vocabulary of these technical processes.[71] Thus, the term hydraulic fracturing in article 1 of the French moratorium must be understood as referring to this particular technological process, and not as meaning shale gas extraction as a whole. Grammatical interpretation hence supports the view that article 1 of the French moratorium outlaws shale gas extraction only insofar as it is done by hydraulic fracturing.

But this is merely the result of grammatical interpretation. It is important to emphasize that the different methods of interpretation cannot be separated from each other.[72] Only a synopsis of different interpretation methods may deliver a reasonable and comprehensive interpretation of the law.[73] Although grammatical interpretation creates strong evidence for a certain way of understanding,[74] the application of only one method of legal interpretation may not suffice. This is particularly the case because semantically clear cases can be legally unclear and semantically unclear cases can be legally clear.[75]

The legislator constitutes the ultimate authority in European democ-

---

67 John Paterson 'Health and Safety at Work Offshore' in Greg Gordon, John Paterson and Emre Üsenmez '*Oil and Gas Law: current practice and emerging trends*' 2nd edition (Dundee University Press, Dundee 2011) paragraphs 8.36–8.69 (hereinafter: Gordon/Paterson/Üsenmez).

68 See Chapter 1 above on 'Shale Gas in a Nutshell'.

69 UK report I, 7/8.

70 Kramer 84; Larenz/Canaris 142/143.

71 Kramer 167/168.

72 Karl Larenz '*Methodenlehre der Rechtswissenschaft*' 2nd edition (Springer, Berlin 1991) 231 (hereinafter: Larenz).

73 Larenz 216 and 231.

74 Kramer 57.

75 Matthias Klatt 'Semantic normativity and the objectivity of legal argumentation' (2004) 90 Archiv für Rechts-und Sozialphilosophie 58.

racies because only the legislator is able to enact a law; his will and the purpose he wished to give to the law is of upmost importance.[76] 'Telos' means objective or aim in Greek[77] and teleological interpretation thus endeavours to liberate the user of the law from slavish and formalistic abidance to words.[78] With the help of systematic, historic and teleological interpretation is it possible to discover what the legislator intended when he created a specific norm, reconstructing in some sense his genuine will.[79] Legal interpretation can then be orientated on the original purpose of the norm and not on the individual views and readings of a single law user.[80]

Historical interpretation of documents on the evolution of the French moratorium seems to suggest that the French legislator wished to react to the emerging societal issue of shale gas extraction as comprehensively as possible.[81] The initiators of the bill wished to ensure that no extraction of unconventional hydrocarbons whatsoever might take place in France.[82] Hence, they proposed a provision that reads: 'unconventional oil and gas extraction in France is illegal'.[83]

However, this proposition never made it into the actual law. Despite

---

76 Hager 54; Krüper § 9 paragraphs 28 et sqq. and 34.

77 Kramer 146.

78 Ibid.

79 Hager 33; Krüper § 9 paragraph 34. Note, however, that Larenz rightly highlighted the difficulties with investigating the 'will' of the legislator in modern democracies, since a huge number of individuals may have come to vote in favour of a certain law. The 'will of the legislator' must hence be confined to the underlying ideas of the initiators of a bill, insofar as they are publicly available, and to the opinions expressed by MPs during parliamentary debates and meetings of parliamentary committees and commissions. Legal interpretation has to orientate itself on what may be deduced from these documents as the perceived 'will' of the legislator, see Larenz 216/217.

80 Krüper § 9 paragraph 17.

81 See for instance: French Senate 'Prohibition of exploration and mining hydrocarbons by hydraulic fracturing. – Discussion Text of the Commission OFFICIAL JOURNAL OF THE FRENCH REPUBLIC REGULAR SESSION 2010–2011 FULL REPORT Meeting on Wednesday, 1 June 2011 (112th sitting day of the session)' (*Interdiction de l'exploration et de l'exploitation des mines d'hydrocarbures par fracturation hydraulique. – Discussion en procédure accélérée de trois propositions de loi dans le texte de la commission SÉNAT JOURNAL OFFICIEL DE LA RÉPUBLIQUE FRANÇAISE SESSION ORDINAIRE DE 2010–2011 COMPTE RENDU INTÉGRAL Séance du mercredi 1er juin 2011 (112e jour de séance de la session)*) 4455, 4459/4460, 4464 available at: http://www.senat.fr/seances/s201106/s20110601/s20110601.pdf [accessed 18 August 2014] (hereinafter: French Senate Minutes 1 June 2011).

82 French Senate Minutes 1 June 2011, 4475–7;

83 Ibid.

the will of the initiators, the reference to all unconventional hydrocarbons in the initial version of article 1 was rejected by a majority of MPs during a meeting of the sustainable development commission on 4 May 2011.[84] Instead, article 1 in its current form was voted in and later resolved by the National Assembly and the Senate.[85] Associated discussions in the French sustainable development commission highlight that the legislator was very aware of the meaning of the initial proposition and consciously voted against the idea of putting a ban on all kinds of unconventional hydrocarbon extraction.[86] Historical interpretation hence does not support the view that the legislator wished to impose a comprehensive prohibition. The explicit will of the legislator, to outlaw only shale gas extraction by hydraulic fracturing, must be respected, since all of the deployed methods of legal interpretation came to a similar conclusion.

However, this outcome of the legal interpretation must be viewed against the backdrop of the status of the industry in France to fully grasp its repercussions. Although article 1 of the French ban might in theory allow shale gas extraction with methods other than hydraulic fracturing, in practice such methods are not yet economically viable.[87] Thus, although a loophole in the law exists, it may not (yet) be exploited in practical terms, as shale gas extraction without hydraulic fracturing is currently not commercially viable.

This assessment has repercussions for the appraisal of the French ban's ability to reconcile environmental protection interests with energy security needs. The domestic extraction of shale gas in France could potentially boost the security of energy supplies by reducing French dependency on gas imports[88] and potentially lower gas prices in the long term.[89] However, because article 1 of the French ban practically excludes shale gas extraction in France, such energy security benefits cannot materialize. The French ban would thus be lopsided towards environmental protection if no other gateways to provide for energy security interests are included in its further articles.

---

[84] French Sustainable Development Commission 2 and 15.

[85] Ibid.

[86] Ibid.

[87] Meiners et al. Bund C 59/C 60.

[88] French National Assembly Minutes 10 May 2011 part 2, 2929, 2932/2933, 2935, 2938; French National Assembly Minutes 10 May 2011 part 1, 2898, 2900.

[89] French National Assembly Minutes 10 May 2011 part 2, 2928 and 2938. The potential energy security effects of shale gas extraction have been discussed in general terms above in Chapter 1.

3.2.2 Article 3 – The Issue of Pre-Existing Licences

A second possible gateway for energy security benefits to come into play is article 3 of the French ban. Prior to the coming into force of the French ban, several licences for gas exploration by hydraulic fracturing had been issued by the French government.[90] Article 3 of the French ban regulates these pre-existing licences. If they could be upheld under the law and the ban only applied to the *future* issuance of licences, shale gas could still be extracted under the old licences. The law would thus provide for environmental protection insofar as it prohibits shale gas extraction in the future. At the same time, however, it would provide for French energy security, as shale gas could be extracted under the pre-existing licences.

However, the grammatical interpretation of article 3 of the French ban does not provide evidence for this. The entire French ban is headed 'Law No. 2011-835 of 13 July 2011 to ban the exploration and exploitation of oil and gas by hydraulic fracturing *and to repeal the exclusive licences of projects using this technique for mining*'.[91] If the law was not intended to target pre-existing licences, the second part of this heading would make no sense. A licence may, logically, only be 'repealed' if it has already been issued by the time that the law that orders its repeal comes into force.

The heading is an expression of conscious intent by the legislator, as grammatical interpretation of the body of the French ban confirms. Paragraph 1 of article 3 prescribes that every holder of an exclusive licence for oil and gas extraction on the French territory must issue a report to the state authorities which explains the techniques used for hydrocarbon extraction. This had to be done within two months after the enactment of the French ban. Article 3 (2) of the French ban stipulates that if the licence holder fails to do so, or his report includes a reference to hydraulic fracturing, his licence will be abrogated. Crucially, neither paragraph limits the number of licence holders who must report to the authorities. Every licence holder must report, which inevitably includes holders of licences that existed prior to the enactment of the French ban. Grammatical interpretation of article 3 of the French ban hence supports the interpretation that this section of the law aims to annul pre-existing licences.

---

[90] Although these licences did not distinguish between unconventional and conventional hydrocarbons, see French National Assembly Minutes 10 May 2011 part 1, 2896/2897 and 2900 and Lauriol paragraph 7.87.

[91] Emphasis added; the French original reads: '*LOI n° 2011-835 du 13 juillet 2011 visant à interdire l'exploration et l'exploitation des mines d'hydrocarbures liquides ou gazeux par fracturation hydraulique et à abroger les permis exclusifs de recherches comportant des projets ayant recours à cette technique*'.

The same conclusion must be drawn from historical and teleological interpretation of article 3. A joint committee of the National Assembly and the Senate moulded the bill that went on to become law no. 2011-835 and it explicitly wished to annul pre-existing licences. This conclusion is based on the mission statement of the committee meeting[92] and the fact that the abrogation of licences was discussed, and explicitly endorsed, during the meeting. In particular, MPs Michel Havard and Pascal Terrasse supported the annulment of pre-existing licences and their demands faced no opposition.[93] It is a recognized method of teleological interpretation to assume that statements of committee members which were not opposed may legitimately be deemed as becoming part of the will of the legislator.[94] The assessment that the legislator deliberately targeted pre-existing licences is further underpinned by the fact that this point has been explicitly discussed in other supporting documents.[95]

These theoretical considerations have been validated by an Order of the French government, dating from 12 October 2011. This order bases itself on article 3 of law no. 2011-835 and revokes three shale gas exploration licences for failing to meet the discussed criteria.[96]

However, the compliance of this action and law no. 2011-835 as a whole with the French constitution was challenged in front of the French Constitutional Court in 2013.[97] The pertaining case was concerned with

---

[92] The heading of the report about the joint committee meeting reads: 'Meeting of the joint committee to propose a text (. . .) and to repeal the exclusive licenses to include projects using this technique' (*Réunie au Sénat le mercredi 15 juin 2011, la commission mixte paritaire chargée de proposer un texte sur les dispositions restant en discussion de la proposition de loi visant à interdire l'exploration et l'exploitation des mines d'hydrocarbures liquides ou gazeux par fracturation hydraulique et à abroger les permis exclusifs de recherches comportant des projets ayant recours à cette technique*), see French Joint Committee on the draft law on the exploration and exploitation of hydrocarbon 'Gathering in the Senate Wednesday, 15 June 2011' (*Commission mixte paritaire sur la proposition de loi relative à l'exploration et l'exploitation des mines d'hydrocarbures Réunie au Sénat le mercredi 15 juin 2011*) available at: http://www.senat.fr/compte-rendu-commissions/20110614/cmp.html#toc4 [accessed 7 May 2014] (hereinafter: France Joint Committee).

[93] France Joint Committee first speeches of Michel Havard and Pascal Terrasse.

[94] Larenz 217; Küper § 9 paragraph 33;

[95] France Report on PP 15.

[96] For details see Lauriol paragraph 7.87.

[97] French Constitutional Court 'Decision no. 2013-346 QPC of 11 October 2013' (no° 2013-346 QPC *Société Schuepbach Energy LLC [Interdiction de la fracturation hydraulique pour l'exploration et l'exploitation des hydrocarbures – Abrogation des permis de recherches]*) available at: http://www.conseil-constitutionnel.fr/

article 3 of the French moratorium and the court had to assess the compliance of this revocation-provision with French constitutional rights and overarching laws.[98]

The French Constitutional Court concluded that, while implementing law no. 2011-835

> (. . .) the legislator has pursued a goal of general interest and environmental protection; that, having regard to current knowledge and the state of the art, the restriction thereby imposed both on the prospecting [for] and exploitation of hydrocarbons resulting from Article 1 of the Law of 13 July 2011 is not disproportionate in nature having regard to the objective pursued (. . .).[99]

As a result, it ruled that the abrogation of pre-existing licences was in line with the French constitution.[100] As the cited statement highlights, the French Constitutional Court merely explored the general interest of environmental protection and whether or not safeguarding that interest by revoking pre-existing licences is disproportionate under 'current knowledge'.[101] It, however, failed to look at 'the bigger picture', especially at two things of upmost relevance for this chapter. First, to consider alternative means of shale gas regulation, which could be better suited to conciliate environmental concerns with energy security interests. Second, it did not elaborate on different interpretations of the precautionary principle as principle of inaction or as principle of cautious action.[102]

The omission of a thorough discussion of alternative shale gas regulation which might conciliate environmental concerns with energy security interests is grounded, first, in procedural reasons. The court actually mentioned the interplay of environmental protection with economic development and, moreover, noted that article 6 of the French Charter for the Environment obliges the state to reconcile both interests.[103] However, the court ruled that this provision does not establish a fundamental right

---

conseil-constitutionnel/english/priority-preliminary-rulings-on-the-issue-of-constitutionality-qpc-/sample-of-decisions-qpc/2013/decision-no-2013-346-qpc-of-11-october-2013.138596.html [accessed 8 November 2013] (hereinafter: French Constitutional Court).

[98] See French Constitutional Court Preface.

[99] French Constitutional Court paragraph 12.

[100] Ibid.

[101] Ibid.

[102] To be discussed further below.

[103] French Constitutional Court paragraph 19.

guaranteed under the Constitution and its violation can therefore not be invoked by the claimant.[104]

Second, although the court recognized the precautionary principle as a legitimate legal principle to steer regulatory decision-making in shale gas cases,[105] it did not discuss how this principle should be interpreted. The court simply deemed any objections on the ground that a ban could conflict with the precautionary principle as 'in any case inoperative'.[106] Unfortunately, the court did not further elaborate on that point, so the reasons for its statement must remain speculative.

3.2.3 Articles 2 and 4 – Shale Gas Research as a 'Back Door' to Extraction?

The preceding result rules out the recognition of energy security interests under article 3 of the French ban. That leaves one last possible gateway for the recognition of these interests: article 2 in conjunction with article 4, that is, the provisions of the French ban which allow for scientific research by means of hydraulic fracturing. These provisions allow for research as long as the research adheres to specific requirements.[107]

Some French MPs feared,[108] others hoped,[109] that this stipulation could

---

[104] Ibid.

[105] French Constitutional Court paragraphs 5 and 20.

[106] French Constitutional Court paragraph 20.

[107] Article 2 in conjunction with article 4 French moratorium.

[108] French National Assembly Minutes 10 May 2011 part 2, 2928, 2933, 2945, 2947; French National Assembly Minutes 11 May 2011 2973, 2975; French National Assembly Minutes 21 June 2011 4380/4381, 4383; French Senate Minutes 1 June 2011 4462, 4468, 4472; French Senate 'Prohibition of exploring for and mining of hydrocarbons by hydraulic fracturing – Further discussion on accelerated procedure of three legislative proposals and adoption of the text of the amended commission OFFICIAL JOURNAL OF THE FRENCH REPUBLIC REGULAR SESSION OF FULL SENATE REPORT 2010-2011 Session Thursday, 9 June 2011 (115th sitting day of the session)' (*Interdiction de l'exploration et de l'exploitation des mines d'hydrocarbures par fracturation hydraulique. – Suite de la discussion en procédure accélérée de trois propositions de loi et adoption du texte de la commission modifié JOURNAL OFFICIEL DE LA RÉPUBLIQUE FRANÇAISE SESSION ORDINAIRE DE 2010–2011 COMPTE RENDU INTÉGRAL SÉNAT Séance du jeudi 9 juin 2011 (115e jour de séance de la session)*) available at: http://www.senat.fr/seances/s201106/s20110609/s20110609.pdf [accessed 7 May 2014] 4715, 4718, 4724, 4733/4734 (hereinafter: French Senate Minutes 9 June 2011).

[109] French National Assembly Minutes 10 May 2011 part 2 2928, 2935; French National Assembly Minutes 21 June 2011 4377, 4386; French Senate Minutes 1 June 2011 4449/4450, 4452, 4478.

serve as a 'back door' to introduce shale gas extraction by hydraulic fracturing in France. According to a number of French MPs, shale gas production could be disguised as scientific experiments and the extracted gas could be used for French electricity production.[110] If large enough, the amounts of shale gas produced in that way could boost French energy security, they argued.[111] Other MPs reckoned the practical result of the research provisions of article 2 in conjunction with article 4 of the French ban would be that shale gas could be extracted, while the environment would not be destroyed.[112] Thus, the research provision of article 2 in conjunction with article 4 of law 2011-835 could provide for an indirect gateway for energy security interests to come into play.

Despite these hopes, grammatical interpretation of articles 2 and 4 of the French ban shows that the legislator attached several conditions to the conduct of research by means of hydraulic fracturing. First of all, the research must comply with guidelines of a national commission.[113] By 21 March 2012 Decree No. 2012-385 set up such a national commission.[114] It assesses and advises on the environmental risks of hydraulic fracturing techniques or alternative techniques.[115]

In addition, scientific experiments may only be conducted under public control, as article 2 stipulates. All of these constraints suggest that the legislator did not wish to allow for a broad interpretation of this possibility, but that a restrictive handling of these research provisions is intended.

Moreover, the research must be explicitly 'scientific' in nature.[116] The production of gas by industry professionals and/or the marketing of extracted shale gas would be difficult to reconcile with this. If the legislator chooses to use expressions of common language, such as 'scientific research', rather than professional terminology, this intention must also be respected.[117] Grammatical interpretation must hence orientate itself

---

110 French National Assembly Minutes 10 May 2011 part 1 2898; French National Assembly Minutes 21 June 2011 4384.

111 French National Assembly Minutes 10 May 2011 part 2 2928, 2935; French National Assembly Minutes 21 June 2011 4377, 4386; French Senate Minutes 1 June 2011 4449/4450, 4452, 4478.

112 Ibid.

113 According to article 2 of the French ban.

114 Decree No. 2012-385 of 21 March 2012 related to the creation of a national commission of guidance, monitoring and evaluation of the techniques used for the exploration and exploitation of liquid and gaseous hydrocarbons.

115 Article 2 in conjunction with article 4 of law no. 2011-835; for more see Lauriol paragraph 7.87.

116 Article 2 of law no. 2011-835.

117 Larenz 208.

on the perception of the general public in these cases.[118] An analysis of the wording of article 2 and article 4 of the French moratorium thus suggests that energy security interests cannot be satisfied by the use of these research provisions for shale gas extraction.

This assessment is also supported by systematic interpretation of the paragraphs in question. Besides grammatical interpretation, a norm may require systematic interpretation to classify it within a certain law and/or area of law.[119] The grouping of the law text and the sequence of paragraphs and single articles might be used to interpret the law's content in these cases.[120] While article 1 of the French ban sets the general rule that oil and gas extraction by hydraulic fracturing should be prohibited, articles 2 and 4 establish an exception to that rule. Legal exceptions, however, may not be interpreted so broadly that their use would foil the general rule.[121] For shale gas to have any perceivable impact on French energy security, its extraction would have to take place at a considerable scale. The amounts of gas that could be produced during scientific experiments would be too small to achieve such impact, except where commercial-scale production is disguised as research. Endorsing such production under article 2 in conjunction with article 4 of the French ban would turn the rule-exception interrelation with article 1 upside down.

Historical and teleological interpretation show that the research provision was hardly ever debated during the legislative procedure.[122] The stipulation appeared relatively late, during a meeting of a joint committee and was thought of as a political compromise to gather sufficient support from French MPs for the bill.[123] It was the task of a joint committee of both chambers of parliament to produce a single, unified bill out of five different proposals that existed prior to the meeting of the joint committee.[124] The Senate initially favoured a proposal that would have allowed

---

[118] Ibid.

[119] Krüper § 9 paragraph 25.

[120] Kramer 91/92.

[121] Larenz 213/214.

[122] See for instance the exhaustive discussions of these stipulations in the French Joint Committee meeting.

[123] France Joint Committee Speeches of MPs Michel Havard, Claude Biwer, Ladislas Poniatkowski, Franck Riester.

[124] French Senate 'Prohibition of exploring for and mining of hydrocarbons by hydraulic fracturing – Discussion in the accelerated procedure on three legislative proposals in the text of the commission OFFICIAL JOURNAL OF THE FRENCH REPUBLIC REGULAR SESSION OF FULL SENATE REPORT 2010–2011 Meeting on Wednesday, 1 June 2011 (112th sitting day of the session)' (*Interdiction de l'exploration et de l'exploitation des mines d'hydrocarbures par*

experiments with hydraulic fracturing. Crucially, this proposition did not define who was allowed to conduct these experiments, so that commercial players would have been free to extract shale gas for 'experimental' purposes.[125] This Senate proposal, however, was rejected and the joint committee decided to adopt the current article about *scientific* research that has to be conducted under public control.[126] The tightening up of the research exemption during the legislative process shows that the legislator wished to implement a very narrow exception, which does not lend itself to any form of disguised commercial shale gas production.

To sum up, the legality of exploiting this research 'loophole' cannot be supported by any means of legal interpretation. The research provisions of article 2 in conjunction with article 4 of the French ban are unsuitable as a gateway for inclusion of energy security interests. As all of the three possibilities to cater indirectly for energy security interests had to be rejected, the French ban does not address energy security interests. The French law no. 2011-835 is hence unable to cater for such interests.

3.2.4 The Precautionary Principle and its French Interpretation as Basis of the Ban

Although this assessment clarifies the actual meaning of the French ban, it does not explain the underlying reasons for its rigour. Such reasons shall now be investigated. They might be found in the particular interpretation of the precautionary principle that is favoured by the French legislator.

A 2010 report of the French parliament established that the precautionary principle should be considered as basis and guideline for any decision-making on shale gas in France.[127] Subsequent parliamentary debates on the French shale gas ban focussed on that point and on the correct implementation of the precautionary principle in the case of shale gas

fracturation hydraulique. – Discussion en procédure accélérée de trois propositions de loi dans le texte de la commission JOURNAL OFFICIEL DE LA RÉPUBLIQUE FRANÇAISE SESSION ORDINAIRE DE 2010–2011 COMPTE RENDU INTÉGRAL Séance du mercredi 1er juin 2011 (112e jour de séance de la session)) 4449, 4463, 4470 available at: http://www.senat.fr/seances/s201106/s20110601/s20110601.pdf [accessed 9 May 2014] (hereinafter: French Senate Minutes 1 June 2011); French Senate Minutes 9 June 2011 4714; French National Assembly Minutes 21 June 2011 4378.

125 France Joint Committee Speech of Michel Havard.

126 Article 4 French moratorium.

127 According to: French National Assembly Minutes 21 June 2011 4378; France Report on PP, paragraph 1) a).

extraction.[128] The most common definition of the precautionary principle requires that when reasonable grounds of concern about a certain activity exist but an actual risk cannot (yet) be scientifically proven, precautionary actions must be taken.[129]

A considerable number of French MPs, however, interpreted the principle slightly differently. From their point of view, precaution means that in case of scientific uncertainties about environmental repercussions of a certain activity, environmental protection interests take precedence over all other interests.[130] Accordingly, only a ban on shale gas extraction can serve the precautionary principle correctly, as it is the only measure giving full precedence to environmental protection interests.[131]

However, a minority of French MPs opposed the construction 'implementation of the precautionary principle equals prohibition': they considered the described interpretation of the precautionary principle as too

---

128 French National Assembly Minutes 10 May 2011 part 1 2901; French National Assembly Minutes 10 May 2011 part 2 2911, 2923, 2927, 2932, 2938; French National Assembly Minutes 21 June 2011 4378; French Senate Minutes 1 June 2011 4450, 4461, 4470, 4475; French Senate Minutes 9 June 2011 4737.

129 European Court of Justice Case C-180/96 *United Kingdom of Great Britain and Northern Ireland v Commission of the European Communities* [1998] ECR I-2265, paragraph 105; Case T-70/99 *Alpharma Inc. v Council of the European Union* [2002] ECR II-03495, paragraph 152; Case C-236/01 *Monsanto Agricoltura Italia Spa and Others v Presidenza del Consiglio dei Ministri and Others* [2003] ECR I-08105, paragraph 106; Case C-192/01 *Commission of the European Communities v Kingdom of Denmark* [2003] ECR I-09693, paragraph 49; Case C-24/00 *Commission of the European Communities v French Republic* [2004] ECR I-01277, paragraph 56. German courts ruled in similar vein: German Federal Administrative Court (*Bundesverwaltungsgericht*): BVerwGE 69, 43; BVerwG NVwZ 1986, 208, paragraph 37 and German Federal Constitutional Court (*Bundesverfassungsgericht*): BVerfG, NJW 2002, 1638 (1639). Furthermore, James Cameron, Will Wade-Gery and Juli Abouchar 'Precautionary Principle and Future Generations' in: Emmanuel Agius '*Future Generations and International Law*' (Earthscan Ltd., London 1998) 99; Astrid Epiney and Martin Scheyli '*Strukturprinzipien des Umweltvölkerrechts*' (Nomos Verlag, Baden-Baden 1998) 109/110 (hereinafter: Epiney/Scheyli).

130 French National Assembly Minutes 10 May 2011 part 1 2901; French National Assembly Minutes 10 May 2011 part 2 2911, 2923, 2927, 2932, 2938; French National Assembly Minutes 21 June 2011 4378; French Senate Minutes 1 June 2011 4450, 4461, 4470, 4475; French Senate Minutes 9 June 2011 4737.

131 French National Assembly Minutes 10 May 2011 part 2 2923, 2927, 2932, 2938; French National Assembly Minutes 21 June 2011 4378; French Senate Minutes 1 June 2011 4450, 4461, 4470, 4475; French Senate Minutes 9 June 2011 4737.

far reaching.[132] One MP feared that the interpretation of environmental interests as soaring above all other interests could produce a 'paralyzed and anxious society' ('*société figée et apeurée*').[133] He argued that the precautionary principle could become a principle of inaction, prohibiting any progressive activity only because it involves potential threats.[134]

Despite these sporadic declarations of discontent, the conclusion that shale gas extraction must be met by prohibitive regulation due to overarching environmental law principles appears to be very widespread among French MPs. This is epitomized by the fact that parliamentary discussions did not focus on the question whether a rigorous interpretation of the precautionary principle would be most appropriate. Instead, the debate was mainly concerned with the question of whether the precautionary principle or the, even stricter,[135] preventive principle should be invoked in shale gas cases.[136]

As opposed to the precautionary principle, the preventive principle applies when a threat has been scientifically proven and its existence may be deemed certain.[137] The preventive principle is stricter than the precautionary principle in the sense that it requires factual proof of the actual existence of a threat. For the invocation of the precautionary principle, reasonable grounds of concern about the existence of a potential threat suffice.[138] A number of MPs considered the potential environmental and health threats of shale gas extraction as proven and consequently wanted to apply the preventive principle.[139] The dispute among French

---

[132] See France Report on PP, paragraph I) A) 1) b); French National Assembly Minutes 21 June 2011 4385, 4390; French National Assembly Minutes 10 May 2011 part 2 2919, 2923; French Senate Minutes 1 June 2011 4472.

[133] French National Assembly Minutes 21 June 2011 4385.

[134] Ibid.

[135] For the terms of the threshold that is set for (potential) threats, see immediately below.

[136] See France Report on PP, paragraph I) A) 1) b); French National Assembly Minutes 21 June 2011 4390; French National Assembly Minutes 10 May 2011 part 2 2919, 2923; French Senate Minutes 1 June 2011 4472.

[137] Nicolas de Sadeleer '*Environmental Principles: From Political Slogans to Legal Rules*' (Oxford University Press, 2002) 74/75 and 222 (hereinafter: de Sadeleer); David Freestone 'International Fisheries Law Since Rio: The Continued Rise of the Precautionary Principle' in: Alan Boyle and David Freestone (eds) '*International Law and Sustainable Development: Past Achievements and Future Challenges*' (Oxford University Press, Oxford 1999) 139 (hereinafter: Freestone 1999).

[138] De Sadeleer 74/75 and 222; Freestone 1999, 139.

[139] France Report on PP, paragraph I) A) 1) b); French National Assembly Minutes 21 June 2011 4390; French National Assembly Minutes 10 May 2011 part 2 2919, 2923; French Senate Minutes 1 June 2011 4472.

MPs became so immanent to the debate that it was only settled by the inclusion of an explicit reference in the actual text of law no. 2011-835. Article 1 of the French law now states that the ban is based equally on the 2004 Environmental Charter (*Charte de l'environnement de 2004*) and on article 1.110-1 of the French Environmental Code.[140] These two articles respectively include the precautionary and the preventive principle.[141] The question of whether anything other than a prohibitive approach to shale gas regulation would be possible was not on the French agenda.

3.2.5 Summary France and Shale Gas

The analysis of the French 'ban by law' revealed that the French legislator wished to stop shale gas extraction as comprehensively as possible with immediate effect. It was not of upmost importance whether energy security interests were taken into account. Three possible gateways in the text of the law lend themselves to facilitating indirect recognition of energy security interests. However, neither of the three 'back doors' sustains closer scrutiny, as the French regulator has sealed off all of the three for the foreseeable future.

As a result, the French 'ban by law' renders every effort to cautiously advance shale gas extraction in France impossible. Instead, the interpretation of the precautionary principle as a principle of inaction by a significant number of French MPs required the putting in place of a comprehensive ban. This strict nature makes a 'ban by law' rather unsuitable for reconciling the interests of environmental protection and energy security in shale gas cases. It is hence not an apt role model for regulation on shale gas extraction in Member States which are obliged by constitutional law to create regulations that take both interests into account.

3.3 'MORATORIUM BY LAW' IN GERMANY

3.3.1 Introduction

Shortly before 12am on Friday 24 June 2016 the lower chamber of the German parliament (hereinafter: *Bundestag)* resolved a package of legal

---

140 See article 1 of law no. 2011-835.

141 Article 1 of the French moratorium now reads: '*En application de la Charte de l'environnement de 2004 et du principe d'action préventive et de correction prévu à l'article L. 110-1 du code de l'environnement*'. For the debate see: French Senate Minutes 1 June 2011 4460.

measures that widely outlawed the use of hydraulic fracturing for gas and oil extraction.[142] The timing of this final resolution was rather peculiar, triggering speculation by some members of the German parliament (hereinafter: MPs) that the fracking law was supposed to be 'rushed' through parliament while public attention was diverted to other topics.[143] Only five hours before that vote on fracking, the British Electoral Commission announced the final result of the so called 'Brexit' referendum – Britain had just voted to leave the European Union.[144] At the same time the 2016 European football championship was well underway in France.[145] On 24 July 2016, the single most important topic on Twitter was 'Brexit', while some users were discussing the European football championship and virtually nobody was talking about the German fracking law.[146] Either the Germans had suddenly lost interest in the topic or the other important 'news items' of that day were much more present in the media.

The so called fracking package[147] outlaws the usage of hydraulic frac-

---

142 Bundestag 'Fracking vote by name' (Fracking Namentliche Abstimmung) available at: http://www.bundestag.de/bundestag/plenum/abstimmung/grafik [accessed 29 June 2016] (hereinafter: Fracking vote by name). The upper chamber (hereinafter: *Bundesrat*) followed suit on 8 July 2016, see Bundesrat 'Drucksache 358/16 (Beschluss des Bundesrates)' available at: http://www.bundesrat.de/SharedDocs/drucksachen/2016/0301-0400/358-16(B).pdf?__blob=publicationFile&v=1 [accessed 13 July 2016] (hereinafter: Bundesrat Drucksache 358/16) and Bundesrat 'Drucksache 353/16 (Beschluss des Bundesrates)' available at: http://www.bundesrat.de/SharedDocs/drucksachen/2016/0301-0400/353-16(B).pdf?__blob=publicationFile&v=1 [accessed 13 July 2016] (hereinafter: Bundesrat Drucksache 353/16).

143 Speech of Hubertus Zdebel Bundestag 'Stenografischer Bericht 180. Sitzung Plenarprotokoll 18/180' at 17795 available at: http://dipbt.bundestag.de/dip21/btp/18/18180.pdf#P.17790 [accessed 30 June 2016] (hereinafter: Stenografischer Bericht 18/180).

144 The Electoral Commission '*EU Referendum count processes and results*' available at: http://www.electoralcommission.org.uk/__data/assets/pdf_file/0013/206113/Media-briefing-EU-Referendum-count-processes-and-results.pdf [accessed 29 June 2016].

145 UEFA 'Uefa Euro Championship 2016' available at: http://www.uefa.com/uefaeuro/index.html [accessed 5 July 2016].

146 Twitter .com 'Google Trends' available at: https://twitter.com/googletrends/status/746303118820937728 [accessed 29 June 2016].

147 Bundestag 'Drucksache 18/4713 Bill of the German Federal Government on the alteration of water and environmental protection norms with the aim of prohibiting and minimizing risks associated with the procedures of the fracking technology' (*Gesetzentwurf der Bundesregierung 'Entwurf eines Gesetzes zur Änderung wasser- und naturschutzrechtlicher Vorschriften zur Untersagung und zur Risikominimierung bei den Verfahren der Fracking-Technologie*) available at: http://dipbt.bundestag.de/doc/btd/18/047/1804713.pdf [accessed 7 April

turing for the extraction of hydrocarbons from certain types of rock, whereas extraction from other rock-types by means of hydraulic fracturing remains licit.[148] The new regulatory framework of the fracking package revolves around six themes:

–a particular German definition of so called 'unconventional' and 'conventional' fracking
–the relationship between the package and the German constitution
–the legal nature of the package
–the prohibition of fracking in water protection areas and the use of Best Available Techniques
–liability and the reversal of the burden of proof
–the obligatory conduct of EIAs for all fracking activities.

These features are going to be assessed individually and the policy processes that led to their emergence will be traced. The section proceeds as follows: it starts off with a short outline of the history of (failed) attempts to install a comprehensive, nationwide regulatory framework on fracking[149] in Germany. Subsequently the section focuses on each of the individual features of the fracking package and scrutinizes these features

---

2015] (hereinafter: Drucksache 18/4713); Bundestag 'Drucksache 18/4714 German Government Draft Bill to expand the Mining Damage Presumption to borehole mining and caverns' (*Bundesregierung Entwurf eines Gesetzes zur Ausdehnung der Bergschadenshaftung auf den Bohrlochbergbau und Kavernen*) available at: http://dipbt.bundestag.de/doc/btd/18/047/1804714.pdf [accessed 7 April 2015] (hereinafter: Drucksache 18/4714); Bundesrat 'Drucksache 144/15 German Ministry of the Economy and Energy Ordinance on the introduction of Environmental Impact Assessments and on mining requirements in deployment of the fracking technology and deep drills' (*Verordnung zur Einführung von Umweltverträglichkeitsprüfungen und über bergbauliche Anforderungen beim Einsatz der Fracking-Technologie und Tiefbohrungen*) available at: http://www.bundesrat.de/SharedDocs/beratungsvorgaenge/2015/0101-0200/0144-15.html;jsessionid=86AAD3036B5C2CDA8A8A2B4B565717AF.2_cid349?cms_templateQueryString=Suchbegriff&cms_fromSearch=true [accessed 7 April 2016] (hereinafter: Bundesrat Drucksache 144/15).

148 See note 141 above. http://www.bundesrat.de/SharedDocs/drucksachen/2016/0301-.

149 The technology will be explained below. The term 'fracking' is inaccurate when it is used to describe hydraulic fracturing generically, since 'fracking' could also mean pneumatic fracturing or other fracturing techniques, which are not in the spotlight of the public debate. However, the German legislator uses the terminology 'fracking'. In order to facilitate ease of reading and coherence this author adopts the lax German terminology and refers to fracking, when discussing hydraulic fracturing.

against the backdrop of guidance documents, studies, etc. As a consequence, preliminary conclusions will be drawn.

3.3.2 Evolution of the German Fracking Package

The first initiatives to develop particular rules for the regulation of fracking and shale gas extraction were launched in the upper chamber of parliament (hereinafter: *Bundesrat*) and the *Bundestag* in 2011.[150] In 2013 the then German government pushed for the passage of a shale gas law but in the end failed to get one adopted and the project was withdrawn from the government's agenda.[151]

Germany is a federal republic and the federal states (hereinafter: *Länder*) are entitled to resolve individual regulations in the field of mining, as long as the nation state has not put in place regulation for particular mining issues.[152] Up until 2016 some of the German *Länder*, namely those

---

[150] Bundestag 'Motion guidance notes on transparency and environmental soundness during unconventional gas production of 8 November 2011' (*Antrag Leitlinien für Transparenz und Umweltverträglichkeit bei der Förderung von unkonventionellem Erdgas*) Bundestagsdrucksache 17/7612 available at: http://dipbt.bundestag.de/dip21/btd/17/076/1707612.pdf [accessed 16 April 2014]; Bundesrat 'Resolution of the *Bundesrat* on the handling of the application of fracking-technologies with environmentally toxic chemicals during the exporation and production of unconventional deposits of 3 December 2012' (*Entschließung des Bundesrates zum Umgang mit dem Einsatz von Fracking-Technologien mit umwelttoxischen Chemikalien bei der Aufsuchung und Gewinnung von Erdgas aus unkonventionellen Lagerstätten*) Bundesratsdrucksache 754/12 available at: http://www.bundesrat.de/SharedDocs/beratungsvorgaenge/2012/0701-0800/0754-12.html [accessed 15 April 2014].

[151] *Die Zeit* 'Fracking-Gesetz scheitert am schwarz-gelben Streit 4 June 2013' available at: http://www.zeit.de/wirtschaft/2013-06/fracking-gesetz-union-fdp [accessed 16 April 2015]; Cairney/Fischer/Ingold; Chailleux/Moyson.

[152] In the German legal system, environmental regulation and regulation pertaining to mining and energy extraction are competences that are shared (*konkurrierende Gesetzgebung*) between the nation state (*Bund*) and the German states (*Länder*). With regard to environmental regulation, this shared competence is prescribed by article 74 No 24, 29 and 32 in conjunction with article 72 (3) No 2 and 5 German constitution for air protection, water protection and general environmental protection, see Hans Dieter Jarass and Bernd Pieroth '*Grundgesetz Kommentar*' 11th edition (C H Beck, München 2011) article 74 paragraphs 69 and 79. With regard to mining activities and the energy industry in general, this shared competence is prescribed by article 74 (1) No 11 in conjunction with article 72 (1) German constitution. Article 74 (1) No 11 not only provides the legislator with the power to introduce general regulations on mining and energy extraction, but also reserves to the legislator the right of bringing in additional legislation to regulate new energy technologies, such as shale gas extraction, see Josef Isensee and Paul

boasting the biggest shale gas reserves[153] (North Rhine-Westphalia,[154] Lower Saxony[155] and Hesse[156]) imposed moratoria on shale gas extraction on their respective territories. But a nationwide German moratorium had not been put in place. Due to the system of 'shared' or 'competing' legislative competence under the German constitution, the *Länder* regulations will now be replaced automatically with the regulations of the 2016 fracking package.[157]

On 1 April 2015 the national government, the so called cabinet (*Bundeskabinett*), decided it was time to take the initiative and put forward a package of legal measures.[158] This fracking package consists of three different proposals, two bills and one draft ordinance.[159]

Kirchhof *'Handbuch des Staatsrechts Band IV'* (C F Müller, Heidelberg 1990) § 100 paragraph 173.

[153] According to Niedersachsen Landesamt für Bergbau, Energie und Geologie '*Erdöl und Erdgas in der Bundesrepublik Deutschland 2012*' (Hannover, 2013) 42/43 available at: http://www.lbeg.niedersachsen.de/portal/live.php?navigation_id=655&article_id=936&_psmand=4 [accessed 15 April 2014].

[154] Ministerium für Klimaschutz, Umwelt, Landwirtschaft, Natur- und Verbraucherschutz des Landes Nordrhein-Westfalen '*Pressemitteilung Umweltministerium und Wirtschaftsministerium legen Risikogutachten zu Fracking vor*' 7 September 2012 available at: http://www.umwelt.nrw.de/ministerium/service_kontakt/archiv/presse2012/presse120907_a.php [accessed 15 April 2014].

[155] Niedersächsisches Ministerium für Umwelt, Energie und Klimaschutz '*Zulassung von Vorhaben zur Aufsuchung und Gewinnung von Erdgas aus konventionellen Lagerstätten mittels hydraulischer Bohrlochbehandlung zur Risserzeugung in einem Verfahren mit Umweltverträglichkeitsprüfung*' available at: http://www.umwelt.niedersachsen.de/aktuelles/pressemitteilungen/strenge-auflagen-bei-der-ergasfoerderung-122495.html [accessed 14 April 2014]; *Hannoversche Allgemeine* 'Die Fracking-Pause dauert bis 2016' available at: http://www.haz.de/Nachrichten/Politik/Niedersachsen/Die-Fracking-Pause-dauert-bis-2016-in-Niedersachsen [accessed 9 January 2015].

[156] Hessisches Ministerium für Umwelt, Klimaschutz, Landwirtschaft und Verbraucherschutz 20.08.2014 'Fracking-Klage gegen das Land Hessen zurückgenommen' available at: https://www.hessen.de/presse/pressemitteilung/fracking-klage-gegen-das-land-hessen-zurueckgenommen-0 [accessed 9 January 2015].

[157] Subject only to the signature of the German president, a rather formal act.

[158] German Government (*Bundesregierung*) 'Kabinettbeschluss Fracking: Mehr Schutz durch strenge Regeln' available at: http://www.bundesregierung.de/Content/DE/Artikel/2015/04/2015-04-01-fracking-gesetz-kabinett.html;jsessionid=7198048376FAAF673F3B194C34B7DE85.s2t2 [accessed 7 April 2015].

[159] Bundestag 'Drucksache 18/4713 Bill of the German Federal Government on the alteration of water- and environmental protection norms with the aim of prohibiting and minimizing risks associated with the procedures of the fracking technology' (*Gesetzentwurf der Bundesregierung 'Entwurf eines Gesetzes zur*

The most important one[160] is the bill on water protection provisions and the prohibition and risk-minimization of the procedures of the fracking technology (hereinafter: Bill on water protection), which proposes certain alterations to the Water Protection Act (*Wasserhaushaltsgesetz* hereinafter: *WHG*).[161] A recent study, commissioned by the German Federal Environmental Ministry, identified two laws as most important for the regulation of shale gas extraction.[162] These are the *WHG* on the one hand and German Mining Law (especially the Federal Mining Act (*Bundesberggesetz (BBergG)*) and several applicable Mining and Environmental Ordinances) on the other hand.[163]

According to § 19 (2) and (3) Water Protection Act, mining authorities are obliged to consult water authorities if an operating schedule requires the usage of water and/or groundwater. As this example highlights, the German Water Protection Act (*Wasserhaushaltsgesetz*) and the German Federal Mining Act include cross-references but are juxtaposed to each other.[164] One

---

Änderung wasser- und naturschutzrechtlicher Vorschriften zur Untersagung und zur Risikominimierung bei den Verfahren der Fracking-Technologie) available at: http://dipbt.bundestag.de/doc/btd/18/047/1804713.pdf [accessed 7 April 2015] (hereinafter: Drucksache 18/4713); Bundestag 'Drucksache 18/4714 German Government Draft Bill to expand the Mining Damage Presumption to borehole mining and caverns' (*Bundesregierung Entwurf eines Gesetzes zur Ausdehnung der Bergschadenshaftung auf den Bohrlochbergbau und Kavernen*) available at: http://dipbt.bundestag.de/doc/btd/18/047/1804714.pdf [accessed 7 April 2015] (hereinafter: Drucksache 18/4714); Bundesrat 'Drucksache 144/15 German Ministry of the Economy and Energy Ordinance on the introduction of Environmental Impact Assessments and on mining requirements in deployment of the fracking technology and deep drills' (*Verordnung zur Einführung von Umweltverträglichkeitsprüfungen und über bergbauliche Anforderungen beim Einsatz der Fracking-Technologie und Tiefbohrungen*) available at: http://www.bundesrat.de/SharedDocs/beratungsvorgaenge/2015/0101-0200/0144-15.html;jsessionid=86AAD3036B5C2CDA8A8A2B4B565717AF.2_cid349?cms_templateQueryString=Suchbegriff&cms_fromSearch=true [accessed 7 April 2016] (hereinafter: Bundesrat Drucksache 144/15).

160 Insofar as it includes the fundamental decisions on fracking in Germany, see Drucksache 18/8916 at 17 and Stenografischer Bericht 18/180 at 17790.

161 Drucksachen 18/4713 and 18/8916. Germany adopted a new Water Protection Act (*Wasserhaushaltsgesetz*) in 2009, with the overall aim of unifying water protection stipulations at the national level, see Manfred Czychowski and Michael Reinhardt '*Wasserhaushaltsgesetz*' 10th edition (Beck, München 2010) Einleitung paragraphs 9 and 12.

162 Meiners et al. Bund B1.

163 Ibid.

164 Meiners et al. Bund B 9.

applies simultaneously alongside the other,[165] as the explanatory memorandum to the Federal Mining Act clarifies.[166]

The hitherto existing Water Protection Act already contained at least seven provisions, which impact upon the regulation of German shale gas extraction.[167] The bill on water protection moves beyond the existing legal framework and includes detailed provisions on the prohibition of certain usages of hydraulic fracturing.

The second bill is concerned with liability and mining damages.[168] The third legislative proposal is a state ordinance, aiming to ensure that Environmental Impact Assessments (hereinafter: EIAs) are conducted for every shale gas extraction project.[169]

The package thus did not propose the establishment of a distinct and stand-alone legal regime. Instead, it operates within the confines of the existing regulatory framework on hydrocarbon extraction and is fine-tuning the regime for the purpose of fracking and shale gas extraction.[170]

Only one month after this package had been introduced to the German parliamentary procedure, the upper chamber, the *Bundesrat*, announced its view on the bills and the German government produced a rebuttal by 20 May 2015.[171] The overall stance of the *Bundesrat* was critical towards the proposal, highlighting that the suggested amendments did not go far enough in some respects.[172] After this 'clash' between the German

---

165 Piens/Schulte/Graf Vitzthum § 1 paragraph 9.

166 Gesetzentwurf der Bundesregierung Entwurf eines Bundesberggesetzes Bundestagsdrucksache 8/1315 at 87 available at: http://suche.bundestag.de/search_bt.do?actualPage=165&resultsSubCategoryFilter=&sort=da&queryAll=Energie&tab=all&datumBis=%20&oneCategoryOnlySearch=false&drucksachennr=false&schlagwort=false&titel=false&resultsPerPage=10&resultsCategoryFilter=&displayCategories=true&language=de&datumVon= [accessed 8 January 2015].

167 For a detailed account, see Meiners et al. Bund B 15 – B 142.

168 Drucksache 18/4714.

169 Bundesrat Drucksache 144/15.

170 The package as such is applicable to all forms of 'unconventional' gas extraction, but targets shale gas extraction specifically.

171 Bundestag 'Drucksache 18/4949 Report by the government concerning Drucksache 18/4713 Position of the Bundesrat and rebuttal of the government' available at http://dipbt.bundestag.de/doc/btd/18/049/1804949.pdf [accessed 5 July 2016] (hereinafter: Drucksache 18/4949); Bundestag '*Drucksache 18/4952 Report by the government according Drucksache 18/4714 Position of the Bundesrat and rebuttal of the government*' available at http://dipbt.bundestag.de/doc/btd/18/049/1804952.pdf [accessed 5 July 2016] (hereinafter: Drucksache 18/4952).

172 Drucksache 18/4949 at 1-10; Drucksache 18/4952 at 1-4.

government and the German *Bundesrat*, discussions on the proposal were stalled.[173]

The slowing down of the legislative process, however, was not due to the different views that prevailed in the *Bundesrat* and the *Bundestag*, but rather to a fierce conflict within the political parties, the CDU/CSU and the SPD, the parliamentary basis of the government in both chambers. A considerable number of MPs from these three governing parties opposed the proposals of their own government, arguing behind closed doors that the government proposal was too lenient.[174] The conflict simmered for a year with individual MPs demanding that the local interests of their particular constituencies needed to be represented in the future national regulation on fracking.[175]

In June 2016, this conflict escalated when rebellious MPs were getting the upper hand. They modified the government bills substantially, turning what might be considered as a well-balanced approach into a strict prohibition on the use of fracking. Final victory for the rebellion was in sight when they launched their considerable amendments to the government bill in the respective committees of the *Bundestag* on 22 June 2016,[176] taking

[173] This is most apparent when looking at the long journey of the bills through the respective parliamentary committees. On 7 May 2015 the two bills were put before the environmental committee and the committee of economic affairs and energy of the *Bundestag* respectively, see Drucksache 18/8916 at 7 and Bundestag Drucksache 18/8907 'Recommendation and Report of the Committee for Economic Affairs and Energy concerning Drucksache 18/4714 (*Beschlussempfehlung und Bericht des Ausschusses für Wirtschaft und Energie zu Drucksache 18/4714*)' available at: http://dipbt.bundestag.de/doc/btd/18/089/1808907.pdf [accessed 5 July 2016] at 7 (hereinafter: Drucksache 18/8907). However, they were not resolved by the committees until 22 June 2016, see below and Drucksache 18/8916 at 1, 7 and 22.

[174] *Der Spiegel* 'Kritik im Bundestag: Dutzende Abgeordnete torpedieren umstrittenes Fracking-Gesetz' available at: http://www.spiegel.de/politik/deutschland/fracking-teile-von-cdu-spd-und-gruenen-gegen-gesetz-a-1026585.html [accessed 7 April 2015].

[175] A good example is provided by the speech of Andreas Jung, MP in the Bundestag on 24 June 2016, who reflects on his efforts to push the particular hydrological interest of the Bodensee region into the limelight, see Stenografischer Bericht 18/180 at 17802/17803.

[176] Bundestag 'Drucksache 18/8916 Recommendation and Report of the Committee for Environment, Nature Protection, Building and Reactor-safety concerning Drucksache 18/4713 (*Beschlussempfehlung und Bericht des Ausschusses für Umwelt, Naturschutz, Bau und Reaktorsicherheit zu Drucksache 18/4713*)' available at: http://dipbt.bundestag.de/doc/btd/18/089/1808916.pdf [accessed 5 July 2016] (hereinafter: Drucksache 18/8916); Bundestag 'Drucksache 18/8907 Recommendation and Report of the Committee for Economic Affairs and Energy concerning Drucksache 18/4714 (*Beschlussempfehlung und Bericht des Ausschusses*

opposition and government equally by surprise.[177] Only two days later the parties resolved the substantially amended bills during the final reading in the *Bundestag* on 24 June 2016 and the *Bundesrat* followed suit on Friday 8 July 2016. The successful attempt to amend the governmental bill was heralded by MPs of these governing parties as a 'gigantic success' and a '*Sternstunde des Parlaments*' (moment of glory for parliament; *Sternstunde* literally meaning sidereal hour).[178] So, what is the precise content of the newly resolved legislative package on fracking?

3.3.3 Main Features of the German Fracking Package

3.3.3.1 'Unconventional' and 'conventional' fracking

The first important feature is a differentiation by the German legislator between so called conventional and unconventional fracking. This differentiation is particularly present in the Bill on water protection.[179] It departs from the view that fracking activities could conflict with the sustainable management of German water resources, particularly with the objective to maintain and safeguard present and future uses of public water supplies.[180] The use of fracking, according to the German government, could lead to the risk of groundwater and drinking water contamination.[181]

To fully understand the regulations that are included in the German fracking package it is important to recall that gas may be produced by hydraulic fracturing from several rock-types, not only from shale. Unconventional gas may also be present in layers of coal and sandstone. However, the production methods for the extraction of all types of unconventional hydrocarbons are quite similar. All of the described 'trapping' rocks are low permeability structures.[182]

---

für Wirtschaft und Energie zu Drucksache 18/4714)' available at: http://dipbt.bundestag.de/doc/btd/18/089/1808907.pdf [accessed 5 July 2016] (hereinafter: Drucksache 18/8907).

177 The opposition criticized the short notice with which the proposals were put before the committees and called the procedure undemocratic, see Drucksache 18/8916 at 17 and Speeches of Hubertus Zdebel and Julia Verlinde, Stenografischer Bericht 18/180 at 17795 and 17798.

178 Speeches of Matthias Miersch, Frank Schwabe and Claudia Roth MPs, Stenografischer Bericht 18/180 at 17790, 17801 and 17807

179 Drucksachen 18/4713 and 18/8916.

180 This is the terminology used in § 6 (1) No 4 WHG.

181 Drucksache 18/4713 at 1.

182 Lars Dietrich and Till Elgeti 'Rechtliche Implikationen der Aufsuchung und Förderung von unkonventionellem Erdgas' (2011) 127 (7–8) *Erdöl Erdgas Kohle* 311; Pearson et al. 56/57.

The three German legislative actions target 'fracking', an inaccurate term that is meant to describe hydraulic fracturing. The European Commission commonly uses the more accurate term 'high volume hydraulic fracturing (HVHF)' (the differentiation between small hydraulic fracturing and HVHF originates from industry use in the 'motherland' of shale gas extraction, the United States).[183] The German government and the parties supporting it, however, did not tune in on that definition, but instead introduced their own definition.

The now resolved bill on water protection differentiates between conventional and unconventional fracking.[184] According to the German Ministry of Economic Affairs and Energy, conventional fracking has been practiced for many years in Germany, whereas unconventional fracking refers to new applications of hydraulic fracturing.[185] The crucial criterion for the differentiation is the type of rock:

> conventional fracking takes place in sandstone (mainly at greater depths) [than those where unconventional fracking is applied]. Unconventional fracking takes place in layers of shale-, argillite and marlstone rock strata, as well as in coal seams.[186] As opposed to the hitherto exploited German sandstone reservoirs, there is no experience or knowledge concerning extraction of natural gas from these 4 types of rock.[187]

Accordingly, the extraction from these four rock-types has been labelled 'unconventional fracking', whereas fracking sandstones, according to the German government, amounts to 'conventional' fracking.[188]

This differentiation has been heavily criticized by the opposition in the German parliament. It was called 'scientifically untenable'[189] and 'arbitrary'[190] as well as 'scandalous'.[191] Indeed, there is no evidence that

---

183 Department of Environmental Quality Michigan 'Hydraulic Fracturing in Michigan' available at: http://www.michigan.gov/deq/0,4561,7-135-3311_4231-262172--,00.html [accessed 30 June 2016].

184 See for instance Drucksache 18/4949 at 11 and Law prohibiting and minimizing risks to water from fracking 1.

185 German Ministry of Economic Affairs and Energy 'Fracking' available at: http://www.bmwi.de/DE/Themen/Industrie/Rohstoffe-und-Ressourcen/fracking,did=653918.html?view=renderPrint [accessed 30 June 2016] (hereinafter: BMWI Fracking).

186 Ibid.

187 Ibid.

188 BMWI Fracking.

189 Drucksache 18/8916 at 17.

190 Drucksache 18/8916 at 15.

191 Drucksache 18/8907 at 10; speech of Hubertus Zdebel, Stenografischer Bericht 18/180 at 17794, 1779.

such a differentiation is made in other countries, among the industry or in geoscience.

Strikingly, this definition is not even used by the government's own agency in charge of geoscience, the Federal Agency of Geoscience and Natural Resources (*Bundesanstalt für Geowissenschaften und Rohstoffe* (hereinafter: BGR)). That agency consistently differentiates between conventional and unconventional *reservoirs* but not between conventional and unconventional *fracking*.[192]

The practical repercussion of the government's odd differentiation is that only 'unconventional fracking' in the four rock formations is addressed by the law, whereas the use of hydraulic fracturing in other rock formations, remains perfectly legal.[193] The government justifies that fundamental differentiation with the following reasons:

> Fracking in shale and argillite rock, marlstone and coalbed seams has almost never been practiced in Germany – as opposed to fracking in sandstone (tight gas); that is why necessary knowledge is missing. In [the former cases] a higher number of drills and drilling pads and a larger volume of frack-fluid per well is required. Moreover, the procedure [fracking] might [in the former cases] also be applied at smaller depths (starting at ca. 1000 metres) and accordingly a smaller gap to groundwater resources and less mighty barriers between the frack-horizon and utilizable groundwater exist. Hydrological barriers are, for instance, saltstones or permian which may prevent upwards migration of fracking fluids from deeper layers (. . .). In order to prevent geological, hydrological and environment-specific dangers, particularly for drinking water, the approval of fracking permits [in the four named types of rock] for commercial purposes shall be prohibited until further sufficient research into possible risks has been conducted.[194]

Tight gas, which is named at the beginning of this statement, is gas produced from tight sand- or limestone formations.[195] In the USA the use of hydraulic fracturing for oil and gas production from sandstones is a well-known and standard practice, and its products are commonly referred

---

[192] Bundesanstalt für Geowissenschaften und Rohstoffe (BGR) '*Wissenswertes über Schieferöl und Schiefergas*' available at: https://www.bgr.bund.de/DE/Themen/Energie/Projekte/laufend/NIKO/FAQ/faq_inhalt.html [accessed 30 June 2016]; --, '*Schieferöl und Schiefergas in Deutschland Potenziale und Umweltaspekte*' (Bundesanstalt für Geowissenschaften und Rohstoffe, Hannover 2016) 13/14 (hereinafter: NIKO).

[193] As pointed out by MPs Hubertus Zdebel, Julia Verlinden and Annalena Baerbock, see Stenografischer Bericht 18/180 at 17795/17796, 17799 and 17801.

[194] Drucksache 18/4713 at 22.

[195] NIKO 196.

to as so called tight oil and tight gas.[196] Fracking has been used for the production of tight gas in Germany since the 1960s.[197] Almost one third of Germany's domestic gas production comes from tight gas reservoirs that are being exploited with the help of hydraulic fracturing.[198] Similarly, the production of geothermal energy often entails the use of hydraulic fracturing.[199] In all of these cases the technological process is similar to that of shale gas extraction. The only difference is the depth at which the respective rocks may be found.

Shale gas may be extracted in Germany from depths spanning 1 000–2 500 metres below surface, whereas tight gas and geothermal energy are often found at depths greater than 3 500 metres below the surface.[200] Tight gas is hence mainly encountered at greater depths than the other source rocks of unconventional gas.[201] Geological and hydrological barriers are assumed to prevent migration of dangerous substances from these deeper sandstone layers, but not from shale, argillite, marlstone or coalbed seam layers, which are often buried closer to the surface.[202]

However, particular spots of Germany shale and tight oil and gas may be encountered at identical depths.[203] As the opposition party, the Left,

---

196 US Energy Information Administration '*Energy in Brief Tight Oil*' available at: https://www.eia.gov/energy_in_brief/article/shale_in_the_united_states.cfm [accessed 13 July 2016]; Lin Sen-Hu et al. 'Status quo of tight oil exploitation in the United States and its implications' (2011) Vol 23 Issue 4 Lithologic Reservoirs 25–30.

197 Bundesrat Drucksache 144/15 at 1.

198 Bundesrat Drucksache 144/15 at 6.

199 Ronan L Hébert and Béatrice Ledésert 'Calcimetry at Soultz-Sous-Forets enhanced geothermal system: Relationships with fracture zones, flow pathways and reservoir chemical stimulation results' in Jianwen Yang '*Geothermal energy, technology and geology*' (Nova Science Publishers, New York 2012) 94.

200 Exxon Mobil Europeunconventionalgas.org 'Tight Gas' available at: http://www.europeunconventionalgas.org/unconventional-gas/types-of-unconventional-gas/tight-gas [accessed 30 June 2016] (hereinafter: Exxon Mobil Tight Gas).

201 This reservation has even been made by the German ministry of economic affairs and energy, see BMWI fracking.

202 Drucksache 18/4713 at 22.

203 See a graphic of the German Federal Agency of Geoscience and Natural Resources available at: http://www.bmwi.de/BMWi/Redaktion/PDF/I/info-grafik-gasforderung,property=pdf,bereich=bmwi2012,sprache=de,rwb=true.pdf [accessed 30 June 2016]. For tight oil, see Landesamt für Bergbau, Energie und Geologie Niedersachsen (LBEG) '*Erdöl und Erdgas in der Bundesrepublik Deutschland 2014*' (Landesamt für Bergbau, Energie und Geologie, Hannover 2015) 14–18 and 35/36 available at: http://www.lbeg.niedersachsen.de/portal//search.php?_psmand=4&q=erdgas+in+deutschland+2014 [accessed 8 July 2016] (hereinafter: LBEG 2014).

aptly pointed out during the decisive session of the environmental committee of the *Bundestag*, fracking in sandstone for oil and gas remains legal.[204] Production of oil and gas from these sandstone layers is one of the most common ways of 'conventional' hydrocarbon extraction.[205] Sandstones differ in terms of brittleness and permeability. A study on oil and gas extraction in Germany found that certain sandstone hydrocarbon reservoirs had not been used for production because the sandstones were not brittle and rather impermeable.[206] Particularly in these sandstones that do not easily give away hydrocarbons, the flow might be stimulated by hydraulic fracturing.

Crucially, oil-bearing sandstone reservoirs are often located close to the German surface, up to a couple of hundred metres in depth.[207] Thus, they are located at the same depth, or occasionally even closer to the German surface, than hydrocarbon-carrying shale, argillite, marlstone or coalbed seam layers.[208] As a result, fracking for oil is now legal in some rock-strata (tight oil from sandstones) whereas it is illegal in others (tight oil from shale), although both types of rock might be found at exactly the same depths.

Accordingly, the upper chamber of the German parliament (*Bundesrat*) and MPs of the opposition in the *Bundestag* argued that potential threats of fracking in tight gas and oil reservoirs might not be different to the potential threats of fracking in the other four types of rock strata.[209] The equation 'greater depth = more geological and hydrological barriers between the point of fracking and groundwater = more safety' hence does not apply in these circumstances.

To sum up, the geological and hydrological circumstances for tight gas extraction/geothermal activities/tight oil on the one hand and shale gas extraction/unconventional gas extraction on the other hand are not always different. There are certain areas where, according to the stipulations of the new fracking package, outlawed and licit activities may take place at similar depths and the very same techniques may be applied in a similar way (directional drilling in combination with hydraulic fracturing).

---

204 Drucksache 18/8916 at 17.

205 Stoneley 27.

206 LBEG 2014 at 14.

207 Ibid. and LBEG 2014 at 14–18 and 35/36.

208 LBEG 2014 at 14–18 and 35/36 and Drucksache 18/8916 at 17.

209 Bundestag Drucksache 18/4949 at 9; Stenografischer Bericht 18/180 at 17794; Annalena Baerbock 'Fracking: Keine Entwarnung' available at: http://www.annalena-baerbock.de/pmfracking-keine-entwarnung/ [accessed 12 July 2016].

Fracking is a cost-intensive technology[210] and the fact that oil prices are not particularly high at the moment might be the reason that this issue has not yet received much attention in Germany. But as the markets pick up and the cost-benefit analyses look better, it might be an option to use fracking, for instance for the production of oil from sandstones, in Germany in the future at larger scale.

The upper chamber of parliament, the *Bundesrat,* shared these concerns. The *Bundesrat* objected that, according to expert evidence, the differentiation with regard to depth was not justifiable.[211] The *Bundesrat* stated clearly in that regard: 'the dangers of fracking for ground and drinking water exist, irrespective of the depths at which the technology is deployed'.[212]

The true reason for the exclusion of certain types of hydrocarbon and geothermal energy production from the fracking-prohibition is probably the fact that Germany hosts Europe's greatest tight gas industry and tight gas production is most advanced here.[213] While protecting the industry might be a legitimate interest of itself, it is highly doubtful whether or not this differentiation complies with the legal requirements concerning equal treatment.

3.3.3.2 Conflict with the German constitution

The fundamentally different legal treatment for tight gas, tight oil and geothermal energy production via fracking on the one hand (allowed) and fracking for shale gas, etc. (prohibited) is problematic. This artificial differentiation might conflict with the principle of equal treatment, enshrined in article 3 of the German constitution. Article 3 German constitution demands that issues, which are essentially the same have to be treated in the same way and issues that are essentially different might be treated differently.[214] Although the legislator has leeway for discretion, this is ultimately limited by the fundamental rights of a person or an entity, as enshrined in the German constitution.[215]

The legality of the German fracking package might be challenged by a

---

[210] Consider the different technologies required, as discussed above.

[211] Drucksache 18/4949 at 2.

[212] Ibid.

[213] According to Exxon Mobil Tight Gas; the link to the fracking law has been made in Drucksache 18/8916 at 17.

[214] Bruno Schmidt-Bleibtreu and Franz Klein, '*Kommentar zum Grundgesetz*' 10th edition (Wolters Kluwer Deutschland GmbH, München 2004) article 3 para 2 (hereinafter: Schmidt-Bleibtreu/Klein).

[215] Schmidt-Bleibtreu/Klein article 3 para 3/4.

company wanting to drill for shale gas in Germany and not allowed to do so. It could litigate against the fracking package by arguing that another company aiming to drill for tight oil or gas at similar depths with the same techniques (hydraulic fracturing) may apply for a licence.

If the former company specializes in shale gas activities, the new law might, in extreme cases, conflict with the fundamental right of the owners of existing oil and gas companies to have and conduct their own business. This fundamental right (*Recht am eingerichteten und ausgeübten Gewerbebetrieb*) is guaranteed under article 14 German constitution and may only be taken away from an individual by a law that provides adequate, effective and timely compensation.[216] The German fracking package does not provide for any compensation.

Moreover, the fracking package might similarly conflict with the right of all Germans to freely choose their occupation under article 12 German constitution, for the very same reasons.[217] The legislator would have to justify the unequal treatment. If he is not able to deliver on that, the German fracking package might not be reconcilable with article 3 and possibly article 12 of the German constitution. This author concludes that, as the law stands, it is prone to be struck down by the Federal Constitutional Court, if challenged.

This, however, is unlikely to happen in the foreseeable future. The reasons are of a practical rather than legal nature: oil and gas companies might not have an interest in launching a legal challenge. Shale gas has been highly controversial in Germany[218] and the debate centred on the clash of strongly held individual beliefs and nationwide campaigns by non-governmental organizations.[219] Polls from April 2015 showed that two thirds of Germans favour a ban on fracking above other means of

---

[216] BVerfGE 1, 264 (276 ff.); 45, 142 (173); BGHZ 23, 157 (162 f.); 30, 338; 57, 359 et sqq; 67, 190 (192); 81, 21 (33); 92, 34 (37); BVerwGE 62, 224 (226); for more see: Schmidt-Bleibtreu/Klein article 14 para 3.; Theodor Maunz and Günter Dürig (eds), '*Grundgesetz Kommentar*' 76. Ergänzungslieferung (C H Beck, München since 1958) article 12 para 95–113 (hereinafter: Maunz/Dürig).

[217] Maunz/Dürig article 12 para 1 et sqq; Schmidt-Bleibtreu/Klein article 12 para 1 et sqq.

[218] Dominik Greinacher and Sebastian 'Revising the Environmental Impact Assessment Thresholds: The Case of Germany' in Cecile Musialski et al. (eds) '*Shale Gas in Europe*' (Claeys & Casteels, Deventer 2013) 508.

[219] Achim Lang and Jale Tosun 'The Politics of Hydraulic Fracturing in Germany: An Analysis of Discourse Networks' in: Christoph Weible, Karin Ingold, Manuel Fischer and Tanya Heikkila (eds) '*Mapping the Political Landscapes of Hydraulic Fracturing*' (Palgrave Macmillan, Basingstoke 2016) (hereinafter: Lang/Tosun).

regulation.[220] However, the industry is still allowed to use fracking for tight gas, sandstone oil and geothermal purposes. Companies are likely to avoid sparking public anger by a legal challenge of the fracking package until prices increase significantly and European shale gas extraction and tight oil extraction via fracking becomes economically viable.[221]

In its reply to the governmental bill, the *Bundesrat* made an interesting proposal which could be used to improve the law and overcome its currently shaky stature. The *Bundesrat* demanded doing away with differentiations between rock types altogether and instead proposed the usage of only one central criterion for the water-permitting procedure of fracking activities: the principle of apprehension in the Water Protection Act (*wasserrechtlicher Besorgnisgrundsatz*).[222]

The principle of apprehension, enshrined in article 48 (1) (2) WHG, is considered to be 'the fundamental norm' of the entire Water Protection Act.[223] A permit for the introduction of substances into groundwater may not be issued if a detrimental change of water quality has to be dreaded (or *apprehended*, the verbatim translation of the German *Besorgnis*).[224] The principle must be adhered to by all water users under article 9 (1) and (2) WHG and also applies to the depositing and storing of substances.[225]

Applying this principle to the issue of fracking would mean that a permit may only be issued if there is no reason to believe the use of fracking could

---

[220] According to figures from April 2015, see Bundestag Drucksache 18/8916 'Recommendation and Report of the Committee for Environment, Nature Protection, Building and Reactor-safety concerning Drucksache 18/4713 (*Beschlussempfehlung und Bericht des Ausschusses für Umwelt, Naturschutz, Bau und Reaktorsicherheit zu Drucksache 18/4713*)' available at: http://dipbt.bundestag.de/doc/btd/18/089/1808916.pdf [accessed 5 July 2016] at 15 (hereinafter: Drucksache 18/8916).

[221] This has also been pointed out by several German MPs, namely in their speeches in the Bundestag, see Stenografischer Bericht 18/180 at 17795/17796 and 17799.

[222] Drucksache 18/4949 at 5/6.

[223] Frank Sieder, Herbert Zeitler et al. 'Wasserhaushaltsgesetz Abwasserabgabengesetz Band 1'(C H Beck, München 2016) § 48 para 1 (hereinafter: Sieder/Zeitler).

[224] § 48 (1) (1) WHG.

[225] Sieder/Zeitler § 48 para 1. An exception applies, according to the most common interpretation of § 48 (1) (2) WHG to 'insignificant' amounts of substances, a term that is highly controversial in Germany, see Sieder/Zeitler § 48 para 3–8. Fracking and the disposal of fracturing fluid, flow back and waste water is now clearly defined as a water use by article 9 (2) No 3 and 4 *WHG* (these have been newly introduced by the fracking package).

bring about detrimental changes to water quality.[226] The German government opposed that proposal of the *Bundesrat*, merely stating that the government does not share the view of the *Bundesrat* that the use of the principle of apprehension of the Water Protection Act could resolve all issues related to fracking and water contamination.[227] However, it gave no further reasoning as to why it arrived at this conclusion.

Despite the misgivings of the government, the proposal of the *Bundesrat* would, indeed, not differentiate between any types of rock. The risk of the law being struck down as not complying with the principle of equal treatment by a constitutional court would, thus, fall away. By altering the law in that respect, the legislator could comply with the demands of the German constitution, establish a legally 'watertight' regulation and provide potential investors in unconventional gas extraction with a clear framework that is easy to apply.

3.3.3.3 Ban or moratorium?

The third important feature of the fracking package is its legal nature. The initial fracking package stated explicitly that its measures are not designed to bring about a general prohibition of fracking.[228] However, the text that did eventually become law maintains that extraction of natural gas (and now also oil) from shale, argillite, marlstone and coalbed seams by hydraulic fracturing is strictly prohibited, irrespective of the depth of the deposit.[229]

It is somewhat difficult to determine whether this regulation actually constitutes a ban or a moratorium. By way of a reminder: a moratorium is a temporary measure with the aim to *suspend* an activity, whereas a ban is of indefinite duration and wants to *suppress and prohibit* an activity in general.

While the German government and the parties supporting it insist that the fracking package includes an indeterminate ban on fracking,[230] opposition MPs even doubted that the law constitutes a moratorium, given that it permits the extraction of tight gas, tight oil and geothermal energy.[231] According to them, the only thing that is going to be outlawed by this law

---

226 Drucksache 18/4949 at 5/6.
227 Drucksache 18/4949 at 12.
228 Drucksache 18/4713 at 2.
229 Drucksache 18/8916 at 2.
230 Drucksache 18/8916 at 16/17; speeches of Matthias Miersch, Karsten Möring, Andreas Lenz and Maik Beermann, Stenografischer Bericht 18/180 at 17790/17791, 17805/17806, 17808, 17872.
231 Drucksache 18/8916 at 17 and Stenografischer Bericht 18/180 at 17790.

is the use of fracking in the four described types of rock and even there no strict ban is put in place.[232]

The name of the most important[233] of the three legislative acts (Bill on water protection provisions and the *prohibition and risk-minimization* of the procedures of the fracking technology (Drucksachen 18/4713 and 18/8916) highlights the ambivalent nature of what has now become law. There is an element of prohibition on the one hand and at the same time risk-minimization on the other. The conjunction 'and' suggests a dual approach. This is, indeed, what the law wants to achieve: prohibiting some activities (fracking in shale, argillite marlstone and coalbed streams), while allowing fracking in sandstone, limestone or for geothermal purposes, as discussed above.[234] But the main question is whether or not the former part, the prohibition, is a time-sensitive measure that suspends the activity or an indeterminate ban that wants to suppress the activity concerned in general and forever.[235]

The new § 13 a (7) WHG establishes that a review of the prohibition of fracking in the four types of rock has to take place by 2021. This, in particular, is central to the characterization of the law. Many of the rebellious MPs argued that this § 13 a (7) merely opens up the possibility for the *Bundestag* to review the prohibition by 2021.[236] In case the *Bundestag* declines to amend the law, it would stay in place beyond 2021, they argued.[237] One MP called § 13 a (7) WHG thus a '*redundant formality*', since every law may anyway be revised by the *Bundestag* at any time.[238] He went on to say 'even if it is written in that law that by 2021 a report shall be issued [. . .] is it up to it [the Bundestag] to take a decision by 2020 or 2025. This is all open.'[239] According to this view the prohibition might only be lifted by a renewed resolution of the *Bundestag,* which makes it an indeterminate ban, according to several MPs.[240]

---

232 Ibid.

233 Insofar as it includes the fundamental decisions on fracking in Germany, see Drucksache 18/8916 at 17 and Stenografischer Bericht 18/180 at 17790.

234 Drucksache 18/4713 at 1 and 2.

235 As discussed above, these definitions are taken from Black's Law Dictionary 154 and 1031.

236 Drucksache 18/8916 at 16/17; speeches of Matthias Miersch, Karsten Möring, Andreas Lenz and Maik Beermann, Stenografischer Bericht 18/180 at 17790/17791, 17805/17806, 17808, 17872.

237 Ibid.

238 Stenografischer Bericht 18/180 Speech of Karsten Möring at 17805/17806.

239 Stenografischer Bericht 18/180 Speech of Karsten Möring at 17805/17806.

240 Drucksache 18/8916 at 16/17; speeches of Matthias Miersch, Karsten Möring, Andreas Lenz and Maik Beermann Stenografischer Bericht 18/180 at 17790/17791, 17805/17806, 17808, 17872.

As opposed to this interpretation of the law, the actual law text shows more characteristics of a moratorium than of a ban. The exact wording of the newly imposed § 13 a (7) WHG is as follows:

> In the year 2021 the German parliament reviews the suitability of the prohibition of section 1 sentence 1 number 1 on the basis of the then existing state of knowledge and technology (. . .).[241]

The first eight words 'In the year 2021 the German parliament reviews (. . .)' places an obligation on the *Bundestag* to review the laws by the end of 2021. There is no leeway for discretion. If the legislator had wanted to leave the decision to review the law to MPs' discretion, a different formulation would have been used.

German administrative law envisages three ways in which discretion might be apportioned by a law: *expressis verbis*, out of an unequivocal context, or by legal description.[242] As the law neither includes an *expressis verbis* section on discretion nor is it based in a context that, unequivocally, provides discretion, only the third group – discretion by legal description – remains as a possibility.

This alludes to the so called '*kann, muss, soll*' provisions of German law.[243] These are particular formulations in a law text, indicating different levels of discretion.[244] A typical formulation would read: 'In the year 2021 the German parliament *may/should/shall* review (. . .).' Such a formulation, however, is not included in the WHG.

The *Bundestag* is, indeed, free to come to the conclusion that the prohibition needs to be prolonged, but it may only come to that conclusion after conducting a review in 2021. The text offers no ambiguity as to the fact that a review is obligatory. Thus, a definite element of time is included in the law text. The prohibition is, hence, more akin to a moratorium than to a ban.

Moreover, the fracking package includes an exception from the general prohibition on fracking in the four types of rock. According to the new § 13 a (2) Water Protection Act, four exploratory trials for scientific

---

241 § 13 a (7) WHG reads in the original: '*Im Jahr 2021 überprüft der Deutsche Bundestag auf der Grundlage des bis dahin vorliegenden Standes von Wissenschaft und Technik die Angemessenheit des Verbotes nach Absatz 1 Satz 1 Nummer 1*'; translation by author.

242 Steffen Detterbeck '*Allgemeines Verwaltungsrecht mit Verwaltungsprozessrecht*' 3rd edition (C H Beck, München 2005) paragraphs 316 et sqq.

243 Detterbeck paragraph 320/321.

244 Ibid.

purposes may be conducted which are defined rather narrowly.[245] These trials need to be approved at national level, but also by the *Länder* where they are supposed to take place.[246] They will be supervised by an expert commission, which assesses the trials, compiles annual reports on progress and submits these reports to parliament and the public.[247] The very fact that an explicit provision to enable scientific research features in the fracking package is further evidence that the technology will not be outlawed indefinitely, but only until scientific knowledge is increased.

Furthermore, a strict ban on hydraulic fracturing has actually been proposed in Germany, but was dismissed during the legislative procedure. The *Bundesrat* proposed an amendment to the Federal Mining Act that would have imposed a strict ban on the use of hydraulic fracturing in shale, argillite and marlstone rock as well as in coalbed seams.[248] In its rebuttal to that proposal of the *Bundesrat,* the German government argued:

> It is, however, not the aim of the government to ban a technology forever that is not yet sufficiently researched. Moreover, it is its task to ensure that human health and the environment are not endangered by application of the technology, as well as, to sustain research possibilities and potential economic perspectives under these prerequisites.[249]

Note, however, that this statement has been made by the government with a view to the original government bill that has been substantially altered since. Nevertheless, the resolved law does not distance itself from this line of reasoning, but rather builds upon it. The rebellious MPs wanted to alter the government bill, but abstained from launching their own proposal in the *Bundestag*.

The Minister of the Environment of Schleswig-Holstein, Robert Habeck, pointed out that a ban, if it had been desired by parliament, could easily have been achieved by introducing a sentence into the Federal Mining Act.[250] Such a sentence could be rather simple and prescribe that

---

[245] Drucksache 18/8916 at 3 and 19.

[246] Drucksache 18/8916 at 3 and 17.

[247] New § 13 a (6) Water Protection Act (hereinafter: WHG), according to Drucksache 18/8916 at 4.

[248] Drucksache 18/4949 at 10.

[249] Drucksache 18/4949 at 14.

[250] Bundesrat 'Plenarprotokoll 947. Sitzung 8.7.2016' at 282, available at: http://www.bundesrat.de/DE/dokumente/plenarprotokolle/plenarprotokolle-node.html [accessed 15 July 2016] (hereinafter: Bundesrat Plenarprotokoll 947. Sitzung).

the breaking of rock by hydraulic pressure for the purpose of exploration for or extraction of hydrocarbons is prohibited.[251]

The fact that MPs adopted neither of these actions suggests that they wanted to stick to the fundamental idea of the government to have a moratorium on certain applications of the fracking technology, but that the terms and conditions of that moratorium should be tightened up.

When the new, amended bill was put before the *Bundesrat* for its final approval, the committees of the *Bundesrat* assessed the amended bill, which now became law, concluding that

> it is with regret that the Bundesrat has to establish that the exploration for and exploitation of hydrocarbons (. . .) by hydraulic fracturing still is not entirely banned. It [the technology] is merely going to be prohibited for commercial use in some areas and rock formations. In all of the areas where [fracking] is not explicitly outlawed by the law, the technology could be used for exploration and production of hydrocarbons for scientific purposes, also in unconventional reservoirs.[252]

Thus, the committees of the *Bundesrat* are not of the opinion that this law constitutes a ban. This is also apparent from the committee's recommendations in which they urged the *Bundestag* 'to put into place a law that entails a non-time-restricted and factually unrestricted ban of the fracking technology (. . .)'.[253]

During the debate on the fracking package in the *Bundesrat* on 8 July 2016, several Prime Ministers and Ministers of different German states made it very clear that they do not consider these laws and the ordinance as putting into place a ban.[254] Rather, some of them pushed for a stricter regulation and considered the moratorium to be a compromise.[255]

This view is finally supported by a statement made by the CDU/CSU faction during discussions on the alterations of the government bill. In view of the new version of that bill (which has now become law) the faction stated 'A general ban on fracking was never the intention. It is also not necessary. The pivotal point is that this extraction method is made more secure compared to the hitherto existing legal framework and that has been achieved.'[256]

---

251 See his proposal, ibid.
252 Bundesrat Drucksache 353/1/16 at 2 No 3.
253 Bundesrat Drucksache 353/1/16 at 4 No 9.
254 Bundesrat Plenarprotokoll 947. Sitzung at 281, 282, 283.
255 Bundesrat Plenarprotokoll 947. Sitzung at 281.
256 Drucksache 18/8916 at 17.

3.3.3.4 Strict prohibition of fracking in certain water protection areas and consistent use of BAT

The fourth important feature of the fracking package is its emphasis on water protection and best available techniques (BAT). Protection of sensitive sites, like water protection areas or their drainage basins, mineral water resources, sources of water used for the production of beverages, national parks, environmental protection areas and Natura 2000 habitats, are particularly important.[257] The fracking package puts an extensive prohibition of fracking in these areas in place which, according to the environmental committee of the *Bundestag*, is required to avoid the occurrence of diverging regulations in individual German *Länder*.[258]

Fracking is now also prohibited underneath water protection areas. This rule has been explicitly incorporated into the law to prevent drilling from outside a protected site into the protected area.[259] When compared to the initial governmental bill, the final law also includes better protection of drainage basins of water sources.[260]

Interestingly, the use of fracking remains licit for the opening up of healing/thermal water springs (*Heilquellen*). This type of fracking has been

---

[257] On the one hand by a respective supplement to § 13 a (1) WHG and on the other hand by alterations to §§ 1, 23, 24 and 33 Nature Protection Act (*Bundesnaturschutzgesetz*).

[258] Drucksache 18/8916 at 18. This, indeed, is a justified fear since, under current legislation, the German states are in charge of establishing water protection zones to safeguard drinking water supplies by individual ordinances, within which certain uses are precluded, see §§ 51 I and 52 I No 1 Water Protection Act. These uses typically comprise of, inter alia, prohibitions on interfering with the soil by drilling. For general information: the German water protection zones are grouped into three categories, from the strongest protection in direct vicinity to a well (zone 1) to a general protection category (zone 3), see Michael Kotulla *'Wasserhaushaltsgesetz Kommentar'* (2nd edition, Kohlhammer, Stuttgart 2011) at § 51 paragraphs 22–7. As opposed to the very strict regulations with regard to zone 1, the stipulations pertaining to zones 2 and 3 do not effectively exclude shale gas extraction. The very same zoning and water protection requirements apply to mineral healing springs, natural springs from which therapeutic waters flow, that form the basis for a network of health-related facilities throughout Germany, according to § 53 (5) in conjunction with § 51 (2) and § 52 Water Protection Act; Meiners et al. Bund B 128. More information on these springs and their importance for German water supplies and recreational purposes may be found at: Germany Travel 'Mineral and thermal springs – harnessing the healing power of water' available at: http://www.germany.travel/en/leisure-and-recreation/health-wellness/spas-and-health-resorts/mineral-and-thermal-springs/mineral-and-thermal-springs.html [accessed 20 July 2015].

[259] Drucksache 18/4713 at 24.

[260] Stenografischer Bericht 18/180 at 17803.

used for quite some time in Germany without causing any issue and it may not be performed by using water-endangering substances.[261] The exception has apparently been introduced to cater particularly for the needs of Bavaria, where thermal water springs are often used.[262]

The quality of all of these water resources in the vicinity of fracking sites has to be controlled via monitoring and benchmarked against baseline studies.[263] The absence of such baseline studies has been identified as a major issue in the US, which hampered adequate investigation into the repercussions of fracking on drinking water because the quality of the water prior to fracking activities was unknown.[264] How may one determine if something has been polluted, if the original status, against which pollution must be benchmarked, is unknown?

Furthermore, all substances that will be used for fracking purposes, as well as their envisaged amounts must be disclosed and published, according to § 13 b (1) (2) Water Protection Act.[265] This regulation is stricter than the REACH regulations at EU level, where certain types of information might be spared from public disclosure if publication would undermine legitimate commercial interests.

The two new laws and the new EIA Ordinance coherently require operators of facilities where fracking is taking place to use Best Available Techniques (BAT) (German term *Stand der Technik)* for extraction.[266] Under German technology regulation, the use of BAT is well-established.[267] A good example for this is the German CCS law, according to which operators of CCS plants have to use the best available technology (BAT) (*Stand der Technik)* during CCS operations to protect humans and the environment.[268] Instructive examples of BAT for the regulation of shale gas extraction include 'green completion' and the three casing system for well insulation.[269]

---

261 Drucksache 18/4713 at 24.

262 Stenografischer Bericht 18/180 at 17808.

263 § 13 b (1) – (5) WHG according to Drucksache 18/4713 at 3 and 11; Drucksache 18/413 at 14; Drucksache 18/8916 at 20.

264 EPA Pavillion 39; US EPA Study Plan 111.

265 Drucksache 18/4713 at 3.

266 See § 13 a (4) No 2, (5), (6) WHG and Drucksache 18/8916 at 19/20.

267 § 3 No 11 Water Protection Act, see also BVerwGE 55, 250; BVerfGE 49, 89 (143) and 53, 30 (58); Meiners et al. Bund B 113.

268 § 13 (1) No 4 CCS law. The law will be discussed in more detail below in Chapter 6.

269 Energy and Climate Change Committee of the House of Commons 'Shale Gas' Fifth Report of Session of the House of Commons 2010–12, Vol. I (Crown 2011) 46; SRU Faulstich 35.

3.3.3.5 The Mining Damage Presumption – liability and the reversal of the burden of proof

The fracking package also contains a second law concerned with the extension of the Mining Damage Presumption to borehole mining and caverns.[270] The so called Mining Damage Presumption (*Bergschadensvermutung*) is a pillar of the German legal framework on mining.[271] Its function is to reverse the burden of proof for mining activities.[272] If damage occurs in a mining area as a result of subsidence, compression or stretching of the surface, it shall be presumed that the damage is caused by the mining company active in that particular area.[273]

Whether or not this Mining Damage Presumption also applies to borehole mining and to caverns had been discussed in the literature, due to the unfortunate wording of § 120 Federal Mining Act.[274] The explicit aim of the current law is to extend the Mining Damage Presumption to these activities generally and to fracking activities in particular.[275]

The bill functions in two ways. First, it enlarges the scope of the Mining Damage Presumption to earth liftings, fissures and tremors,[276] which also encompasses the issue of induced seismicity.[277] Second, it amends the applicable Impact Area Mining Ordinance (*Einwirkungsbereichs-Bergverordnung*), which regulates what might be considered to be the geographical extent of an area that has been damaged by mining activities.[278]

---

270 *Gesetz zur Ausdehnung der Bergschadensvermutung auf den Bohlochbergbau und Kavernen* see Bundestag Drucksache 18/4714 and Drucksache 18/8907 as well as Drucksache 18/4952.

271 Included in § 120 German Mining Act; for details, see Gerhard Boldt and Herbert Weller '*Bundesberggesetz*' (de Gruyter, Berlin 1994) § 120 RN 2 et sqq. (hereinafter: Boldt/Weller); Reinhart Piens and Hans-Wolfgang Schulte and Stephan Graf Vitzthum, '*Bundesberggesetz*' 2nd edition (Kohlhammer, Stuttgart 2013) § 120 RN 3et sqq. (hereinafter: Piens/Schulte/Graf Vitzthum).

272 Ibid.

273 § 120 (1) (2) No 2, See Drucksache 18/4714 at 7 and Drucksache 18/8907 at 3; for more see: Boldt/Weller § 120 Rn 2 et sqq.; Piens/Schulte/Graf Vitzthum § 120 RN 3 et sqq.

274 For the discussion see Boldt/Weller § 120 Rn 8.; Piens/Schulte/Graf Vitzthum § 120 RN 14; Drucksache 18/4714 at 11.

275 Ibid.

276 By amending § 120 (1) (2) No 2, see Drucksache 18/4714 at 7 and Drucksache 18/8907 at 3.

277 Drucksache 18/4952 at 2 and 5/6.

278 The issue here was how to establish in advance the area that could potentially be affected by tremors caused by mining (and particularly fracking) and where affected citizens could claim compensation. The government argued that it would be impossible to know the area affected in advance and, thus, the area in which the Mining Damage Presumption applies would be impossible to determine,

3.3.3.6 The fracking EIA ordinance – obligatory EIAs for fracking activities

Besides the two laws the legislative package also consists of a third measure, the issuance of an ordinance on EIAs for fracking activities.[279] The situation in Germany with regard to EIAs had not previously been much different to the situation in Europe as a whole: the threshold for obligatory EIAs stood at 500 000 cubic metres daily gas production.[280] Thus, fracking operations were subjected only to a scoping procedure under the German EIA-law (*Gesetz über die Umweltverträglichkeitsprüfun g*).[281] This scoping procedure assessed the need to carry out a full-fledged EIA on a case-by-case basis.[282]

The new EIA ordinance makes EIAs obligatory for fracking activities.[283] This pertains particularly to the exploration for and production of oil and gas via hydraulic fracturing, including scientific trials,[284] exploration for oil and gas via hydraulic fracturing and disposal of waste water via deep drillings (*Verpressen*).[285] Moreover, hydraulic fracturing for geothermal purposes is also subjected to an EIA in cases where the geothermal drillings involve water-endangering substances, according to the new § 1 No 8 and 8a of the Ordinance on EIAs for Mining Activities (*Verordnung über die Umweltverträglichkeitsprüfung bergbaulicher Vorhaben* (hereinafter: *UVP-V Bergbau*)) *UVP-P Bergbau*.[286]

On a final note, the new EIA fracking ordinance also amends part of the Mining Ordinance for all Mining Plants (*ABBergV*) to ensure adequate monitoring and that disposal of fracking waste water via deep ground injection may not take place in areas that are located close to the surface.[287] This is mainly achieved by the introduction of a new § 22 b (monitoring)

---

see Drucksache 18/4952 at 7. The interesting solution to this issue is that the impact area for tremors is to be determined after the seismic activity takes place, see Drucksache 18/4952.

279 Bundesrat Drucksache 144/15.

280 Annex I German EIA law in conjunction with § 1 No 2 (a) of the German Ordinance concerning Environmental Impact Assessment in Mining Projects (*Verordnung über die Umweltverträglichkeitsprüfung bergbaulicher Vorhaben (UVP-V Bergbau)*).

281 Meiners et al. Bund B 138; Philippe & Partners paragraph 145.

282 Ibid.

283 Bundesrat Drucksache 144/15 at 1.

284 The latter has been introduced by the Bundesrat, see Bundesrat Drucksache 358/16 at 2 No 2.

285 Bundesrat Drucksache 144/15 at 1/2 and 6/7 with amendments in Bundesrat Drucksache 358/16 at 2/3 No 3.

286 Ibid.

287 Ibid.

and § 22 c (treatment and disposal of flow back and *Lagerstättenwasser*) to the *ABBergV*.[288]

The effects of the ordinance will be evaluated after four years.[289] Although the ordinance is thought of as supplementing the two laws on fracking, it currently has some inadequacies, which suggest that it does not fall perfectly into line with those two laws.[290] These inadequacies are mainly due to the fact that the ordinance was modelled upon the original government bills, which have now been substantially altered. It remains to be seen to what extent the *Bundesrat* is going to alter the ordinance to make it more fitting to the two resolved laws.

3.3.4 Summary Germany and Shale Gas

The most striking point of the German prohibition of shale gas extraction is that scientists deem it unnecessary. Although a number of studies on the specific German situation regarding shale gas extraction have been released,[291] not one asked for a strict prohibition. This fact has been acknowledged during the parliamentary debate in the German *Bundestag*, where Andreas Jung, MP said: 'We are doing much more than just implementing the suggestions of the experts of the German Federal Environmental Agency. We are a quantum leap ahead.'[292]

The last sentence might prove to be controversial. In fact, it is rather alarming to see how little the expert opinion of scientists seems to have mattered to the German legislator. As a result of that approach the core-regulation of the package is a fundamentally different legal treatment of so called 'unconventional' and 'conventional' fracking, which does not sit well with the German constitution.

Very small amendments could improve and fortify the fracking package and take it out of constitutional criticism. But this requires trust and a willingness to listen amongst the three major institutions of the legislative and the executive power. The fierce debate over the package between the

---

288 Bundesrat Drucksache 144/15 at 3–5.

289 Bundesrat Drucksache 144/15 at 13.

290 The most striking example is the fact that the ordinance adopts the depth criterion (3000 metres), which was included in the initial draft package, but was removed in the actual laws.

291 Meiners et al. Bund; Dannwolf et al.; Ewen; Meiners et al. NRW; Fritsche/Herling; SRU Faulstich; Andruleit et al.; NIKO and many more, as discussed throughout this book.

292 Stenografischer Bericht 18/180 at 17803.

Bundesrat, the government and the *Bundestag*, however, provide little hope for this to happen in the future.

3.4 'POLITICAL MORATORIUM' IN THE UK

As opposed to the outright ban on shale gas extraction by hydraulic fracturing in France, and the German moratorium with its long duration, the United Kingdom took an incremental approach towards shale gas regulation. In essence, this approach is characterized by step-by-step advancement of shale gas extraction in England and Wales, while asserting that appropriate precautionary measures are in place at each step along the way.[293] If existing precautionary measures prove to be insufficient, shale gas extraction is halted, additional precautionary measures are put in place and shale gas extraction resumes with the new, additional precautionary measured attached.[294] Interestingly, Scotland and Northern Ireland adopted different action on shale gas regulation. In these two countries moratoria are currently in place, as will be discussed further below.

When companies began to push for shale gas extraction in the UK, the House of Commons initially conducted an investigation and provided the government with a report that was broadly supportive of shale gas extraction in the whole of Britain.[295] At that time, the House of Commons report stated that 'induced seismicity'[296] was not thought to be a significant risk

---

[293] Good overviews on the approach in the UK as a whole and the individual regions may be found at: Jill Morgan 'Sustainability and stakeholder participation: shale gas extraction in the United Kingdom' in John C Dernbach and James R May (eds) '*Shale Gas and the Future of Energy Law and Policy for Sustainability*' (Edward Elgar Publishing, Cheltenham, 2016) 143–64 (hereinafter: Morgan); Musialski UK chapter 517–36; Hunter/Usenmez/Paterson 383–90; Gordon/McHarg/Paterson paragraphs 14.30 and 14.33–14.35.

[294] Ibid.

[295] Energy and Climate Change Committee of the House of Commons '*Shale Gas*' Fifth Report of Session 2010–2012, Vol. I (Crown 2011) 55 (hereinafter: UK report I); Energy and Climate Change Committee '*Shale Gas*' Fifth Report of Session of the House of Commons 2010–12, Vol. II (Crown 2011) (hereinafter: UK report II).

[296] 'Induced seismicity' refers to seismic events which are triggered by the injection of fluid into shale plays, see Cuadrilla Resources Ltd, 'Geomechanical Study of Bowland Shale Seismicity Synthesis Report' (2011) 53 available at: http://www.cuadrillaresources.com/wp-content/uploads/2012/02/Geomechanical-Study-of-Bowland-Shale-Seismicity_02-11-11.pdf [accessed 24 April 2016] (hereinafter:

in the UK.[297] While this report was still being printed,[298] seismic activities occurred at the first shale gas exploratory drilling site in the UK, in the area of Blackpool.[299] These earth tremors made it obvious that the initial assessment of shale gas extraction threats was insufficient.

Due to these shortcomings, the UK government imposed a moratorium on shale gas extraction for the entire UK.[300] This moratorium was not implemented by a formal law, but was rather the result of an internal governmental agreement that reached the public in the guise of a decree.[301] Since the existence and efficiency of this moratorium does not depend on a law, but on the will of politicians, this kind of moratorium has been dubbed 'political moratorium'.[302]

While this 'political moratorium' was in place, the UK government pushed for a thorough investigation into the causes of the earth tremors.[303] By 2012, scientific surveys into the Blackpool tremor concluded that the 'induced seismicity' was mainly due to the specific geological structure of the area surrounding the site.[304] Shale gas extraction could be resumed if

---

Cuadrilla report). A technically precise definition may be found at Cuadrilla report 36.

297 UK report I, Ev 94.

298 The committee had heard its last evidence by April 2011, see UK report I, Ev. 109, and went to print on 10 May 2011, see UK report I, 1.

299 *BBC* 'Fracking tests near Blackpool "likely cause" of tremors' available at: http://www.bbc.co.uk/news/uk-england-lancashire-15550458 [accessed 4 April 2014]; Musialski UK chapters 519 and 529.

300 Written Statement Davey; Morgan 143; Musialski UK chapter 529; *BBC* Blackpool; *Telegraph* 31 May 2011.

301 The moratorium was imposed by the governmental body in charge at that time, the Department of Energy and Climate Change (DECC), according to Written Statement Davey.

302 Dąborowski/Groszkowski 23.

303 Written Statement Davey; The Royal Society and the Royal Academy of Engineering '*Shale gas extraction in the UK: a review of hydraulic fracturing*' (London, 2012) available at: https://royalsociety.org/policy/projects/shale-gas-extraction/report/ [accessed 11 April 2014] (hereinafter: Royal Society); Christopher A Green et al. 'Preese Hall Shale Gas Fracturing Review and Recommendations for Induced Seismic Mitigation Report on behalf of the UK Department of Energy and Climate Change' available at: https://www.gov.uk/government/uploads/system/uploads/attachment_data/file/15745/5075-preese-hall-shale-gas-fracturing-review.pdf [accessed 11 April 2014] (hereinafter: Green et al.); Government response to Royal Academy of Engineering and Royal Society report on 'Shale Gas Extraction in the UK: a review of hydraulic fracturing' (10 December 2012) available at: https://www.gov.uk/government/uploads/system/uploads/attachment_data/file/49541/7269-government-response-sg-report-.pdf [accessed 11 April 2014] (hereinafter: Government response).

304 Green et al., ii/iii and 13/14; Royal Society 4–7.

additional precautionary measures to avoid 'induced seismicity' would be implemented.[305]

Shortly after the publication of these scientific reports the UK government incorporated these recommendations into a technical guidance document on prudent shale gas extraction.[306] These additional precautionary measures are intended to safeguard against earth tremors and include the obligation to review information on pre-existing geological faults before the commencement of drilling activities and to implement continuous monitoring when drilling is taking place.[307] Moreover, the operator has to submit a 'fracking plan', specifying volumes of fluids used for hydraulic fracturing and the timing of the 'fracks'.[308] The operator is also asked to monitor the growth in height of the fracture away from the borehole.[309] Finally, independent experts must be present at the drilling site to ensure compliance with all required precautionary measures.[310]

The UK government subsequently decided to lift the 'political moratorium'.[311] Ever since then, shale gas extraction in England and Wales has been possible, although with the additional requirement for operators to utilize the tighter precautionary measures. The UK government is currently perceived of as being 'strongly in favour' of shale gas extraction,[312] wanting to see shale gas extraction rolled out quickly' in England and Wales.[313]

Indeed, the government is actively engaged in stimulating shale gas extraction activities in the UK, mainly with the help of two means: changes to the access rights to land (including changes to the rules on trespass) and tax incentives.[314] With regard to access rights, the Infrastructure Act 2015 has been passed. S. 43 Infrastructure Act, in effect, grants shale gas

---

[305] Ibid and Morgan 143/144.

[306] United Kingdom Onshore Operators Group (UKOOG) 'UK Onshore Shale Gas Well Guidelines' (Issue 1, February 2013) available at: https://www.gov.uk/government/uploads/system/uploads/attachment_data/file/185935/UKOOG ShaleGasWellGuidelines.pdf [accessed 11 April 2014]. For explanations on the background of the British solution to use legally non-binding guidance documents of an industry body and the underlying 'goal-setting approach' to regulation, see Gordon/Paterson/Üsenmez paragraphs 8.36–8.69.

[307] Written Statement Davey.

[308] Ibid and Green at al. iii; Musialski UK chapter 532.

[309] Ibid.

[310] Ibid.

[311] Written Statement Davey; Morgan 143; Musialski UK chapter 533.

[312] Gordon/McHarg/Paterson paragraph 14.30.

[313] Morgan 150.

[314] Morgan 151–5.

operators the right to drill in deep-level land (at least 300 metres below surface level) to access energy resources. Remarkably, such action may be taken without having to engage in any form of negotiations over rights of access with the landowner.[315] This constitutes a considerable departure from the current system for Petroleum Exploration and Development Licenses (hereinafter: PEDL, the UK onshore petroleum licence),[316] according to which unpermitted entry onto another person's land is prohibited.[317]

In view of tax incentives, the UK government introduced a tax allowance, specific to shale gas activities, in 2013.[318] This measure has been supplemented by a regime which wants to stimulate local planning authorities to permit shale gas operations. According to the newest regulations, councils may keep 100 per cent (as opposed to the hitherto 50 per cent) of business rates (in effect property taxes) that they collect from shale gas sites.[319] This right could be worth up to £1.7 million per year, per site.[320] Gordon, McHarg and Paterson labelled the measure as an 'apparent attempt to encourage English authorities to approve fracking applications'.[321] These authors, however, deemed the attempt unsuccessful because local authorities continued to hesitate before approving fracking applications (. . .)'.[322]

As opposed to this pessimistic assessment there are, in fact, signs that the policy is working: North Yorkshire Council just gave the green light for the first exploratory works to be carried out at a shale gas site in England after the lifting of the 'political moratorium'.[323]

---

315 S.43 Infrastructure Act 2015; Morgan 151; Gordon/McHarg/Paterson paragraph 14.34.

316 More on PEDLs can be found at: Greg Gordon 'Petroleum Licensing' in Greg Gordon, John Paterson and Emre Üsenmez '*Oil and Gas Law: current practice and emerging trends*' 2nd edition (Dundee University Press, Dundee 2011) paragraph 4.71–4.73.

317 See s. 9 (2) Petroleum Act 1998; Gordon/McHarg/Paterson paragraph 14.32; Morgan 152.

318 HM Treasury '*Budget 2013*' pages 4, 36 and 82 available at: https://www.gov.uk/government/uploads/system/uploads/attachment_data/file/221885/budget2013_complete.pdf [accessed 22 September 2016]; for a discussion see Hunter/Usenmez/Paterson 391–4.

319 Morgan 150; Gordon/McHarg/Paterson paragraph 14.34.

320 UK Government 'Local Councils to Receive Millions in Business Rates from Shale Gas Developments' 13 January 2014 available at: http://www.gov.uk/government/news/local-councils-to-receive-millions-in-business-rates-from-shale-gas-developments [accessed 10 June 2016].

321 Gordon/McHarg/Paterson paragraph 14.34.

322 Ibid.

323 Josh Halliday, *The Guardian* 'North Yorkshire Council backs First UK Fracking Test for Five Years' available at http://www.theguardian.com/environ

The United Kingdom is not the only European country or region that opted for a 'political moratorium' on shale gas extraction. Others include the Netherlands, Romania, the Czech Republic, Denmark, the German states of North Rhine-Westphalia, Lower Saxony and Hesse, as well as the Belgian region of Flanders.[324] While the Netherlands and the Czech Republic currently operate a 'political moratorium', Romania, like the UK, already lifted its moratorium.[325] All cases have in common that the decision to install a 'political moratorium' was based on an internal agreement between government ministers. The public has only been informed on the outcome of consultations and, accordingly, very few official documents on the 'political moratoria' exist. A 'traditional' systematic legal interpretation of 'political moratoria' by Savigny's methods is hence not possible.

However, indirect sources can be utilized to trace legislative and governmental reasoning. In the Czech Republic and in Romania, legislative documents illuminate the official position. In these documents the government and supporting Senators explained themselves to the public.[326] In the Netherlands, a scientific report on shale gas, commissioned and endorsed

---

ment/2016/may/23/north-yorkshire-council-backs-first-uk-fracking-tests-for-five-years?CMP=fb_gu [accessed 10 June 2016].

324 See Introduction to this chapter for references.

325 Daborowski/Groszkowski 25; Savu/Timu; Irina Savu *Bloomberg News* 'Romania Ends Moratorium on Shale Gas Exploration, Premier Says' available at: http://www.bloomberg.com/news/2013-03-19/romania-ends-moratorium-on-shale-gas-exploration-premier-says.html [accessed 1 April 2014]; Neil Buckley *Financial Times* 'Romania and Lithuania back fracking' available at: http://www.ft.com/cms/s/fa2812bc-6fa6-11e2-956b-00144feab49a,Authorised=false.html?_i_location=http%3A%2F%2Fwww.ft.com%2Fcms%2Fs%2F0%2Ffa2812bc-6fa6-11e2-956b-00144feab49a.html%3Fsiteedition%3Duk&siteedition=uk&_i_referer=http%3A%2F%2Fsearch.ft.com%2Fsearch%3FqueryText%3DRomania%2Band%2BLithuania%2Bback%2Bfracking#axzz2xo4pyAZd [accessed 3 April 2014].

326 Czech Republic, see immediately below; Romania: Romanian Senate 2012 Legislative proposal prohibiting exploration and exploitation perimeters with liquid or gaseous hydrocarbon by hydraulic fracture (cracking), exclusive exploration licences, and cancellation of all projects that use this technique (L228/2012)) (*2012 Propunere legislativă privind interzicerea explorărilor şi exploatărilor perimetrelor cu zăcăminte de hidrocarburi lichide sau gazoase prin fracturarea (fisurarea) hidraulică şi anularea licenţelor exclusive de explorare a tuturor proiectelor care recurg la această tehnică (L228/2012)*) (hereinafter: Romanian proposal); Romanian Senate 'Minutes of Senate meeting on 5 November 2013 Summary' (*Stenograma şedinţei Senatului din 5 noiembrie 2013 Sumar*) 59/60 and 62–7 available at: http://www.senat.ro/PAGINI/Stenograme/Stenograme2013/13.11.05.pdf [accessed 26 May 2014] (hereinafter: Romanian Senate Minutes 5 November 2013).

by the Dutch government, exists,[327] alongside a short governmental reasoning on the establishment of a moratorium.[328]

When assessing these 'political moratoria' the unresolved questions that have been identified as crucial in the French ban and the German moratorium will build the focus of the investigation into the situation in the UK. First, it will be established whether the 'political moratorium' targeted hydraulic fracturing specifically or shale gas extraction in general. Second, it needs to be assessed how the regulation catered for the interplay of environmental protection with energy security and third, if and to what extent the regulation was based upon the precautionary principle.

3.4.1 Outlawing Shale Gas Extraction or Hydraulic Fracturing?

The implementation of a 'political moratorium' allowed the UK government to avoid the pitfalls of ambiguous wording that became obvious in the assessment of the French moratorium. The UK government did not need to engage in semantic discussions on the differences between 'hydraulic fracturing' and 'shale gas extraction'. Since no abstract law had to be elaborated, the government adopted a rather pragmatic stance on the issue and approached the only company that was undertaking shale gas extraction in the UK at that time, ordering it to halt extraction.[329]

Moreover, the government decided internally that it would not consent to any requests for permission to extract shale gas for the duration of the UK moratorium.[330] The responsible government minister stressed, however, that hydraulic fracturing as a technique would remain licit for purposes other than shale gas extraction.[331] As a result, the UK government established a moratorium that efficiently targeted shale gas extrac-

---

[327] Ministry of Economic Affairs Directorate Energy Market 'Additional studies on potential risks and impacts of the exploration and production of shale and coal in the Netherlands Final report research '(*Ministerie van Economische Zaken Directie Energiemarkt Aanvullend onderzoek naar mogelike risico's en gevolgen van de opsporing en winning van schalie- en steenkoolgas in Nederland Eindrapport onderzoeksvragen*)' available at: http://www.rijksoverheid.nl/ministeries/ez/documenten-en-publicaties/rapporten/2013/08/26/aanvullend-onderzoek-naar-mogelijke-risico-s-en-gevolgen-van-de-opsporing-en-winning-van-schalie-en-steenkoolgas-in-nederland-eindrapport-onderzoeksvragen-a-en-b.html [accessed 14 April 2014].

[328] Netherlands Statement Minister of Economic Affairs; Schavemaker.

[329] *BBC* Blackpool; *Telegraph* 31 May 2011; Written Statement Davey.

[330] Written Statement Davey.

[331] Ibid.

tion, but not hydraulic fracturing. Even if other means to extract shale gas became available in the UK during the period of the moratorium,[332] it would have been a breach of the moratorium to apply them.

3.4.2 Balancing Environmental Concerns with Energy Security Interests in a 'Political Moratorium'

While the 'political moratoria' in the Czech Republic and Romania are silent on the balancing of environmental concerns and energy security interests,[333] the weighting has been a central point in UK discussions. The House of Commons report identified the balance between environmental risks and potential energy security gains as crucial for the appraisal and handling of shale gas extraction.[334] It dismissed a moratorium on the

---

332 A possibility that was explicitly dismissed by the UK government, which states that hydraulic fracturing is the only means currently available, see Written Statement Davey.

333 In the case of the Czech Republic for instance, parliamentary discussions were dominated by commonly encountered environmental concerns about shale gas extraction, see Czech Senate 'Proposal for a Senate Bill by Senators Pakosty Peter, George Oberfalzer and Paul Trpák' (*Návrh senátního návrhu zákona, senátorů Petra Pakosty, Jiřího Oberfalzera a Pavla Trpáka*) 8/9 available at: http://www.senat.cz/xqw/webdav/pssenat/original/64897/54755 [accessed 23 May 2014] (hereinafter: Czech Explanatory Memorandum); Czech Senate 'Minutes of the 23. Senate meeting (2nd day of meeting – 14.06.2012)' (*Těsnopisecká zpráva z 23. schůze Senátu (2. den schůze – 14.06.2012)*) Speeches of Petr Pakosta and Jiří Bis available at: http://www.senat.cz/xqw/xervlet/pssenat/hlasovani?action=steno&O=8&IS=4817&D=14.06.2012#b12986 [accessed 23 May 2014] (hereinafter: Czech Minutes 14 June 2012). Energy security aspects by contrast, are missing entirely from the 'political moratorium' and have only been briefly addressed in the debate, see: Czech Explanatory Memorandum 9; Czech Senate 'Minutes of the fourth Senate meeting (2nd day of meeting – 31 January 2013)' (*9. funkční období Těsnopisecká zpráva z 4. schůze Senátu (2. den schůze – 31.01.2013)*) Speech of Jiří Bis available at: http://www.senat.cz/xqw/webdav/pssenat/original/67114/56512 [accessed 27 March 2014] (hereinafter: Czech Minutes 31 January 2013).

334 UK report I, 3. The same is true for the Netherlands, see Ministry of Economic Affairs Directorate Energy Market 'Additional studies on potential risks and impacts of the exploration and production of shale and coal in the Netherlands Final report research '(*Ministerie van Economische Zaken Directie Energiemarkt Aanvullend onderzoek naar mogelijke risico's en gevolgen van de opsporing en winning van schalie- en steenkoolgas in Nederland Eindrapport onderzoeksvragen*) available at: http://www.rijksoverheid.nl/ministeries/ez/documenten-en-publicaties/rapporten/2013/08/26/aanvullend-onderzoek-naar-mogelijke-risico-s-en-gevolgen-van-de-opsporing-en-winning-van-schalie-en-steenkoolgas-in-nederland-eindrapport-onderzoeksvragen-a-en-b.html [accessed 14 April 2014].

basis that, from the Committee's point of view, environmental threats of shale gas extraction are hypothetical and unproven.[335] When weighing these potential threats against possible energy security benefits, the House of Commons Committee concluded that the case for environmental concerns was not strong enough to justify a moratorium.[336] Another UK study on shale gas agreed that environmentally sound extraction could be achieved without a moratorium, if certain precautionary measures were implemented.[337]

Despite these scientific recommendations, the UK government stopped shale gas extraction for a short period after the Blackpool tremors (May 2011–December 2012).[338] The efficacy of the 'political moratorium' meant that, as no shale gas could be extracted at that point in time, energy security interests were not furthered for the duration of the 'political moratorium'. The responsible UK minister justified the lifting of the UK moratorium with the argument that the additional precautionary measures, implemented after the Blackpool tremor, made shale gas extraction safe. They would now allow the regulator to cater for both environmental concerns and energy security interests.[339]

To sum up, the 'political moratorium' can only deliver a reasonable balance between environmental protection and energy security if it is embedded in the broader context of an incremental approach to shale gas regulation. It was this incremental approach which allowed for the simultaneous consideration of environmental concerns and energy security interests before and after the moratorium was in place. The 'political moratorium' properly speaking, however, did not make any concessions to energy security interests and aimed to provide the strongest environmental protection possible. For the duration of the 'political moratorium', conciliation of both interests was hence not possible.

3.4.3 The British Interpretation of the Precautionary Principle as a Principle of Cautious Action

Prior to the earth tremors in Blackpool, the precautionary principle had not been explicitly invoked by the UK regulator with regard to shale gas extraction. The words 'precaution' or 'Precautionary Principle' do not

---

335 UK report I, 10.
336 Ibid.
337 Green et al., ii/iii and 13/14; Royal Society 4–7.
338 Written Statement Davey.
339 Ibid.; the same argument was invoked in Romania, see Truth Live Ponta.

feature in the House of Commons report.[340] Similarly, no reference to precaution was made in the government's response to the report.[341]

After the tremors, discussions on the precautionary principle became more visible in official documents. Although still not being explicitly referenced as a principle, the governmental inquiry into the causes of the tremors suggested a 'cautious continuation' of hydraulic fracturing operations in the UK.[342] The responsible government minister accordingly stressed that the government will move forward on the shale gas issue with appropriate caution.[343] Such statements typically constitute an implicit invocation of the precautionary principle.[344]

The UK government's interpretation of the precautionary principle as a principle of cautious action can be traced in regulatory actions and statements regarding the future conduct of the government and the industry. Not only did the government pursue the described sequence of regulatory actions, the responsible minister also emphasized that the new precautionary measures, put into place after the tremors to mitigate possible threats in the future, are not to be regarded as the final word on the matter. He said

> The controls are not at this stage to be regarded as definitive, but as appropriate precautionary measures for our present state of knowledge. Initial operations under these controls will be subject to careful scrutiny to ensure the effectiveness of the controls. And they will be reviewed, as experience develops, to ensure that they are proportionate to the risks.[345]

This statement is indicative of an understanding of the precautionary principle as principle of cautious action. The minister reckons that the current precautions are suitable. However, if this assessment should turn out to be wrong, he promises to review them to ensure that proportionate measures are in place to counter potential threats.[346] This incremental approach to future regulations would be hard to reconcile with an understanding of the

---

340 Although this is true for the main report, a very brief reference to precaution was made during evidence-hearing sessions, see: UK report I, Ev 12 and 83.

341 Energy and Climate Change Committee of the House of Commons '*Shale Gas: Government Response to the Committee's Fifth Report of Session 2010–12*' (Crown 2011).

342 Green et al., iii.

343 Written Statement Davey.

344 Trouwborst 2006, 59 and 62.

345 Written Statement Davey.

346 More on this can be found below in the chapter on precautionary measures.

precautionary principle as a principle of inaction, because the latter would not allow for any advancement of an activity in the face of uncertainty.[347]

3.4.4 Specifics in Scotland and Northern Ireland

As opposed to this governmental support for shale gas extraction in England and Wales, the devolved governments of Northern Ireland[348] and Scotland[349] maintained or implemented moratoria to outlaw shale gas extraction. Although the UK remains a unitary state which does not operate a federal structure in the strict sense of the word, certain powers have been devolved by the central government to devolved governments.[350] Whether or not this is also the case for energy policy and law-making differs from country to country, with Northern Ireland enjoying the greatest devolved energy policy competence of all British nations.[351]

3.4.4.1 Northern Ireland

According to schedule 2 paras 9 and 18 of the Northern Ireland Act 1998, only nuclear power regulation and energy taxation are subject to 'London rule'. All other areas of energy policy and law-making fall into the competence of the Northern Ireland Assembly and the Northern Ireland Executive.[352] As Gordon, McHarg and Paterson aptly point out, 'there is also potential for unconventional gas production, but the NI [Northern Ireland] Executive currently opposes it'.[353]

However, it is not only the Executive, but also the Northern Ireland Assembly, which opposes shale gas extraction. On 6 December 2011, the Assembly of Northern Ireland resolved a 'moratorium by law' on shale gas extraction, the text of which is rather short, reading:

> Resolved: That this Assembly believes that a moratorium should be placed on the onshore and offshore exploration, development and production of shale gas by withdrawing licences for hydraulic fracturing (fracking), at least until the publication of a detailed environmental impact assessment into the practice; notes that hydraulic fracturing can put local water sources at risk of contamination; further notes that, amongst a variety of adverse environmental impacts,

347 As above.
348 Northern Ireland Minutes.
349 Gordon/McHarg/Paterson paragraph 14.30.
350 Russell Deacon '*Devolution in the United Kingdom*' 2nd edition (Edinburgh University Press, Edinburgh 2012) 248.
351 Gordon/McHarg/Paterson paragraph 14.11.
352 Ibid.
353 Gordon/McHarg/Paterson paragraph 14.11 footnote 21.

> the process of fracking can cause serious well blowouts, which put both workers and local communities at risk; considers that the production of hard-to-reach fossil fuels is not compatible with efforts to achieve carbon reduction targets; and urges the Minister of Enterprise, Trade and Investment to give greater support to the generation of energy from renewable sources instead.[354]

The analysis of that prohibition should start with the question of whether it constitutes a ban or a moratorium. According to the text, the prohibition shall stay in place at least until the publication of a detailed environmental impact assessment, thus, making explicit that the law is only a temporary measure. Moreover, the text itself clearly speaks of a 'moratorium'. There are no indications to the contrary.

Besides that clear categorization as moratorium, a second aspect should be discussed, namely the extent to which the moratorium caters for the balancing of environmental protection interests with energy security needs. The short text of the moratorium includes a number of references to environmental concerns, including water protection, climate change issues and renewable energy generation, which are explicitly invoked as reasons for the imposition of the moratorium.

As distinct from other moratoria, it is argued that 'the process of fracking can cause serious well blowouts, which put both workers and local communities at risk.' The debate in the Northern Ireland Assembly on the moratorium clarified that the risk of a 'blowout' is associated with insufficient well casing and well integrity, which could lead to damage of water reserves.[355] Critical Members of the Assembly, however, pointed out that the wording is very imprecise, since well-blowouts may occur during any kind of oil and gas extraction and to single out shale gas extraction would not do justice to the activity.[356]

While the text of the moratorium is rather explicit on potential threats, it does not elaborate on potential energy security benefits. However, during the debate on the moratorium in the Northern Ireland Assembly some of its Members contemplated the balance between environmental protection and energy security. They concluded that 'at this stage no economic argument can overcome the concerns regarding the environmental and health risks that have been raised about the impact of hydraulic fracturing'.[357] Moreover, the Members noted that 'it cannot be determined that the

---

[354] Northern Ireland Assembly Deb 6 December 2011, Vol 69 No 6, cols 300 and 336 (hereinafter: Northern Ireland minutes).

[355] Northern Ireland minutes col. 301.

[356] Northern Ireland minutes cols. 331/332.

[357] Northern Ireland minutes col. 302.

possible economic benefits would present a public interest that would override the potential detrimental environmental and health damages.'[358]

A third point of interest is the relation between the Northern Irish moratorium and the precautionary principle. As a point of departure, it is interesting to note that Northern Ireland resolved a moratorium on shale gas extraction, despite not having issued any permits for shale gas extraction in the first place. The Minister for Enterprise, Trade and Investment accordingly pointed out:

> We can hardly impose a moratorium on hydraulic fracturing, because no permit has yet been issued, and it is unlikely that an application will come to the Department for at least another 12 to 18 months. During that period, we will reap the benefit of several in-depth scientific and engineering studies currently in progress, notably in the United States.[359]

As opposed to that view, the imposition of the precautionary principle prior to the issuance of permits might well fall in line with the logic that underpins the precautionary principle, since precautionary measures must be taken by governments actively and as early as practicable.[360]

But although the withdrawal of licences might be an exercise that remains to be of mere theoretical importance to Northern Ireland, it is worthwhile scrutinizing the Northern Irish position on this for comparison purposes. The withdrawal might also be read as a precautionary measure, as discussed in the context of the French moratorium above. However, the text of the Northern Irish moratorium states that hydraulic fracturing should be stopped 'at least until the publication of a detailed environmental impact assessment into the practice' has been carried out.[361] The withdrawal of licences, however, would not be a temporary measure.

Suspicions were rife among Members of the Assembly that the intention behind the law was to ban hydraulic fracturing altogether, not only during the time required to conduct research.[362] If the aim of the moratorium was to install a temporary prohibition of shale gas extraction, it would have been more appropriate to provide the possibility to suspend licences until the research has been carried out, instead of withdrawing them straight away. Once a licence is withdrawn a complete new licence would have to be issued if the scientific inquiry established the safety of the technology.

358 Ibid.
359 Northern Ireland minutes col. 333.
360 Trouwborst 2006, 181.
361 Text of the Northern Ireland Resolution.
362 Ibid. 305.

But given the fact that there were no pre-existing licences, this problem is likely to remain of theoretical nature.

3.4.4.2 Scotland

Compared to the situation in Northern Ireland, Scotland was rather late in putting into place a moratorium on shale gas extraction (2015 in Scotland compared to 2011 in Northern Ireland). Initially, the Scottish government did not follow the Northern Irish example to impose a moratorium in the early days (2011) of the shale gas debate.[363] Instead, it opted for obstructing and impeding the implementation of the, arguably, lenient line of the UK government towards shale gas regulation.

This point may be illustrated by an announcement of the Scottish government of 19 October 2013, in which it pledged to strengthen the position of local councils in planning policy related to unconventional oil and gas activities.[364] This meant that the approval of local councils was now needed prior to the commencement of works that had been, in principle, licensed under a PEDL by the UK government.[365] The Scottish government at that time must have been all too aware of the fact that local councils were broadly opposed to any form of unconventional hydrocarbon extraction.

But there is another crucial difference between shale gas regulation in Northern Ireland and Scotland, namely the fact that the reservation of energy policy competences to Westminster is much more extensive in Scotland.[366] Under schedule 5 Head D of the Scotland Act 1998 large parts of energy policy, including natural gas-ownership in situ and exploration and production of natural gas, fall outside of Scottish competence.[367] This, however, might change in the future, as a legislative proposal on further devolution has entered the legislative process in the wake of the Scottish 'No' vote in the 2014 Scottish Independence Referendum.[368] Moreover, questions concerning the future relationship between Scotland and the rest of the UK are abounding after the Brexit referendum.

For now the Scottish government has only two competences that

---

[363] The Northern Irish moratorium was resolved by the Northern Ireland Assembly in 2011, see Northern Ireland minutes.

[364] Scottish government 'Policy on unconventional gas' 19 October 2013 available at: http://news.scotland.gov.uk/News/Policy-on-unconventional-gas-547.aspx [accessed 10 June 2016].

[365] Ibid.

[366] Ibid.

[367] Schedule 5 Head D section D 2 a) Scotland Act 1998; Hunter/Usenmez/Paterson 386/387.

[368] Hunter/Usenmez/Paterson 387/388.

relate to energy, the right to regulate renewables and the environment in Scotland.[369] These latter competences have been sufficient to allow the Scottish government to pursue distinct energy policy goals, for instance a ban on new nuclear plants and hydraulic fracturing.[370] Scotland did not base its moratorium on the energy competence, but instead the (devolved) environmental competence on water protection instead as a gateway to prohibit shale gas extraction.

On 8 October 2015, the Scottish Parliament resolved the Water Environment (Controlled Activities) (Unconventional Oil or Gas Development) (Scotland) (No. 2) Direction 2015, which constitutes the legal basis of the Scottish moratorium. The Direction uses the mechanisms provided by the Water Environment (Controlled Activities) (Scotland) Regulations 2011, which itself is based on the Water Environment and Water Services (Scotland) Act 2003.[371] Article 20 (1) Water Environment and Water Services (Scotland) Act 2003 apportions to the Scottish Ministers the right to regulate activities for the purpose of water protection.

Contentwise, article 2 Water Environment (Controlled Activities) (Unconventional Oil or Gas Development) (Scotland) (No. 2) Direction orders the Scottish Environment Protection Agency (SEPA) to refer to the Scottish Ministers any application under the Water Environment (Controlled Activities) (Scotland) Regulations 2011 for an authorization to carry on any controlled activity in connection with unconventional oil or gas development. This complicated mechanism is nothing more (or less) than a transfer of the power to decide upon applications for shale gas extraction from SEPA to the Scottish Ministers.

But that was only the formal law part which gave the Ministers the necessary leeway for discretion. How they intended to use this leeway for discretion to approve or disapprove of applications for shale gas extraction was already clear prior to the resolution of this formal law. A political moratorium has already been proclaimed on 28 January 2015 by the then Scottish Energy Minister Fergus Ewing. He announced via the webpage of the Scottish government

> a moratorium on the granting of planning consents for all unconventional oil and gas developments, including fracking. This moratorium will continue until

---

[369] Ruven Zeuschner 'United Kingdom: The Development of the Consenting Regime for Offshore Renewable Energy Projects in Scotland' (2011) 4 *International Energy Law Review* 150/151.

[370] Gordon/McHarg/Paterson paragraph 14.11.

[371] According to the preamble of the Water Environment (Controlled Activities) (Scotland) Regulations 2011.

> such time as the work I have set out to Parliament today, including a full public consultation, is completed.[372]

The formal law was thus used to reinforce the 'political moratorium'. Moreover, the announcement highlighted the need for further research, which it phrased in the following terms: 'the independent Expert Scientific Panel published their report in July 2014 saying that more evidence is needed into the effects of unconventional extraction oil and gas.'[373] The Minister added: 'We recognize that local communities are likely to bear the brunt of any unconventional oil and gas developments, particularly in terms of increased traffic and related emissions and noise impacts. These are issues that must be researched further.'[374]

This statement refers to the fact that an Independent Expert Scientific Panel, commissioned by the Scottish government to assess unconventional oil and gas extraction in Scotland, issued its final report on 28 July 2014.[375] Indeed, the report[376] flagged up that further research is currently on its way.[377]

However, this research was associated with the estimation of reserves and the further understanding of public opinion towards shale gas extraction.[378] The executive summary does not include a substantial call for further investigations, but concludes that, by and large, the issues are rather well-researched and do not appear to be insurmountable.[379]

One and a half years(!) after the publication of this report, the Scottish government decided that, in spite of the conclusion that no substantial new research was needed, it had to put a moratorium in place. The evident misrepresentation of the report's findings highlights that the need for further research is rather unlikely to have been the true reason behind the

---

372 The Scottish Government 'Moratorium called on Fracking' available at: http://news.scotland.gov.uk/News/Moratorium-called-on-fracking-1555.aspx[accessed 8 September 2016] (hereinafter: Scottish Moratorium Announcement).

373 Ibid.

374 Ibid.

375 The Scottish Government 'Issued on behalf of the Expert Scientific Panel' available at: http://news.scotland.gov.uk/News/Issued-on-behalf-of-the-Expert-Scientific-Panel-f2a.aspx [accessed 8 September 2016] (hereinafter: Announcement Independent Panel).

376 Independent Expert Scientific Panel 'Report on Unconventional Oil And Gas' (Crown 2014) available at: http://www.gov.scot/Resource/0045/00456579.pdf [accessed 8 September 2016] (hereinafter: Independent Panel Report).

377 Paragraphs 4.8, 8.4 and 8.11 Independent Panel Report.

378 Ibid.

379 Page v Executive Summary Independent Panel Report.

introduction of the moratorium. An arguably much bigger role was played by the public resistance to shale gas extraction.[380]

Moreover, it is worth noting that the Scottish Moratorium does not outlaw hydraulic fracturing.[381] The Scottish government took a broader approach and put in place a moratorium on 'all unconventional oil and gas developments, including fracking'. In light of the discussions on the phrasing of the French ban and the German moratorium, this must be understood as covering all unconventional operations and all types of rock.

Furthermore, it is interesting to investigate the nature of the prohibition – namely whether it constitutes a ban or a moratorium. Scholars who did write on the subject were apparently undecided or did not pay attention to the matter, as they used both terms at the same time to describe the same prohibition.[382] It is, indeed, interesting to give some thought to that issue, but not so much in terms of ban or moratorium (looking at the explicit reference to further research in the proclamation it is rather clear that this is thought to be a time-restricted measure, thus a moratorium). Rather, the question is whether the Scottish prohibition is a 'moratorium by law' or a 'political moratorium', given that it is based on a formal legal act in conjunction with a statement of the government.

Although there are reasons to view the Scottish prohibition as a mixture of both, there are better arguments for classifying it as a 'political moratorium'. For one, the political announcement was made *prior* to the coming into force of the formal act. Second, the formal act does not address the question of prohibition of shale gas extraction, but rather reserves this decision to the political sphere.

Finally, some attention should be paid to the balance of environmental protection with energy security interests in the Scottish moratorium. While it is clear from several passages in the text of the proclamation that the Scottish government was attentive to potential environmental threats of shale gas extraction,[383] this is less obvious with energy security. Although the announcement as such does not include a direct reference to energy security, the government concluded its statement with the words:

> We have listened to legitimate concerns about potential negative impacts. However, we must also acknowledge that some take a different view and see

---

380 Compare the text of the Scottish Moratorium Announcement.

381 As Gordon/McHarg/Paterson paragraph 14.11 wrongly claimed.

382 Compare Gordon/McHarg/Paterson paragraph 14.11 with 14.30.

383 See for instance the explicit references in the Scottish Moratorium Announcement to 'increased traffic and related emissions and noise impacts'.

> opportunities in unconventional oil and gas extraction. We should never close our minds to the potential opportunities of new technologies – but we must also ensure that community, environmental and health concerns are never simply brushed aside. This government will not allow that to happen and I hope the actions I have announced today will be widely welcomed as proportionate and responsible.[384]

It is apparent from that statement that the government is aware of the potential benefits of shale gas extraction for Scotland, but does not want to give the impression that it actually believes in great opportunities. The reference in the last sentence to proportionality could be viewed as an attempt to come to some kind of balance. However, the document does not pursue the discussion of the interplay between environmental protection and energy security any further.

3.4.5 Summary Great Britain and Shale Gas

To sum up, the UK example of incremental shale gas regulation shows that 'political moratoria' can be a flexible regulatory tool. The UK circumvented the pitfalls of prescriptive regulation; it ensured that shale gas extraction cannot take place, irrespective of the techniques used. In addition, the British example showcased that 'political moratoria' are flexible in the sense that they may be annulled at governmental discretion, whenever the lifting of a moratorium is deemed appropriate.

The main reasoning behind the use of an incremental approach to shale gas regulation, which might include a 'political moratorium', is the possibility it offers to cater for environmental protection interests and energy security needs simultaneously. Prior to the instalment of the 'political moratorium', the UK regulator advanced shale gas extraction, but required operators to put precautionary means in place, according to the hitherto existing state of knowledge. During that initial period it would have been possible to extract shale gas in the UK. The extracted gas would not have had to be imported from other countries and, thus, could have contributed to the security of energy supplies in the UK.[385]

At the same time, however, the implemented precautions prevented considerable environmental harm: as soon as minor earth tremors were detected at the first shale gas drilling site near Blackpool, operations were shut down and a 'political moratorium' was imposed. The precautionary measures were re-assessed and, once new measures were adopted in line

---

[384] Scottish Moratorium Announcement.
[385] A connection described in general terms above in Chapter 1.

with the new state of knowledge, the moratorium was lifted and shale gas extraction could resume. This approach ensures that the environment is protected with the most up-to-date techniques, while future possible positive effects on energy security are safeguarded.

One should, however, be cautious in concluding that a 'political moratorium' inevitably strikes a balance between environmental protection and energy security. The Scottish example aptly highlights that a 'political moratorium' might provide for little more than commonplace-statements and abstain from engaging in a genuine balancing exercise.

Nonetheless, at least the *potential* of the described incremental approach to shale gas regulation for balancing the interests of environmental protection and energy security on a flexible basis is evidenced by the developments in the UK. With regard to the precautionary principle the incremental approach can be viewed as a first practical example of how the proposition of this book, an interpretation of environmental law principles as principles of cautious action, may play out in practice.

3.5 CONCLUSION

As far as prohibitive shale gas regulation in European countries is concerned, moratoria are currently the preferred measure of choice. However, not all prohibitions are quite the same. This chapter revealed that prohibitive shale gas regulation in Europe tends to fall into one of three categories, 'ban by law', 'moratorium by law' and 'political moratorium'. Although all three represent examples of prohibitive shale gas regulation, their main features are quite distinct from each other.

The French 'ban by law' as well as the German 'moratorium by law' are rather rigid and inflexible tools that do not regulate shale gas extraction in a very efficient manner. Probably the most obvious example is German shale gas legislation. The descriptive approach to shale gas regulation taken there and its strict nature leads to inconsistencies, which leave the prohibition law exposed to substantial challenges in front of the German Federal Constitutional Court.

Turning to the interplay of environmental protection with energy security interests in shale gas extraction, the French ban arguably gives rise to the greatest imbalance. Environmental protection has been given clear preference, not only in the text of the ban, but also in the underlying reasoning and in the balance of different interests.

The reason for this lopsided approach is the French regulator's view on the precautionary principle. The French regulator interprets the environmental law principle of precaution rather narrowly, as a command not to advance a potentially risky technology unless scientific research has shown that no potential environmental threat exists.[386] Under this interpretation, new technologies cannot be advanced on a confined 'trial and error' basis. Only if their security has been proven by scientific experiments beyond any doubt may such technologies be deployed in France.

This conceptual interpretation of environmental law principles as principles of inaction has three main setbacks. For one, the strong focus on environmental protection overshadows the existence of other societal interests, which are also relevant to the issue of shale gas extraction.

Second, the use of environmental law principles as principles of inaction comes with a heavy 'price-tag' attached to it, the so called 'opportunity costs'.[387] 'Opportunity cost' is a term used in economics, which refers to the fact that omitting an opportunity may create costs in the form of foregone benefits and chances.[388] In the case of the French moratorium, these 'opportunity costs' relate to the omission of the chance to improve the security of energy supplies of the country by domestic production.

Third, the use of the precautionary principle as principle of inaction could lead to a 'fossilization' of knowledge. In the debate about the French shale gas moratorium, MPs pointed out that the immediate and outright ban on shale gas in France could cripple innovative industrial research.[389] If research and exploration are not allowed a country could lose its ability to adapt to changing circumstances by developing new technologies and as a result, its state of knowledge could 'fossilize'.[390]

The application and interpretation of the precautionary principle can be very different, as the UK example highlights. Under the English and Welsh interpretation, precaution is deemed to be a principle of cautious action, demanding an incremental approach to shale gas regulation. A shale gas moratorium is seen as a *possible*, not a *compelling* consequence of precautionary logic.

This interpretation led to the English and Welsh incremental advancement of shale gas extraction, which initially required operators to put

---

[386] French National Assembly Minutes 10 May 2011 part 2, 2921, 2923, 2931; French National Assembly Minutes 11 May 2011, 2973/2974.

[387] More on 'opportunity costs' may be found at: Nigel Haigh 'The Introduction of the Precautionary Principle into the UK' in: Timothy O'Riordan and James Cameron (eds), '*Interpreting the Precautionary Principle*' (Earthscan Publications Ltd., London 1994) 231 (hereinafter: Haigh).

[388] Haigh 231.

[389] French National Assembly Minutes 10 May 2011 part 2, 2931.

[390] French National Assembly Minutes 11 May 2011, 2973/2974; French National Assembly Minutes 10th May 2011 part 2, 2931.

precautionary means in place, in line with the relevant state of knowledge at that time. Shale gas could be extracted in the UK and had not to be imported from other countries, contributing to the security of energy supplies. When issues arose, the applicable regulatory framework was revised. This cautious 'trial and error' approach ensures that the environment is protected by the most up-to-date techniques, while shale gas extraction can resume with a view to enhancing energy security. The advantage of the cautious action-approach lies in its ability to cater for both interests, environmental protection and energy security, at the same time.

Under a 'political moratorium' the government can target shale gas operations directly and at its discretion, without the need for very abstract laws. Quarrels surrounding the definition of terms like 'hydraulic fracturing' and 'shale gas extraction' that occurred in France and Germany are circumvented. Precautionary measures may be handled flexibly and adjusted in one or the other direction, whenever there are indications to do so. This flexibility allows for confined shale gas extraction and the reaping of energy security benefits. Thus, very few 'opportunity costs' are being incurred.

At a conceptual level, this interpretation of the precautionary principle as principle of cautious action enables a learning society. Technological knowledge is extended in an ongoing manner, which avoids a freezing of the status quo and the fossilization of knowledge.

PART II

A new methodology for energy regulation: the trias of objectives, principles, rules

4. Objectives and shale gas

4.1 INTRODUCTION

The EU's 2014 Shale Gas Recommendation[1] and the pertaining 2014 Shale Gas Communication[2] require Member States to make up their own minds on the approach they want to take towards shale gas regulation.[3] Having explored the cautious but permissive approach to shale gas regulation on the one hand and the strictly prohibitive approach on the other hand a more conceptual question arises: which cornerstones exist for the development of regulatory energy policies in EU Member States?

Whether to allow a controversial way of gas extraction or restrict it is a choice that has to be made by society as a whole and its political representatives. The different approaches to shale gas regulation that Member States are thus adopting are based on political choices. The legislator has leeway for discretion in taking such decisions.

Having said that, national legislators may not simply adopt *any* law they see fit. These new laws must comply with certain legal requirements which may, for instance, be found in constitutional law. The natural starting point for the development of new regulation in many Member States is the constitution, just as regulations at EU level have to be developed from a primary law basis.[4]

---

[1] Commission Recommendation 2014/70/EU of 22 January 2014 on minimum principles for the exploration and production of hydrocarbons (such as shale gas) using high-volume hydraulic fracturing [2014] OJ L 39/72. The Recommendation targets not only shale gas extraction, but every kind of unconventional gas extraction that uses 'high-volume hydraulic fracturing', a term defined in Commission Shale Gas Recommendation article 2 (a); for reasons of coherence and readability, the author will use the generic term 'shale gas extraction' when referring to 'high-volume hydraulic fracturing'.

[2] Commission 'Communication on the exploration and production of hydrocarbons (such as shale gas) using high volume hydraulic fracturing in the EU' (Communication) *COM (2014) 23 final/2* (hereinafter: 2014 Commission Shale Gas Communication).

[3] See Chapter 3 above.

[4] Although the legislator is generally entitled to adopt new laws according to perceived needs, this right is not unlimited. Take the example of the biggest EU

In that regard, environmental protection and energy security, being the most salient aspects in the case of shale gas extraction,[5] are not mere societal interests. The current chapter highlights that they are constitutional objectives and hence amount to legally binding norms, which have to be taken into account when developing new regulations on shale gas extraction.

Constitutional objectives are norms that oblige the state to further a given aim by all possible means and to align state actions with that objective.[6] Crucially, objectives do not convey subjective rights to individuals,[7] but they are legally binding upon the state, which has to permanently regard and fulfil them, when implementing new laws.[8]

Environmental regulation should contribute to the achievement of the objective of environmental protection, whereas a measure related to energy should ensure energy security.[9] Constitutional objectives thus have a bearing on the make-up of concrete shale gas regulation.

The essential question is what if the legislator is asked to achieve two constitutional objectives which compete with each other, as is the case with shale gas regulation?[10] To find a solution to this conflict, the current book puts forward a new methodology, the so called trias, which builds upon a constitutional rights theory that was developed by the modern

---

Member State, Germany: here the leeway for discretion is curtailed by article 20 (3) German constitution (*Grundgesetz*), which obliges German decision-makers to comply with the stipulations of the constitution.

5 See Introduction and Chapter 1 above.

6 Nicolas de Sadeleer '*Environmental Principles: From Political Slogans to Legal Rules*' (Oxford University Press, 2002) 310 (hereinafter: De Sadeleer); BVerfGE 14, 263 (275); 59, 57 (108); Hermann von Mangoldt and Friedrich Klein et al. *'Das Bonner Grundgesetz Kommentar Band 2: Artikel 20 bis 78'* 4th edition (Verlag Franz Vahlen, München 2000) article 20 a pararaph 33 (hereinafter: V Mangoldt); Michael Sachs '*Grundgesetz Kommentar*' 6th edition (Beck, München 2011) Sachs article 20a paragraph 17 (hereinafter: Sachs).

7 Klaus Stern '*Das Staatsrecht der Bundesrepublik Deutschland Band III/2*' (Beck, München 1994) 1486 and 1714 (hereinafter: Stern III/2); Kloepfer 388.

8 See for instance the rulings of the German Federal Constitutional Court in BVerfGE 14, 263 (275); 59, 57 (108); indirectly BVerfGE 102, 1 (18); Bericht der Sachverständigenkommission 'Staatszielbestimmungen/Gesetzgebungsaufträge' 1983 in Bundesminister des Innern/Bundesminister der Justiz (eds) '*Staatszielbestimmungen/Gesetzesbestimmungen*' (Konkordia, Bonn, 1983) paragraphs 7 et sqq.

9 Astrid Epiney '*Umweltrecht in der Europäischen Union*' 3rd edition (Nomos, Baden-Baden 2013) 141 (hereinafter: Epiney Umweltrecht).

10 See Chapter 1 above.

'founding fathers' of this discipline, Ronald Dworkin and Robert Alexy.[11]

According to the trias three levels of norms exist: objectives, principles and rules which interact with each other. New energy technologies can be regulated via translation of all three types of norms into concrete measures, as will be demonstrated further below. The first level of the trias, in case of shale gas extraction the objectives/interests of environmental protection and energy security, is the subject of this chapter.

The meaning and legal function of environmental protection and energy security reaches beyond Member States. Both interests have been heavily influenced by EU primary law. In the context of EU law, environmental protection and energy security have been called 'quasi-constitutional objectives'.[12] The chapter discusses EU 'quasi-constitutional' law first and then turns to the constitutions of Member States. After some initial explanations, both will be discussed together. This helps to obtain a holistic understanding of the interplay of both objectives and their influence on shale gas specific regulation. To start with, the meaning of the term 'quasi-constitutional' has to be explained.

Although the Treaty on the Functioning of the European Union (hereinafter: TFEU) and the Treaty on the European Union (hereinafter: TEU) are not actually called 'constitution', their evolution into the current form is the result of endeavours to draw up a constitutional document for Europe.[13] TEU and TFEU contain the provisions of the failed European Constitution of 2005, with minor modifications only, and are therefore considered to be quasi-constitutional norms.[14]

It would lie beyond the scope of this book to run through the constitutional set-up of 28 EU Member States individually. Instead, this chapter

---

[11] Ronald Dworkin '*Taking Rights Seriously*' (Harvard University Press, Cambridge 1977) 24 (hereinafter: Dworkin); Robert Alexy '*A Theory of Constitutional Rights*' (Oxford University Press, 2004) 44–7 (hereinafter: Alexy). Both acknowledge themselves, however, that the ancient roots of the discipline run very deep.

[12] Craig/De Burca 25 and 75.

[13] Craig/De Burca 23 and 25.

[14] More on the term quasi-constitutional and the discussions surrounding it may be found at: Lucia Serena Rossi 'A New Revision of the EU Treaties After Lisbon?' in Lucia Serena Rossi and Federico Casolari '*The EU after Lisbon – Amending or Coping with the Existing Treaties?*' (Springer International, Cham 2014) 6–9; Edward Best '*Understanding EU Decision-Making*' (Springer International, Cham 2016) 38; Craig/DeBurca 25 and 75.

resorts to a generic approach.[15] It engages in an in-depth analysis of the constitutional objectives of environmental protection and energy security in Germany and France.[16] The findings, however, are exemplary for other EU Member States and broader conclusions about the interplay of environmental protection with energy security in shale gas cases will be drawn.

Finally, a word on the limited scope of this chapter is required. The present work is exclusively concerned with the interplay of the constitutional objectives of environmental protection and energy security in the case of shale gas extraction. It does not discuss fundamental rights to any extent. The outcome of this chapter must, therefore, be understood in the broader legal context. The chapter merely assesses which type of shale gas regulation is the legally most sound, *from the perspective of constitutional objectives*. This picture might or might not change if concrete measures are assessed for their potential to interfere with fundamental rights. Making that appraisal, however, lies beyond the scope of the present work.[17]

The chapter proceeds as follows. After a discussion of the anchorage of environmental protection and energy security as (quasi-) constitutional objectives in EU primary law, the role of both state objectives in national constitutions is illuminated by case studies on Germany and France. Having established the precise legal basis of these two state objectives,

---

15 This approach has also been deployed in Chapter 3 above. For details, see there.

16 These jurisdictions have been selected for the following reasons: both are major European countries with considerable shale gas potential and boast the longest traditions of applying the objectives and principles that are relevant to shale gas extraction, see Nicolas de Sadeleer '*Environmental Principles: From Political Slogans to Legal Rules*' (Oxford University Press, 2002) 93; Philippe Sands and Jacqueline Peel '*Principles of International Environmental Law*' 3rd edition (Cambridge University Press, Cambridge 2012) 267. Germany, in particular, has been characterized as the 'environmental frontrunner' in Europe, having in place one of the most advanced environmental regulatory systems, see Jale Tosun and Christoph Knill 'The Extraction of Unconventional Gas in France and Germany: Explaining Cross-Country Variation in Regulation' (2012) Presentation in the workshop '*Intra-institutional Politics of EU Energy Policy-Making: Landscape-Level Shifts, Discourses and Organizational Cultures*' pages 16/17 Robinson College, Cambridge, England, 27–28 April 2012 (hereinafter: Tosun/Knill). Most importantly, however, the legal mechanism that is going to be put forward to strike a balance between environmental protection and energy security interests in shale gas cases has been developed by German scholars looking at the German constitution, as explained further below in this chapter.

17 More on this can be found below in Chapter 7.

the chapter situates their role within the context of constitutional rights theory. In that part the so called trias is introduced. Subsequently, the chapter focusses on the first level of the trias, objectives, and discusses potential solutions to the issue of competing objectives. The chapter uses two 'meta'-principles to align the environmental protection and energy security objectives and concludes that these mechanisms may be used beyond the context of shale gas extraction.

4.2 ENERGY SECURITY AND ENVIRONMENTAL PROTECTION AS STATE OBJECTIVES

4.2.1 Anchorage of Both State Objectives in EU Law

Environmental protection and energy security are objectives that are not only enshrined in national constitutions, but also in EU primary law.[18] To start with environmental protection this position is mainly apparent from article 3 (3) TEU and articles 11 and 191 (2) TFEU.[19] These articles oblige the EU to integrate environmental protection considerations into all EU policies.[20] Environmental protection has also been addressed by the TEU and TFEU in several sector-specific stipulations.[21] Due to this firm positioning the ECJ ruled that environmental protection amounts to one of the '*essential objectives*' of the EU, lying in the '*general interest*' of European citizens.[22]

Energy security is handled notably differently by the TEU and the TFEU. While the TEU does not mention energy security directly, it is generally accepted that the interest represents a specific means to

---

18 Case 240/83 *Procureur de la République v Association de défense des brûleurs d'huiles usagées* [1985] ECR 531 paragraph 13 (hereinafter: ADBHU); Case 302/86 *EC Commission v Denmark* [1989] 1 CMLR 619 paragraphs 8 and 9; Case 72/83 *Campus Oil Ltd v Minister for Industry and Energy* [1984] ECR 2727 paragraph 35 (hereinafter: *Campus Oil)*; Case C-503/99 *Commission v Belgium* [2002] ECR I-4809 paragraphs 23 and 46 (hereinafter: *Commission v Belgium 2002*).

19 V Mangoldt article 20 a paragraph 112; Christian Calliess and Matthias Ruffert '*EUV/AEUV Kommentar*' 4th edition (Beck, München 2011) article 3 EUV paragraph 4 (hereinafter: Calliess/Ruffert).

20 ADBHU paragraph 13; Case 302/86 *EC Commission v Denmark* [1989] 1 CMLR 619 paragraphs 8 and 9.

21 Preface and article 21 (2) (f) TEU; articles 4(2) (e), 114 (3)–(5) and 191–4 TFEU.

22 ADBHU paragraphs 13 and 15; Calliess/Ruffert article 191 AEUV paragraph 6.

promote other overarching objectives that are included in article 3 (3) TEU.[23] One of these objectives is economic and social development, the achievement of which is influenced by energy security.[24] Without a steady and secure supply of energy, economies cannot develop and social welfare is endangered.[25]

In the TFEU by contrast, the security of energy supplies is addressed head on by several articles, including article 194 (1) (b) and 194 (2) TFEU. But the TFEU also includes indirect references in specific sectorial laws, like articles 122 (1) TFEU and 192 (2) (c) TFEU.[26] The energy provision of article 194 (1) TFEU[27] in particular is considered as having a legal quality that is similar to article 11 TFEU, one of the core provisions on environmental protection.[28]

In summary, the ECJ deduced from the above named treaty provisions that energy security constitutes a full-blown objective of European law that exists in its own right,[29] a position which is endorsed by the legal literature.[30] The ECJ recently reiterated its position on energy security in a case concerned with a Dutch law that, inter alia, prohibited privatization of particular parts of generation/production, supply and trade of electricity and gas in the Netherlands.[31] The question posed in that case was whether or not such a prohibition on particular privatizations complies with EU law. The court ruled that certain restrictions on the free movement of

---

23 Beate Sjafjell *'Towards a Sustainable European Company Law'* (Kluwer Law International, Alphen aan den Rijn 2009) 173–9; Calliess/Ruffert article 3 EUV paragraphs 2 and 7.

24 Bjornebye 67/68.

25 Ibid.

26 Rudolf Geiger, Daniel-Erasmus Khan and Markus Kotzur, '*European Union Treaties*' (Beck, München 2015) article 192 paragraphs 6–9 (hereinafter: Geiger/Khan/Kotzur); Eberhard Grabitz, Meinhard Hilf and Martin Nettesheim '*Das Recht der Europäischen Union Loseblattsammlung*' (Beck, München 2012) article 194 paragraph 21 (hereinafter: Grabitz/Hilf/Nettesheim).

27 The energy article 194 TFEU has been discussed in Chapter 2 above.

28 Grabitz/Hilf/Nettesheim article 194 paragraph 14.

29 *Campus Oil* paragraphs 34/35; *Commission v Belgium 2002* paragraphs 23 and 46; Case C-174/04 *Commission v Italy* [2005] ECR I-4933 paragraph 40 (hereinafter: Commission v Italy).

30 See for instance Grabitz/Hilf/Nettesheim article 194 paragraph 21.

31 Joined Cases C-105/12 to C-107/*12 Staat der Nederlanden v Essent NV (C-105/12), Essent Nederland BV (C-105/12), Eneco Holding NV (C-106/12), Delta NV (C-107/12)* available at EurLex: http://curia.europa.eu/juris/document/document.jsf?text=&docid=143343&pageIndex=0&doclang=EN&mode=lst&dir=&occ=first&part=1&cid=573312 [accessed 23April2015] paragraph 2 (hereinafter: Essent II).

capital can be justified by energy security considerations,[32] citing its long history of case law on the special importance of this objective of EU law:

> It must also be observed that the objective of guaranteeing adequate investment in the electricity and gas distribution systems is designed to ensure, inter alia, security of energy supply, an objective which the Court has also recognised as being an overriding reason in the public interest (Case 72/83 Campus Oil and Others [1984] ECR 2727, paragraphs 34 and 35; Case C-503/99 Commission v Belgium [2002] ECR I-4809, paragraph 46; and Case C-174/04 Commission v Italy, paragraph 40).[33]

4.2.2 Germany and France – State Objectives under National Constitutional Law

Turning now to national law, the constitutions of those countries that have been examined in Chapter 3 above shall now be briefly assessed for their take on energy security and environmental protection. The examination will, however, be confined to Germany and France, as they are representative of Member States with a written constitution. The UK has been omitted because it has no written constitution and because of recent political developments.[34]

In Germany, environmental protection and energy security both feature in the constitution as state objectives (so called *Staatszielbestimmungen*).[35] Article 20a of the German constitution stipulates that the state shall protect 'the natural bases of life'.[36] This stipulation has been interpreted by German scholars as referring to environmental protection, elevating the interest to the level of a state objective (*Staatszielbestimmung*).[37] The stipulation effectively obliges all state organs to preserve the environment and to take protective measures against environmental impairments.[38]

Energy security is not addressed explicitly by the German constitution, but it is covered indirectly by constitutional stipulations. Article 20 (1) of

---

[32] Essent II paragraph 69.

[33] Essent II paragraph 59.

[34] Following recent political developments and the so called 'Brexit' referendum, the United Kingdom is about to leave the European Union at the time of writing.

[35] Calliess/Ruffert article 3 EUV paragraph 2.

[36] Article 20a German constitution.

[37] V Mangoldt article 20a paragraphs 14 and 32; Sachs article 20a paragraph 27.

[38] V Mangoldt article 20a paragraph 35.

the German constitution establishes Germany as a 'social state', giving rise to the so called principle of the social state (*Sozialstaatsprinzip*).

According to German courts, this *Sozialstaatsprinzip* obliges the state to safeguard certain 'public services in the general interest' (*Daseinsvorsorge*) under articles 20 (1) and 28 (1) of the German constitution.[39] One of these services is providing a secure and steady supply of energy to German customers.[40] This constitutional obligation, however, is not a subjective right, as it is legally binding only upon the state, making it a constitutional state objective (*Staatszielbestimmung*).[41]

The French Constitution addresses environmental protection and energy security in a comparable manner. However, French constitutional history is rather complicated since France has had a total of fifteen constitutional instruments since the French Revolution in 1789.[42] The current constitution originates from 1958.[43]

In 2004 the French Constitution was supplemented by a Charter for the Environment (*La Charte de l'environnement de 2004*), which is also part of what the French call 'the constitutional bloc',[44] which is the constitution

---

[39] Daniel Hahn '*Staatszielbestimmungen im integrierten Bundesstaat*' (Duncker & Humblot, Berlin 2010) (hereinafter: Hahn) 114; Klaus Stern '*Das Staatsrecht der Bundesrepublik Deutschland Band I*' 2nd edition (Beck, München 1984) 910/911 (hereinafter: Stern I).

[40] BVerfGE 66, 248 (258); BVerfG JZ 1990, 335; BVerwGE 98, 275 et sqq; 122 162/163; Horst Dreier '*Grundgesetz Band II*' (Mohr Siebeck, Tübingen 1998) article 20 (Sozialstaat) paragraph 54; article 28 paragraph 137 et sqq (hereinafter: Dreier).

[41] V Mangoldt article 20 paragraph 116; Hans Dieter Jarass and Bernd Pieroth '*Grundgesetz Kommentar*' 11th edition (C H Beck, München 2011) article 20 paragraphs 122 and 125 (hereinafter: Jarass/Pieroth); Dreier article 20 (Sozialstaat) paragraph 58.

[42] Martin A Rogoff 'French Constitutional Law Cases and Materials' (Carolina Academic Press, Durham (North Carolina) 2014) 1 (hereinafter: Rogoff); John Bell '*French Constitutional Law*' (Oxford University Press 1992) 1 (hereinafter: Bell French Constitution).

[43] An English translation may be obtained From the French Conseil Constitutionnel at http://www.conseil-constitutionnel.fr/conseil-constitutionnel/root/bank_mm/anglais/constiution_anglais_oct2009.pdf [accessed 15 December 2016].

[44] See preamble French Constitution of 1958 and French Constitutional Council 'Juin 2014: La Charte de l'environnement de 2004' available at http://www.conseil-constitutionnel.fr/conseil-constitutionnel/francais/a-la-une/juin-2014-la-charte-de-l-environnement-de-2004.141685.html [accessed 15 December 2016]; furthermore, Rogoff 7; a translation into English is available at Rogoff 509/510.

itself and other sources of law mentioned in its preamble, as well as parts of former constitutions and other laws.[45]

Rogoff commented on the legal character of the Charter for the Environment in the following terms: 'Its provisions are, for the most part didactic and programmatic, although there is no doubt that it has normative content by virtue of reference to it in the preamble of the constitution.'[46]

Recently the Constitutional Council of France even used provisions of the charter to invalidate significant legislative measures.[47] In Decision 2008-564 DC on Genetically Modified Organisms[48] the Council explicitly stated that 'all the rights and duties defined in the Charter for the Environment have constitutional status'.[49]

The Charter consists of 10 articles. Article 6 reads: 'Public policies shall promote sustainable development. To this end they shall reconcile the protection and enhancement of the environment with economic development and social progress.'[50] The Charter for the Environment also includes an article 2, which says that 'everyone' is obliged to participate in preserving and enhancing the environment.[51] Leaving aside the discussion of whether or not the state is 'everyone', both articles together make clear that environmental protection is an objective of constitutional rank and that public policies have to be oriented on this constitutional objective.

With regard to energy security, as in the German case, the French constitution does not include an explicit reference. However, the laws that are building the French constitutional bloc include law no. 48-1268 of 17 August 1948.[52] It was passed in an attempt to restrict the scope of delegation of legislative powers to the government and to define the appropriate areas in which parliament should focus its legislative activities.[53]

In its article 7, law no. 48-1268 prescribes public service and social resources as matters belonging to the regulatory sphere of parliament.[54]

---

45 Rogoff 7.

46 Rogoff 221.

47 Ibid.

48 Constitutional Council Decision 2008-564 DC, Genetically Modified Organisms, of 19 June 2008, Rec. 313 (hereinafter: Constitutional Council GMO).

49 Constitutional Council GMO paragraph 18.

50 Translation by author.

51 See Rogoff 509.

52 Loi n° 48-1268 du 17 août 1948 tendant au redressement économique et financier.

53 Bell French Constitution 281.

54 For the history of that inclusion, see Bell French Constitution 86.

Energy is clustered under '*social resources*' and the parliament is responsible for the control and use of energy as well as the rationing of raw materials and industrial products under article 7. Energy security is linking in with certain aspects of this provision, particularly the control and rationing of energy resources.

Moreover, article 7 law no. 48-1268 obliges the French parliament to provide public services and to regulate them. The legal role of public services and its content has been shaped by the legal practice of the Constitutional Council. Sometimes, the Council refers to 'objects', 'principles' or 'rules' with constitutional status as part of the constitutional bloc.[55] In its Decision 79-105 DC on the right to strike on radio and television[56] the Council stated that 'the continuity of public service (. . .) has the character of an objective of constitutional status'.[57]

Although this decision was not concerned with energy security, energy security can be considered as a public service in the general interest. Viewed together with the provisions of article 7 of law no. 48-1268, this leads to the conclusion that French legislators are obliged to orientate their decisions in relevant areas on the objective of energy security.

To sum up, energy security and environmental protection both feature in European as well as in national law in the form of constitutional and quasi-constitutional objectives. A slight difference in the way in which these objectives are presented at the European and the national level, however, exists. While the European treaties recognize both interests explicitly, their recognition in the national constitutions is often indirect, as the examples of Germany and France highlight. Nevertheless, national courts and the legal literature agree that both interests amount to objectives of constitutional law.

4.2.3 The Trias of Objectives, Principles, Rules

Having established that environmental protection and energy security are (quasi-) constitutional objectives, it is now necessary to discuss the legal function of these objectives. According to constitutional law theory, constitutional objectives belong to an abstract category of norms.[58]

---

[55] Rogoff 223.

[56] Constitutional Council Decision 79-105 DC, Right to Strike on Radio and Television, 25 July 1979, Rec. 33 (hereinafter: Constitutional Council Radio).

[57] Ibid.; Rogoff 223.

[58] Alexy 44; De Sadeleer 308/309; Bernd Rüthers, Christian Fischer and Axel Birk '*Rechtstheorie*' 8th edition (Beck, München 2015) paragraphs 491a–491c (hereinafter: Rüthers/Fischer/Birk); Sommermann 361/362; Hahn 389/390.

Dworkin and Alexy established that every norm can be categorized either as a rule or a principle[59] and that a fundamental distinction between these two categories exists.[60] A rule sets forth a precise solution for specific facts.[61] Once its conditions of application have been fulfilled it leads to a legal solution.[62] A principle, by contrast, provides general orientation and direction for rules.[63] Principles are, thus, guiding policy measures, which can only be implemented by rules.[64] Owing to these differences in nature, principles allow for more discretion than rules.[65]

Alexy also discusses objectives. According to him, objectives fall into the category of principles because they are abstract in nature and their application does not automatically lead to a fixed legal result.[66] An objective thus provides general orientation for the decision-maker and cannot be regarded as fulfilled or unfulfilled, like rules.[67]

Although this statement is correct in terms of the abstract nature of both norms, it has to be taken into account that Alexy developed this point many decades ago. Since then, particularly during the 1980s and 1990s, a trend in constitutional law emerged to increasingly enrich constitutional documents with state objectives.[68] These state objectives have real legal leverage[69] and their quantitative increase means that they are also becoming more influential in shaping constitutional law.

---

[59] Dworkin 22 et sqq; Alexy 48.

[60] Dworkin 24; Alexy 44/45.

[61] Dworkin 24; De Sadeleer 307; Rüthers/Fischer/Birk 491a–491c.

[62] Rüthers/Fischer/Birk 491a–491c; Sommermann 361/362; Hahn 389/390; Dworkin 24.

[63] Dworkin 24; Alexy 48.

[64] Rüthers/Fischer/Birk 491a–491c; Sommermann 361/362; Hahn 389/390; De Sadeleer 309.

[65] De Sadeleer 307.

[66] Alexy 44.

[67] Karl-Peter Sommermann *'Staatsziele und Staatszielbestimmungen'* (Mohr Siebeck, 1997 Tübingen) 361/362 (hereinafter: Sommermann). However, this traditional view has recently been challenged, see Hahn 389/390.

[68] State objectives were included in constitutions way before that time, but during the named period a significant increase in the number of state objectives was perceivable. For the reasons and a good introduction to the development of state objectives see Hahn 1 et sqq.

[69] This is essentially the case because of two things: first, according to the principle of unity of the constitution, all legal interests and objectives, directly or indirectly protected in the constitution, build an ensemble without contradictions. No part of the constitution may be deemed irrelevant. Second, courts have ruled in the past that state objectives have legal leverage. Both points will be discussed in detail immediately below in this chapter.

It is time to recognize this development in constitutional law theory and take the necessary consequences. Objectives often provide an even more abstract command than principles. Take the environment as an example. The (quasi-) constitutional[70] objective of environmental protection requires the state to orientate its relevant policies on environmental protection. One tool to enhance the level of environmental protection lies with environmental law principles, for instance the precautionary principle.[71] The precautionary principle is thus a more concrete embodiment of the objective of environmental protection. Objectives should now be treated as a third, separate category of norms. This has particular consequences in the context of energy and environmental law.

A constitutional objective, formulated in a very general manner, may take on a more concrete form through gradual modifications.[72] The objective (environmental protection) might be translated into a principle that is set out in framework legislation (for instance the precautionary principle).[73] This principle in turn forms a legal rule that is complete and precise (environmental act).[74] In this sense, rules, principles and objectives

[70] This terminology is broadly used when discussing European constitutional law. Although the Treaty on the Functioning of the European Union (hereinafter: TFEU) and the Treaty on the European Union (hereinafter: TEU) are not actually being called 'constitution', their evolution into the current form is the result of endeavours to draw up a constitutional document for Europe. TEU and TFEU contain the provisions of the failed European Constitution of 2005 with only minor modifications and are therefore considered to be quasi-constitutional norms, see Craig/De Burca 23 and 25. More on the term quasi-constitutional and the discussions surrounding it may be found at: Lucia Serena Rossi 'A New Revision of the EU Treaties After Lisbon?' in Lucia Serena Rossi and Federico Casolari 'The EU after Lisbon – Amending or Coping with the Existing Treaties?' (Springer International, Cham 2014) 6–9; Edward Best '*Understanding EU Decision-Making*' (Springer International, Cham 2016) 38; Craig/DeBurca 25 and 75.

[71] The precautionary principle asks the legislator to take cost-effective measures to prevent environmental degradation, even if it is not (yet) proven that a particular activity is leading to serious or irreversible damage (principle 15 of the 1992 Rio Declaration on Environment and Development (hereinafter: Rio Declaration)). An in-depth discussion of the precautionary principle and shale gas extraction can be found in Chapter 5 below. For more on the precautionary principle and its thresholds see: Ruven Fleming and Leonie Reins 'Shale gas extraction, precaution and prevention: A conversation on regulatory responses' (2016) 20 *Energy Research & Social Science* 132 et sqq.

[72] De Sadeleer 310.

[73] De Sadeleer 310; Sommermann 411; Rüthers/Fischer/Birk 491a–491c.

[74] Ibid.

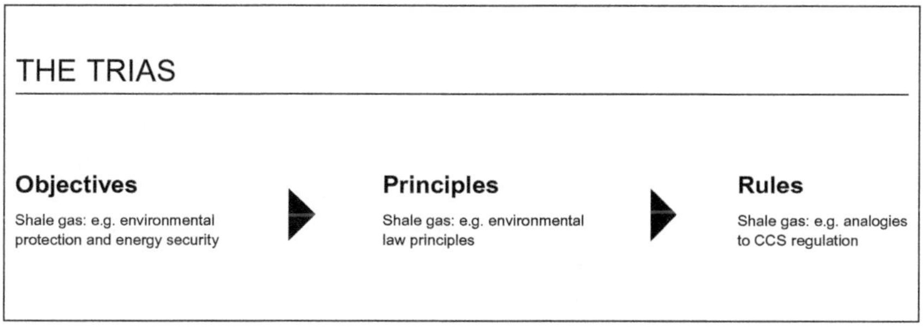

Source: Author's own work.

Figure 4.1 The trias of objectives, principles and rules for the example of shale gas regulation

overlap.[75] All three represent different points on the same sliding scale.[76] This three-fold system constitutes the trias.

However, in the case of shale gas extraction a fundamental issue arises already at the first level of the trias: *both,* environmental protection and energy security objectives must be translated into law principles, which in turn steer the implementation of concrete shale gas regulation.[77] In shale gas regulation thus not one but two objectives apply at the same time and compete with each other.

4.3 THE ISSUE OF COMPETING OBJECTIVES – SOLUTIONS TO THE CONFLICT IN EUROPEAN AND NATIONAL LAW

The two constitutional objectives of environmental protection and energy security pull the legislator into very different directions in the case of shale gas regulation.[78] The reason being that every constitutional objective

[75] Alexy 45; Rüthers/Fischer/Birk 491a–491c; Sommermann 361/362; Hahn 389/390.

[76] De Sadeleer 310; Rüthers/Fischer/Birk 491a–491c.

[77] The latter can possibly be achieved by analogies to other rules, if the applicable objectives and principles do not cover the activity sufficiently, see Chapter 6 below.

[78] See Chapter 1 above.

requires the legislator to realize its individual content to the greatest extent possible.[79]

There are mainly[80] three positions on how to resolve the competition between environmental protection and energy security.[81] According to one opinion, the constitutional objective of energy security generally prevails over the objective of environmental protection, a solution that is often called the 'critical chain' approach.[82] From a second perspective the opposite is correct and environmental protection generally prevails over energy security.[83] According to a third opinion both objectives are equal

---

79 Alexy 47.

80 There are other positions. However, these other positions are merely more differentiated descriptions of the three main positions discussed here. For more details see: Rudolf Streinz '*EUV/AEUV*' 2nd edition (Beck, München 2012) article 191 AEUV paragraph 32–36 (hereinafter: Streinz).

81 Jürgen Basedow 'Zielkonflikte und Zielhierarchien im Vertrag über die Europäische Gemeinschaft' in Ole Due (ed.) '*Festschrift für Ulrich Everling*' (Nomos, Baden-Baden 1995) 62, 65, 67/68 (hereinafter: Basedow); Ivo Schwartz 'EG-Kompetenz für das Verbot der Tabakwerbung?' (1998) 6 *Zeitschrift für Medien- und Kommunikationsrecht AfP* 562 (hereinafter: I. Schwartz); Carl-Otto Lenz and Klaus Dieter Borchardt '*EU Verträge Kommentar*' 5th edition (Bundesanzeiger Verlag, Köln 2010) article 191 AEUV paragraph 10 (hereinafter: Lenz/Borchardt); Epiney Querschnittsklausel 500; Manfred Zuleeg 'Vorbehaltene Kompetenzen der Mitgliedsstaaten der EG' (1987) 4 NVwZ 283 et seq (hereinafter: Zuleeg). See as well: Epiney Umweltrecht 107/108 and 110/111.

82 For the 'critical chain approach' in the context of shale gas extraction see: Maurin/Vivoda 370. Others who have taken this position include: Werner Thieme '*Umweltschutz im Recht*' (Duncker & Humblot, Berlin 1988) 206 (hereinafter: Thieme); Ulrich Battis 'Vereinheitlichung des Umweltrechts im europäischen Binnenmarkt? '(1989) 9 *Natur und Recht* 366 (hereinafter: Battis); Lenz/Borchardt article 191 AEUV paragraph 10; Winfried Haneklaus 'Zur Verankerung umweltpolitischer Ziele im EWG-Vertrag' (1990) 21 *Deutsches Verwaltungsblatt* 1137 (hereinafter: Haneklaus); H-J Glaesner 'Die Einheitliche Europäische Akte' (1986) 2 *Europarecht* 140/141 (hereinafter: Glaesner); Basedow 62, 65, 67/68; I Schwartz 562; Streinz paragraph 32. This opinion bases itself upon ECJ case law, in particular: Case 72/83 *Campus Oil* paragraphs 34/35; Case C-393/92 *Municipality of Almelo and others v NV Energiebedrijf Ijsselmij* [1994] ECR I-1477 paragraph 48 (hereinafter: *Almelo*); Case C-159/94 *Commission v France* [1997] ECR I-5815 paragraph 57; Case C-483/99 *Commission v France* [2002] ECR I-4781 paragraphs 50/51 (hereinafter: *Commission v France 2002*); Case C-503/99 *Commission v Belgium* [2002] ECR I-4809 paragraph 32 (hereinafter: *Commission v Belgium 2002*) and Case C-367/98 *Commission v Portugal* [2002] ECR I-4731 paragraph 52 (hereinafter: *Commission v Portugal 2002*); *Commission v Italy* paragraph 40; *Essent II* paragraph 68.

83 Wolfgang Kahl '*Umweltprinzip und Gemeinschaftsrecht*' (C F Müller, Heidelberg 1993) 204 et seq and 211 et seq (hereinafter: Kahl); Astrid Epiney 'Umweltrechtliche Querschnittsklausel und freier Warenverkehr' (1995) 10/11

in abstract rank[84] and the competing objectives must be reconciled by an individual rule, elaborated against the backdrop of the circumstances of the case in question.[85]

The first and the second possibility may be discounted as rather impractical, due to the results of their application to the creation of shale gas regulation. If the legislator were to pursue the objective of environmental protection in a strict manner, shale gas extraction would never be possible, as this strict measure guarantees the highest possible level of environmental protection. Such a regime would conflict with EU law,[86] as potential energy security effects could not materialize. Vice versa, if the legislator strictly adheres to the objective of energy security, legislation would have to be put into place that could incite a rush to extract lots of gas very quickly, risking severe damage to the environment.

The TFEU takes the third approach and assigns similar weight to all treaty objectives, leaving the concrete balance to the individual case.[87] Coequal ranking of environmental protection with other EU objectives, including energy security, can be inferred from EU primary law.[88] The

---

NuR 500 (hereinafter: Epiney Querschnittsklausel); Epiney Umweltrecht 107/108 and 110/111; Streinz article 191 paragraph 35; Zuleeg 283 et sqq. This opinion also refers to ECJ case law: ADBHU 531; Case 302/86 *Commission v Denmark (Danish Bottle case)* [1988] ECR 4627 paragraphs 8/9 and 21; Case C-2/90 *Commission of the European Communities v Kingdom of Belgium* [1992] ECR I-4431 paragraphs 29 and 32; Case C-379/98 *Preussen Elektra* [2001] ECR I-2099 paragraphs 76, 83 and operative part 2; Case C-573/12 *Alands Vindkraft v Energimyndigheten* [2012] OJ L 140/16 paragraphs 76–82; Case C-204/12 through to C 208/12 *Essent Belgium NV v Vlaamse Reguleringsinstantie voor de Elektriciteits* available at EurLex: http://curia.europa.eu/juris/celex.jsf?celex=62012CJ0204&lang1=de&type=TXT&ancre= [accessed 27 October 2014] paragraphs 92/93.

84 Krämer Focus 12; V Mangoldt article 20 a paragraph 47); Michael Kloepfer *'Verfassungsrecht Band I'* (C H Beck, München 2011) 390 (hereinafter: Kloepfer Verfassungsrecht I).

85 Krämer Focus 12; V Mangoldt article 20a paragraph 47; Kloepfer Verfassungsrecht I 390.

86 See chapter 2 above as well as immediately below.

87 V Mangoldt article 20a paragraph 43; Grabitz/Hilf/Nettesheim article 194 AEUV paragraph 26; Krämer Focus 12; Hans-Werner Ringeling (ed.) '*Handbuch zum europäischen und deutschen Umweltrecht Band I*' (Heymanns, Köln 1998) § 9 paragraph 27; Ulrich Beyerlin 'Die "neue" Umweltpolitik der Europäischen Gemeinschaften' (1989) 9 Umwelt- und Planungsrecht 362.

88 Geiger/Khan/Kotzur article 194 TFEU paragraph 5; Krämer Focus 12; V Mangoldt article 20a paragraph 47; Kloepfer Verfassungsrecht I 390; Johannes Caspar '§ 2 Europäisches und nationales Umweltverfassungsrecht' in Hans-Joachim Koch *'Umweltrecht'* 3rd edition (Franz Vahlen, München 2010) paragraph 16 et sqq; Battis 366.

EU is thus legally obliged to respect all treaty objectives as equal and one objective may not prevail entirely over the other.[89]

At the level of national law, the Federal Constitutional Court of Germany (*Bundesverfassungsgericht*) established a further fundamental distinction. It differentiates between the question of *whether* the legislator should implement constitutional objectives in a certain case and the issue of *how* he should implement them.[90] The court asserted that the legislator has large discretion regarding the decision *whether* he wants to concretize a constitutional objective with a certain law.[91] Owing to the separation of powers which flows from the rule of law,[92] this decision may not be reviewed by the judicial power.[93]

Notwithstanding this discretion, the court has the right to review the law that is *resulting* from the process of deliberation, in other words the *how*.[94] This review may also be based upon objectives which the legislator did not mention or did not even think of, if these objectives are relevant to the regulated area.[95] The Federal Constitutional Court measures a norm not only against a constitutional objective that has been named in the explanatory memorandum to the law. Instead, it reviews the resulting law against the backdrop of all constitutional objectives which it considers to be relevant in a given case.[96]

As a result, the possibilities for the legislator to base shale gas specific regulation solely on the objective of environmental protection or solely on that of energy security are of a rather theoretical nature. Environmental protection and energy security are both of practical relevance in the case of shale gas extraction.[97] Thus, the resulting shale gas regulations are likely

---

89 Calliess/Ruffert article 3 EUV paragraph 11; Stern III/2 1701; Sachs Einführung paragraph 50; Hesse paragraph 318; Haneklaus 1137.

90 BVerfG (1998) NJW 1776 (1777); BVerfGE 21, 292 (299); 33, 171 (186).

91 Ibid.

92 As included in articles 20 (3) and 28 (1) German constitution.

93 Horst Dreier *'Grundgesetz Band II'* (Mohr Siebeck, Tübingen 1998) article 20 Rechtsstaat paragraphs 62 and 66/67 (hereinafter: Dreier); Rainer Wernsmann 'Wer bestimmt den Zweck einer grundrechtseinschränkenden Norm-BVerfG oder Gesetzgeber?' (2000) NVwZ 1363 (hereinafter: Wernsmann).

94 BVerfG (1998) NJW 1776 (1777); BVerfGE 21, 292 (299); 33, 171 (186); but also Wernsmann 1361 et sqq.; Hahn 393.

95 Ibid.

96 BVerfG (1998) NJW 1776 (1777). See Gerrit Manssen *'Staatsrecht II Grundrechte'* 11th edition (Beck, München 2014) paragraph 182 (hereinafter: Manssen) for pertaining arguments. This position, however, is not undisputed, see Wernsmann 1361 et sqq.

97 See Chapters 1 and 2 above.

to be measured against both objectives, irrespective of the intention of the legislator.

4.4 THE 'META' PRINCIPLES OF UNITY OF THE CONSTITUTION AND PRACTICAL CONCORDANCE

European and national laws prescribe that a solution to the competition of constitutional objectives must be sought in the light of what is legally and factually possible.[98] The scope of the legally and factually possible is determined by certain 'meta' principles of the constitution.[99]

'Meta' principles are not legal principles, strictly speaking, as they can be used to review legal principles and objectives.[100] Their function is to provide a benchmark for the review of legal norms.[101] For the interplay of constitutional objectives, such as environmental protection and energy security, two 'meta' principles are decisive: the principle of the unity of the constitution and the principle of practical concordance.[102]

4.4.1 Unity of the Constitution and 'Harmonization'

The principle of the unity of the constitution is a presumption of modern constitutional law theory that all legal interests and objectives, directly or indirectly protected in the constitution, build an ensemble without contradictions.[103] Because of this principle, the constitution has to be interpreted in the light of its very nature as a unitary ensemble of norms. These norms cannot contradict each other.[104] Constitutional objectives, such as environmental protection and energy security, are not simply juxtaposed to each other, but interact. When the legislator decides to regulate an issue

---

98 Alexy xxxi; Rudolf Streinz et al. '*EUV/EGV*' (Beck, München 2003) article 2 EGV paragaph 37 (hereinafter: Streinz EGV).

99 Hesse paragraph 72; Michael Kloepfer '*Verfassungsrecht Band II*' (C H Beck, München 2010) 93/94 and 100 (hereinafter: Kloepfer Verfassungsrecht II); Dreier article 20 Einführung paragraph 10; Alexy 48.

100 Hesse paragraph 72.

101 Ibid.

102 Dreier article 20 Einführung paragraph 10; Alexy 66 footnote 84.

103 Josef Isensee and Paul Kirchhof '*Handbuch des Staatsrechts Band V*' (C F Müller, Heidelberg 1992) § 122 paragraph 6 (hereinafter: Isensee/Kirchhof V); Stern III/2 1701; Sachs Einführung paragraph 50.

104 BVerfG, DVBl 2005, 1458; Stern III/2 1701; Sachs Einführung paragraph 50.

like shale gas extraction that is associated with more than one of them he has to take this interaction into account.[105] If constitutional objectives are in apparent conflict they have to be 'harmonized', as the ECJ put it.[106]

'Harmonization' should occur by implementation of *both* competing constitutional interests into one concrete law.[107] 'Harmonization' requires as a bare minimum that environmental protection and energy security are *both* represented in shale gas regulation. But this does not of itself explain *how* the conflicting objectives of environmental protection and energy security should be assembled in such a law to achieve 'harmonization'. For that purpose, a second 'meta' principle must be applied to shale gas regulation, the principle of practical concordance.

4.4.2 Practical Concordance and 'Optimization'

Practical concordance is a legal concept originating from German constitutional law.[108] Without being named specifically, its basic components have also been introduced in proceedings before the ECJ.[109] Practical concordance can be used to resolve any kind of conflict between constitutional interests, including constitutional objectives, provided they are of equal abstract rank.[110]

Practical concordance requires that the interplay of environmental protection and energy security in shale gas specific regulation is as 'optimal'

---

105 Blasberg 37.

106 Case 9/56 *Meroni & Co., Industrie Metallurgiche, S.p.A. v High Authority of the European Coal and Steel Community* [1957-1958] ECR 135 pages 151/152 (hereinafter: Meroni) still speaks of 'permanent reconciliation'; the term 'harmonization' first appears in: Case 5/73 *Balkan-Import-Export v Hauptzollamt Berlin-Packhof* [1973] ECR 1091 paragraph 24 (hereinafter: *Balkan Import*) and was later reiterated in Case 29/77 *S.A. Roquette Frères v French State – Administration des Douanes* [1977] ECR 1835 paragraphs 3 and 29–31; Case C-44/99 *The Queen v Minister of Agriculture, Fisheries and Food, ex parte Fishermen's Organisations and others* [1995] ECR I-3115 paragraph 37. For Germany see: BVerfG, DVBl 2005, 1458; Stern III/2 1701; Sachs Einführung paragraph 50; Hahn 395.

107 Hahn 390.

108 Endorsed by the Federal Constitutional Court in established law practice, see: BVerfGE 41, 29 (51); 77, 240 (255); 81, 278 (292/293); 81, 298 (308); 83, 130 (143).

109 See for instance Opinion of Advocate General Lagrange in Case 13/57 *Wirtschaftsvereinigung v Hohe Behörde* [1958] ECR 288 at 301 and 372 (hereinafter: Opinion Lagrange); Case 5/73 *Balkan-Import-Export v Hauptzollamt Berlin-Packhof* [1973] ECR 1091 paragraph 24 (hereinafter: *Balkan Import*).

110 BVerfGE 28, 243 (261); Isensee/Kirchhof Band V § 109 paragraph 82; Gellermann 359.

as possible. Where 'harmonization' demands the representation of both competing interests, 'optimization' allows for the *greatest possible* impact on shale gas regulation. Only in exceptional cases, where 'optimization' is entirely impossible, may the legislator apportion precedence to one constitutional objective over the other.[111]

Konrad Hesse, who became the first scholar to define practical concordance as a fixed principle of law,[112] deduced the concept from the principle of the unity of the constitution:[113]

> the principle of the unity of the constitution requires 'optimization': both constitutional interests must be curtailed, in order for both to reach an optimal impact. The delimitation must, hence, be proportional in the individual case; it must not go beyond what is necessary to reach concordance of both constitutional interests.[114]

Four criteria of practical concordance have been established by the German Federal Constitutional Court[115] which are mirrored in the case law of the ECJ.[116] First, practical concordance demands *conciliatory* interpretation and balancing of the respectively competing constitutional objectives in an individual case.[117] Second, only in exceptional cases, where the legislator *strove for* the best possible conciliation of competing objectives and failed, might one interest be given precedence over the other.[118] Third, however, even then the other objective, in the current case environmental protection or energy security, may not be disregarded entirely[119] and precedence must be confined in time and space.[120] This means that infringements of the very core of an objective are prohibited.[121] Finally, weighting of the competing objectives, here environmental protection and

---

111 Hesse paragraph 72; Jürgen Schwarze *Europäisches Verwaltungsrecht* 2nd edition (Nomos, Baden-Baden 2005) 673 (hereinafter: Schwarze); Humberto Bergmann Àvila *Theorie der Rechtsprinzipien* (Duncker & Humblot, Berlin 2006) 97 (hereinafter: Bergmann Avila).

112 Stern III/2 626; Jarass/Pieroth Vorbemerkung vor article 1 paragraph 41.

113 Hesse paragraph 72.

114 Ibid.

115 BVerfGE 35, 202 (225); 39, 1 (43); 77, 240 (255); 81, 278 (292/293); 83, 130 (143).

116 Opinion Lagrange 301 and 372; Balkan Import paragraph 24.

117 BVerfGE 39, 1 (43); 93, 1 (21); BVerfG, DVBl 2005, 1458; Stern III/1 930.

118 BVerfGE 35, 202 (225).

119 V Mangoldt article 20 paragraph 314.

120 BVerfGE 35, 202 (225/226).

121 BVerfGE 119, 1 (29/30).

energy security, must aim for an 'optimal' interplay in a concrete law on shale gas.[122]

These four criteria have subsequently been translated into a commonly accepted formula on the appraisal of the interplay of competing constitutional interests, the so called 'Je-desto-formula'.[123] It asserts that: 'the graver the impact of an intervention on one constitutionally protected interest, the more is the intruding force obliged to avoid conflict with the other competing constitutionally protected interest' ('Je-desto-Formel').[124]

The two approaches to shale gas regulation,[125] a cautious but permissive approach, as envisaged by the European Commission,[126] and a prohibitive approach, as adopted by several Member States,[127] must now be checked for compliance with this formula. Either of the two approaches is, in principle, open to the legislator, but the question is which one of them is the legally soundest, the one adhering best to the principle of practical concordance.

Prohibitive shale gas regulation, for instance in the form of a shale gas moratorium, would run the danger of disregarding one relevant objective, energy security. For the duration of a moratorium, the extraction of shale gas would be illegal and domestic shale gas sources could not be exploited to boost a country's energy security.[128]

Cautious, but permissive shale gas regulation, by contrast, would cater for environmental protection and energy security interests at the same time. It would protect the environment through the imposition of constraints and restrictions on shale gas extraction.[129] Thus, shale gas could potentially be extracted, albeit at an environmentally tenable scale.

---

122 BVerfGE 81, 278 (292); BVerfGE 83, 130 (143).

123 Meinhard Schröder Die Je-desto-Formel des Bundesverfassungsgerichts in der Esra-Entscheidung und ihre Bedeutung für Grundrechtsabwägungen (2008) *DVBl* 148/149 (hereinafter: Schröder DVBl); Dreier article 2 II GG paragraph 48; Isensee/Kirchhof § 109 paragraph 80.

124 Schröder DVBl 148/149; Dreier article 2 II GG paragraph 48; Isensee/Kirchhof § 109 paragraphs 80 and 84/85; Mizdalski 125.

125 Chapters 2 and 3.

126 Chapter 2.

127 Chapter 3.

128 Admittedly, before and after the moratorium shale gas could still be produced and energy security would not be disregarded entirely. However, a moratorium could be implemented without clearly specifying the period of time for which it stayed in place. The practical effect was then that of a ban on shale gas extraction.

129 See Chapter 2 above.

Such regulation would impact less gravely on the energy security interest and avoid, to the extent possible, a conflict with environmental protection. Cautious, but permissive shale gas regulation, hence, complies with the 'Je-desto-Formel', whereas bans might be viewed as failing this test.

However, there is a potential issue with the 'Je-desto-formula', namely the pre-emption of legislator decision-making on the interplay of constitutional objectives. The legislator has the right to adopt the solution to shale gas regulation that he deems appropriate. An obligation to align competing constitutional objectives only at the 'highest point' of maximal impact for both objectives could reduce the legislator's leeway for discretion and unduly restrict his choices.[130]

To maintain the discretion of the legislator a review of legislative action must be restricted to the question: has one relevant objective been disregarded arbitrarily and/or did the legislator commit a manifest error of judgement?[131] In other words: the obligation to strive for 'optimization', under the principle of practical concordance finds its ultimate limit in the principle of proportionality.[132] 'Optimization' might only occur to the extent that legislator discretion is not curtailed in a disproportionate manner.[133]

Meaningful shale gas regulation should *strive for* an 'optimal' solution, wherever this is practical,[134] as the term *practical* concordance already suggests, but must not be misunderstood as restricting options only to *the* 'optimal' solution. Thus, in the case of shale gas regulation, the legislator is not obliged to reach the 'highest point' of conciliation of environmental protection interests with energy security demands. Any approach to shale gas regulation is legitimate, as long as the concrete law delivers on both, environmental protection and energy security.

To sum up, the analysis suggests that shale gas extraction should be

---

130 Rüdiger Konradin Albrecht '*Zumutbarkeit als Verfassungsmassstab*' (Duncker & Humblot, Berlin 1995) 118 paragraph 440 (hereinafter: Konradin Albrecht); Peter Lerche 'Die Verfassung als Quelle von Optimierungsgeboten?' in Joachim Burmeister et al. *'Verfassungsstaatlichkeit. Festschrift für Klaus Stern'* (Beck, München 1997) 205 (hereinafter: Lerche); Rainer Wahl 'Der Vorrang der Verfassung' (1981) 20 Der Staat 504 (hereinafter: Wahl); Gellermann 359.

131 Case C-280/93 *Federal Republic of Germany v Council of the European Union- Bananas* [1994] ECR I-4973 paragraphs 89–91 (hereinafter: *Bananas market*); Opinion Lagrange 301; Case 139/79 *Maizena GmbH and the European Parliament v Council of the European Communities and the European Commission* [1980] ECR 3393 paragraph 23.

132 Alexy 397.

133 Alexy 411.

134 Ibid.

regulated via a cautious, but permissive regulatory approach, as this is the legally soundest option. Legally sound, in this context, means complying to the greatest extent possible with the applicable (quasi-) constitutional objectives.[135] Cautious, but permissive shale gas regulation complies best with the 'meta' principles of unity of the constitution and practical concordance and allows for 'harmonization' and 'optimization' of the relevant objectives of environmental protection and energy security.

However, the legislator is not *obliged* to take that approach. If decision-makers find that the wide array of possibilities for encompassing environmental protection and energy security in one law[136] does not in any way live up to their ambitions, they may implement a shale gas moratorium instead. However, in this case the legislator should take great care to explain the reasoning and make sure that this prohibitive law does not arbitrarily disregard energy security interests. Therefore, the legislator should, at the very least, abstain from indefinite bans to avoid the danger of having shale gas specific regulation struck down by a court.

4.5 CONCLUSION

This chapter showcased why cautious, but permissive shale gas regulation is the legally soundest way to regulate shale gas extraction from the perspective of the two most salient constitutional objectives, environmental protection and energy security. By taking a cautious, but permissive approach both objectives could be reconciled in concrete shale gas regulation. However, the legislator is free to pick a different approach, for instance imposing a moratorium, as long as the resulting law simultaneously caters for both objectives.

The trias of objectives, principles and rules, explained at the beginning of the current chapter, will be used as a template for the remaining chapters of this book. The interplay of the constitutional objectives of environmental protection and energy security has just been scrutinized. The next chapter deals with relevant environmental law principles and their impact on shale gas specific regulation. Finally, the book goes to the third level of concrete rules in Chapter 6 to come up with suggestions for shale gas specific regulation.

---

135 More on this can be found below in Chapters 3 and 4.

136 This could be done by different measures, ranging from lenient to very strict, as outlined in the framework of the European Commission on shale gas regulation, discussed above in Chapter 2.

5. Principles and shale gas

5.1 INTRODUCTION

The interplay of the objectives environmental protection and energy security[1] in shale gas cases is considerably influenced by law principles. Within the trias[2] and the hierarchy of norms,[3] legal principles occupy the second tier.[4] Because legal principles have been initially developed to guide decision-making on environmental issues they are often associated with environmental law.[5] However, their application extends beyond that narrow area.[6]

The particular applicability of five (environmental) law principles to shale gas extraction in the EU has long been emphasized by scientists,[7] lawyers[8] and politicians.[9] These principles are the precautionary principle,

---

1 This has been the subject of Chapter 4 above.

2 Ibid.

3 See beginning of Chapter 5 for explanations on the hierarchy of norms, as used in this work.

4 Paul Craig and Grainne de Burca '*EU Law Texts Cases and Materials*' 5th edition (Oxford University Press, Oxford 2011) 109 (hereinafter: Craig/De Burca). This hierarchy has been discussed above at Chapter 5.

5 Craig/De Burca 109.

6 To name just two examples: first, the precautionary principle also covers potential threats to human health and has even been applied in the context of social security and crime, see Arie Trouwborst '*Precautionary rights and duties of states*' (Martinus Nijhoff Publishers, Leiden 2006) 12–16 (hereinafter: Trouwborst 2006). Second, the principle of sustainable development transcends the environmental context and extends to economic development, see Patricia Birnie, Alan Boyle and Catherine Redgwell, '*International Law and the Environment*' 3rd edition (Oxford University Press, Oxford 2009) 116–18.

7 SRU Faulstich 39 et sqq.; Milieu Ltd. '*Regulatory provisions governing key aspects of unconventional gas extraction in selected Member States*' (2013) available at: http://ec.europa.eu/environment/integration/energy/uff_studies_en.htm [accessed 4 September 2014] 16.

8 Stokes 44; Reins Minimum Principles 16 et sqq.

9 European Parliament debate of 20 November 2012 on 'Environmental Impacts of Shale Gas and Shale Oil Extraction Activities – Industrial, Energy and Other Aspects of Shale Gas and Oil' CRE 20/11/2012 – 11 available at: http://www.europarl.europa.eu/sides/getDoc.do?type=CRE&reference=20121120

the polluter pays principle and the principles of sustainable development, public participation and rectification at source.[10]

The current chapter demonstrates how these environmental law principles can be translated into a set of concrete measures in the case of shale gas extraction to tackle the four most salient potential environmental threats.[11] These measures may be recommended for the regulation of shale gas extraction in EU Member States.

However, many of the law principles could be translated into more than one concrete measure. The principles, and following from that also the measures, might overlap in certain areas. Where that is the case, it will be highlighted in the course of this chapter. To improve readability the author decided to match one specific law principle with one measure only.

The chapter analyses each law principle in the following three steps. First, a short definition of the respective principle and a description of its anchorage in European primary law is provided. Second, the principle is applied to the potential environmental threats of shale gas extraction. Third, the way in which the law principle can be translated into a concrete measure that could form part of concrete shale gas regulation is explained.

5.2 PRECAUTION

Article 191 (2) TFEU lists a number of (environmental) law principles which steer European environmental policy.[12] First in line is the precau-

---

&secondRef=ITEM-011&language=EN [accessed 2 September 2014] Speeches of Josefa Andrés Barea, Niki Tzavela, Mairead McGuinness, Bogusław Sonik, Carl Schlyter, Ana Miranda, Corinne Lepage.

[10] These principles are the most relevant in the case of shale gas extraction, according to SRU Faulstich 6, Stokes 44. However, some of these environmental law principles still lack precise or uniformly accepted definitions, see Daniel Bodanski et al. '*The Oxford Handbook of International Environmental Law*' (Oxford University Press, Oxford 2008) 381/382 (hereinafter: Bodanski et al.); Simon Marr '*The Precautionary Principle in the Law of the Sea: Modern Decision Making in International Law*' (Kluwer Law International, The Hague 2003) 5 (hereinafter: Marr); Trouwborst 2002 at 51. In the quest for such definitions, where European or national law lacking them, environmental lawyers resort to international law, see Stuart Bell and Donald McGillivray '*Environmental Law*' 7th edition (Oxford University Press, Oxford 2008) 56/57 (hereinafter: Bell/McGillivray); Davies 28. This chapter takes a similar approach.

[11] Described in Chapter 2 above.

[12] Nicolas de Sadeleer '*Environmental Principles: From Political Slogans to Legal Rules*' (Oxford University Press, Oxford 2002) 110 (hereinafter: De Sadeleer); see also Chapter 2 above on the European Union. With regard to

tionary principle. Its position as first principle to be named in this article is not due to coincidence. The precautionary principle has such an important role to play that it influences the interpretation of all other environmental law principles and the way in which they may be translated into environmental regulation.[13]

An issue from the outset with the precautionary principle is the fact that jurisprudence and literature have, so far, failed to provide one uniform[14] and imperative definition of the principle.[15] At EU level, the European Commission elaborated a 'Communication on the Precautionary Principle' in February 2000.[16] This Communication does not provide a definition, but aims at harmonizing the application of the precautionary principle in Europe.[17]

---

national law, the example of Germany will be used once again: in German law, it is included in article 20a German constitution, see Hans Dieter Jarass and Bernd Pieroth '*Grundgesetz Kommentar*' 11th edition (C H Beck, München 2011) article 20a paragraph 8 (hereinafter: Jarass/Pieroth); Hermann von Mangoldt and Friedrich Klein et al. *'Das Bonner Grundgesetz Kommentar Band 2: Artikel 20 bis 78'* 4th edition (Verlag Franz Vahlen, München 2000) article 20a paragraphs 69/70 (hereinafter: V Mangoldt); Horst Dreier *'Grundgesetz Band II'* (Mohr Siebeck, Tübingen 1998) article 20a paragraph 49 (hereinafter: Dreier).

13 Precaution is not confined to environmental protection but encompasses sustainable development, the ideas of polluter pays, public participation and rectification at source, see Sonja Boehmer-Christiansen 'The Precautionary Principle in Germany – enabling government' in Timothy O'Riordan and James Cameron '*Interpreting the Precautionary Principle*' (Earthscan Publications Ltd., London 1994) 31–3 (hereinafter: Boehmer-Christiansen); Arie Trouwborst '*Precautionary rights and duties of states*' (Martinus Nijhoff Publishers, Leiden 2006) 254 (hereinafter: Trouwborst 2006). Due to this intricate interplay, the potential environmental threats of shale gas extraction will be assessed comprehensively in the section on the precautionary principle. The subsequent sections only resort to explanations on individual potential threats, where it is necessary to highlight the specific function of the respective principle for shale gas regulation.

14 As many as 19 different formulations are scattered across one article alone: Per Sandin 'Dimensions of the precautionary principle' (1999) 5 *Human and Ecological Risk Assessment* 889–907.

15 Bodanski et al. 381/382; Laurence Boisson de Chazournes 'The Precautionary Principle' in UNEP '*Precaution from Rio to Johannesburg: Proceedings of a Geneva Environment Network Roundtable*' (International Environment House, Geneva 2002) 10; Marr 5; Trouwborst 2002 51.

16 European Commission 'Communication on the Precautionary Principle' (Communication) *COM (2000)1* (hereinafter: Communication on precaution).

17 Although it is not a legally binding document, see Ivo Appel 'Europas Sorge um die Vorsorge. Zur Mitteilung der Europäischen Kommission über die Anwendbarkeit des Vorsorgeprinzips, Neue Zeitschrift für Verwaltungsrecht' (2001) 4 *NVwZ* 395/396. The European Court of Justice ruled that it represents a

For the purpose of definition, legal scholars accordingly point towards international law and the best-known and most widely used[18] explanation of precaution in principle 15 of the so called Rio Declaration.[19] Although the Rio Declaration is not legally binding, it received wide endorsement by European and Member State decision-makers alike.[20] Principle 15 reads:

> In order to protect the environment, the precautionary approach shall be widely applied by states according to their capabilities. Where there are threats of serious or irreversible damage, lack of full scientific certainty shall not be used as a reason for postponing cost-effective measures to prevent environmental degradation.[21]

From this definition, two main pre-conditions for the application of the precautionary principle can be deduced: first, there must be scientific uncertainty about the existence of a threat and second, this potential threat must have the ability to cause serious or irreversible damage to the environment.[22]

Case law from the ECJ and Member State courts suggests that the likelihood for potential environmental threats and pertaining damages must fall into a certain corridor, to be relevant under the precautionary principle.[23] The lower margins of this corridor are purely hypothetical/theoretical

---

codification of the law and may therefore be deemed as a valid legal position, see Case T-70/99 *Alpharma Inc. v Council of the European Union* [2002] ECR II-03495, paragraphs 162 and 144 (hereinafter: *Alpharma*); Case T-13/99 '*Pfizer Animal Health SA v Council of the European Union*' [2002] ECR II-3305 paragraphs 123 and 149. The Communication suggests a two-step approach to determining its applicability in a particular case. First, the plausibility of a threat must be established by assessing and categorizing all issues and potential threats which a certain activity could bring about and second, the identified potential threats should then be combated by appropriate protective measures, see Communication on precaution 4, 13 and 16; Harald Hohmann '*Precautionary Legal Duties and Principles of Modem International Environmental Law*' (Kluwer Law International, The Hague 1994) 334 (hereinafter: Hohmann). This approach will be adopted and used in this chapter.

[18] The European Policy Centre '*Occasional Paper April 2001 Towards a Proportionate Implementation of the Precautionary Principle*' (The European Policy Centre, Brussels 2001) 9 (hereinafter: EPC Occasional Paper); Trouwborst 2006 23 and 147.

[19] 1992 Rio Declaration on Environment and Development (hereinafter: Rio Declaration).

[20] For details see Davies 28.

[21] Principle 15 Rio Declaration.

[22] Marie-Claire Cordonier Segger and Ashfaq Khalfan '*Sustainable Development Law*' (Oxford University Press, 2004) 144 (hereinafter: Cordonier Segger/Khalfan).

[23] *Alpharma* paragraph 159; BVerwGE 69, 37.

threats and pertaining 'marginal' damage.[24] Its upper boundary is a threat that has been proven to exist by consistent scientific evidence and that could cause very serious damage.[25] The potential threats of shale gas extraction are subsequently benchmarked against the lower and the upper boundaries of this corridor.

5.2.1 Lower Margin: Plausibility of Potential Threats of Shale Gas Extraction

To establish the lower boundary in a given case, plausibility of an assumption is used as a criterion.[26] Plausibility helps to differentiate between a purely theoretical threat (outside of the corridor) and a threat that needs to be tackled by precautionary action.[27] In order to determine plausibility of a hypothesis, scientists developed a set of objective epistemic criteria. These criteria gauge whether or not the potential environmental threat fall within the scope of the precautionary principle. These criteria are coherence, explanatory power, analogy, precedence, precision and parsimoniousness of a hypothesis.[28]

The most important potential threats of shale gas extraction that have been pinpointed for Europe will now be assessed against the named six criteria of plausibility. These threats have been discussed above in Chapter 1 and include potential contamination of groundwater and soil by a range of processes involved in the extraction of shale gas, the emission of greenhouse gases (CO_2 from flaring, methane from venting) and competition for land use.[29]

With regard to the first criterion, coherence, the relationship between deep-ground injection of chemicals and the deterioration of soil and water quality is well known.[30] The same is true for the fact that the release of

---

24 *Alpharma* paragraph 159; BVerfG of 28 February 2002 -1 BvR 1676/01 – paragraph 12; BVerfG, NJW 2002, 1638 (1639); BVerwGE 72, 300 (322).

25 *Alpharma* paragraph 159; de Sadeleer 119 et sqq.

26 David B Resnik 'Is the precautionary principle unscientific?' (2003) 34 *Studies in History and Philosophy of Biological and Biomedical Sciences* 339 (hereinafter: Resnik). A multitude of other criteria have been put forward in the past, but none of the others has similar explanatory power and stringency.

27 Ibid.

28 Thomas Samuel Kuhn '*The essential tension*' (University of Chicago Press, Chicago 1977) 321/322; Resnik 339.

29 SRU Faulstich 44/45; see also Chapter 1 above.

30 Donald L Sparks '*Environmental Soil Chemistry*' 2nd edition (Academic Press, San Diego 2003) 1/2; Kim H. Tan '*Principles of Soil Chemistry*' 4th edition (CRC Press; Boca Raton 2011) 5.

CO_2 and methane increases the greenhouse effect in the atmosphere which raises the risk of catastrophic climate change phenomena.[31] In other words, current knowledge suggests that the process of shale gas extraction could lead to environmental damage.[32] The potential threat that shale gas extraction poses to the soil, groundwater and the atmosphere is hence coherent with existing knowledge.

Second, the hypothesis that shale gas extraction could cause groundwater contamination/issues with well integrity, massive land use and the release of harmful greenhouse gas emissions, must be able to explain real-life phenomena. Elevated levels of groundwater pollutants have been found in groundwater samples from US drinking water wells that were situated in close proximity to shale gas sites.[33]

Moreover, water and soil contamination in the past occurred in places where waste water from shale gas sites had been disposed of.[34] Studies established elevated levels of powerful greenhouse gases, like CO_2 and methane, in the ambient air of shale gas wells and shale gas combustion plants as well as an increase in the competition for land.[35] All of these phenomena could be coherently explained if shale gas extraction were the source of the contaminations and emissions.[36]

Third, analogies from related fields should exist, which help to verify

---

31 Ibid.

32 This is not necessarily the case, as will be explained below in Chapter 8.

33 Agency for Toxic Substances and Disease Registry (ATSDR) of the U.S. Department of Health and Human Services 'Health Consultation-Chesapeake ATGAS 2H Well Site Leroy Hill Road, Leroy, Leroy Township, Bradford County, PA November 4 2011' (2011) Conclusion 1 at page iii, available at: http://www.atsdr.cdc.gov/hac/pha/ChesapeakeATGASWellSite/ChesapeakeATGASWellSiteHC110411Final.pdf [accessed 14 March 2013]; Stephen G Osborn et al. 'Methane contamination of drinking water accompanying gas-well drilling and hydraulic fracturing' (2011) Volume 108 No 20 *Proceedings of the National Academy of Science of the United States of America* (PNAS) 8172 and 8175; US Environmental Protection Agency (EPA) '*Draft Investigation of Ground Water Contamination near Pavillion, Wyoming*' (2011) at 37–9 and xiii, available at: http://www.epa.gov/region8/superfund/wy/pavillion/EPA_ReportOnPavillion_Dec-8-2011.pdf [accessed 04 March 2013] (hereinafter: EPA Pavillion).

34 Stefan Lechtenböhmer et al. '*Impacts of shale gas and shale oil extraction on the environment and on human health*' (European Parliament, Brussels 2011) 29/30 (hereinafter: Lechtenböhmer).

35 Howarth/Santoro/Ingraffea 680; SRU Faulstich 35; Christopher L Weber and Christopher Clavin 'Life Cycle Carbon Footprint of Shale Gas: Review of Evidence and Implications' (2012) Vol 46 No 11 *Environmental Science & Technology* 5693.

36 See Chapter 2 above.

that mechanisms or processes, suggested by the hypothesis, are not unparalleled. Carbon Capture and Storage (hereinafter: CCS) could be used as such an analogy. CCS includes the gathering of CO_2 from polluting sources, its injection into underground geological formations and its permanent storage in those formations.[37] The potential environmental issues, such as leakage of gas into water or the atmosphere[38] and the threat of small seismic events,[39] are comparable to those potential threats that are associated with shale gas extraction.[40] Moreover, recent American and German studies found that groundwater contamination by CCS[41] might occur via exposure pathways that are similar to those to be expected in shale gas cases.[42] Because of these similar mechanisms CCS may be used as an analogy.

Fourth, events posited by the hypothesis should be similar to precedents that have been observed previously.[43] Precedence means a decided case that furnishes a basis for determining later cases, involving similar issues or facts.[44] This is rather difficult to establish, given that shale gas *production* never actually started in Europe and so no precedence from past experiences exists. What we have are analogies from other countries and technologies. With regard to the latter, the analogy between shale gas extraction and CCS has just been discussed. Nonetheless, a *decided* case which can be taken as a basis for determining later cases does not exist and the reader may only be referred to the discussions on analogies above.

Fifth, the hypothesis about potential threats should be reasonably precise and not too vague.[45] 'Shale gas extraction' is a clearly defined

---

37 For detailed explanations of the technological process see: Intergovernmental Panel on Climate Change '*IPCC Special Report on Carbon Dioxide Capture and Storage*' (Cambridge University Press 2005) 17/18 (hereinafter: IPCC Special Report).

38 IPCC Special Report 12.

39 IPCC Special Report 13; more can be found in Chapter 2 above.

40 See above Chapter 2 for details on the threats of shale gas extraction.

41 Mark G Little and Robert B Jackson 'Potential Impacts of Leakage from Deep CO_2 Geosequestration on Overlying Freshwater Aquifers' (2010) 44 *Environmental Science & Technology* 9225 and 9230; Ralf E Krupp '*Gutachten zur geplanten Kohlendioxid-Einlagerung (CCS) in der Antiklinal-Struktur Neutrebbin, Ostbrandenburg*' (2011) available at: http://www.co2bombe.de/joomla/images/stories/co2/krupp_gutachten_1_neutrebbin_final.pdf [accessed 5 March 2013] 21, 40 and 48.

42 Compare the pathways for CCS, described in IPCC Special Report 34/35 to those for shale gas, described at Osborn et al. 8175 and above in Chapter 2.

43 Resnik 339.

44 Black's Law Dictionary 156.

45 Resnik 339.

method of gas extraction.[46] The potential threats of groundwater and soil contamination, the emission of greenhouse gases (CO_2 from flaring, methane from venting), and competition for land use are also rather concrete and precise.

Finally, the hypothesis about potential threats should be parsimonious and simple.[47] In other words it should rely on existing knowledge and relationships of cause and effect and posit a simple pathway between the technology and the potential harm. The problems discussed above involve established relationships of cause and effect (this does not apply to the fourth criterion). They are parsimonious and fulfil the sixth and final criterion of plausibility.

To sum up, the existence of environmental threats arising from shale gas extraction is plausible. All plausibility criteria, except for one, were met readily. Although a purely quantitative argument, this result rather supports the plausibility of a hypothesis than its dismissal.[48] The plausibility of the potential threat of water, soil and air contaminations by shale gas extraction has, hence, been established.

5.2.2 Upper Margin Precaution v Prevention in the Case of Shale Gas

The upper margin of the corridor of precaution is determined by the question of whether the potential threats of shale gas extraction are actually not (yet) known. Beyond the upper margin of precaution lies another environmental law principle, the preventive principle.[49] The preventive principle applies when scientific proof of a threat appears and its source can be reliably identified, so that the existence of the threat is certain.[50] The

---

[46] See Chapter 2 above.

[47] Kuhn 322.

[48] Resnik 339; Hohmann 334.

[49] In contrast to the precautionary principle, the preventive principle is an undisputed principle of public international law. It is also a customary rule of international law. Its status is established and well recognized, as opposed to the principle of precaution, see Kingdom of Belgium v Kingdom of the Netherlands, Iron Rhine ('IJZEREN RIJN') Railway arbitration tribunal award of 24 May 2005, available at: http://www.pca-cpa.org/showpage.asp?pag_id=1155 [accessed 21 June 2012] para. 59 and 222. See, for a discussion of the relationship of the principles in international law, A. Trouwborst, 'Prevention, Precaution, Logic and Law: The Relationship between the Precautionary Principle and the Preventative Principle in International Law and Associated Questions' 2 *Erasmus Law Review* 2 (2009) 105–27.

[50] De Sadeleer 74/75 and 222; David Freestone 'International Fisheries Law Since Rio: The Continued Rise of the Precautionary Principle' in Alan Boyle and David Freestone (eds) '*International Law and Sustainable Development: Past*

precautionary principle, by contrast, applies where reasonable grounds of concern for the existence of an environmental threat can be established, but the existence of the threat is yet uncertain.[51] Thus, the 'watershed' criterion to distinguish the precautionary from the preventive principle is scientific uncertainty.[52]

It has been argued that a differentiation between the precautionary and the preventive principle is no longer required as both have become so blurred that they should actually be considered as one.[53] While this argument has been predominantly advanced in the context of international law,[54] it would actually constitute an interpretation *contra legem* with respect to EU law.

Achievements and Future Challenges' (Oxford University Press, Oxford 1999) 139 (hereinafter: Freestone 1999); Epiney/Scheyli 91.

[51] *Britain v Commission* paragraph 105; *Alpharma* paragraph 152; *Monsanto Italia* paragraph 106; Case C-192/01 *Commission of the European Communities v Kingdom of Denmark* [2003] ECR I-09693, paragraph 49; Case C-24/00 *Commission of the European Communities v French Republic* [2004] ECR I-01277, paragraph 56. German courts ruled in similar vein: German Federal Administrative Court (*Bundesverwaltungsgericht*): BVerwGE 69, 43; BVerwG NVwZ 1986, 208, paragraph 37 and German Federal Constitutional Court (*Bundesverfassungsgericht*): BVerfG, NJW 2002, 1638 (1639). Furthermore, the idea is invoked by Communication on precaution 1 and 10; James Cameron and Juli Abouchar 'The Status of the Precautionary Principle in International Law', in David Freestone and Ellen Hey (eds) '*The Precautionary Principle and International Law*' (Kluwer Law International, The Hague 1996) 45; James Cameron, Will Wade-Gery and Juli Abouchar 'Precautionary Principle and Future Generations' in Emmanuel Agius '*Future Generations and International Law*' (Earthscan Ltd., London 1998) 99; Epiney/Scheyli 109/110.

[52] A similar argument is put forward by BVerwGE 69, 37 (43); Trouwborst 2002, 37–9; Trouwborst 2006, 94; Freestone 1999, 139; Charmian Barton 'Status of the Precautionary Principle in Australia: Its Emergence in Legislation and as a Common Law Doctrine' (1998) Vol. 22 Issue 2 *Harvard Environmental Law Review* 535; Arie Trouwborst 'Prevention, Precaution, Logic and Law' (2009) 2 *Erasmus Law Review* 119. For a more detailed discussion of the application of the preventive and precautionary principle to shale gas extraction, see Ruven Fleming and Leonie Reins 'Shale gas extraction, precaution and prevention: a conversation on regulatory responses' (2016) 20 *Energy Research and Social Science* 131–41 (hereinafter: Fleming/Reins).

[53] For examples, see '*Precautionary rights and duties of states*' (Martinus Nijhoff Publishers, Leiden 2006) 94 (hereinafter: Trouwborst 2006). To give just one example from contemporary legal documents, see article 7 of the International Law Association (ILA) Resolution2/2014 on the Legal Principles Relating to Climate Change available at: http://www.ila-hq.org/en/committees/index.cfm/cid/1029 [accessed 7 November 2016].

[54] See ibid. Even there it is, however, not entirely convincing, since a number

The preventive principle is included alongside the precautionary principle in article 191 (2) Treaty on the Functioning of the European Union (TFEU). This article, however, clearly differentiates between the two principles. As the TFEU constitutes one of the two constitutional law documents of the EU,[55] this is a fundamental decision made by the European legislator.

It could, nonetheless, be argued that the treaty is outdated and the situation changed so dramatically that an interpretation of article 191 (2) TFEU *contra legem* would be justified. However, the TFEU is a relatively young treaty (2009), building on pre-existing EU treaties, such as its direct predecessor the Treaty establishing the European Community (hereinafter: TEC). When the TEC was revised and turned into the TFEU in 2009, the debate on merging the principles of prevention and precaution was already rife. In spite of that debate, the legislator took the decision to distinguish between both principles in the TFEU. This decision of the legislator must be respected by courts, governments and legal scholars under the principle of separation of powers. Thus, it is necessary to categorize potential threats of shale gas extraction as falling under either the precautionary or the preventive principle.[56]

Some scholars argued for the application of the preventive principle to the potential threats of shale gas extraction.[57] Their line of reasoning is as follows: what is unknown in the case of shale gas extraction is the degree to which measures must be taken, but not the impacts of the process as such.[58] There is, for example, no doubt that the injection of chemicals into the ground can have environmental impacts.[59] It has further been argued that shale gas has been drilled in the US for up to ten years now, so that a point of reference does exist, even if the American experiences cannot be entirely transferred to Europe.[60] The only uncertainty remaining is if and how exactly impacts might occur and how to address them.[61] Therefore, preventive rather than precautionary measures should be the focus of attention.[62]

---

of international law treaties can be found that still differentiate between the two principles.,

55 See Chapter 2 above.

56 See also on the distinction between precaution and prevention, Kraemer 39; G Van Calster, 'Risk Regulation, EU Law and Emerging Technologies: Smother or Smooth?' (2008) 2 *NanoEthics* at 66.

57 For details on that position, see Fleming/Reins 137.

58 Ibid.

59 Ibid.

60 Reins Conference Paper page 10 of the paper accompanying the presentation.

61 Reins Minimum Principles 26.

62 Ibid.

As opposed to these arguments, a closer look at the four most salient potential threats of shale gas extraction reveals that scientific uncertainty currently prevails and the precautionary principle should be used for the regulation of shale gas extraction.

Concerning the issue of groundwater contamination/issues with well integrity, some US evidence *suggests*, but does not conclusively *prove*, that a link between shale gas extraction and groundwater contamination exists.[63] More recent studies reviewed early investigations and scrutinized a considerably higher number of wells; these studies established that shale gas extraction is in fact unlikely to pose a significant threat to the groundwater.[64] Arguably the most authoritative and comprehensive study has been conducted by the US Environmental Protection Agency (EPA),[65] which looked into the hydraulic fracturing of Coal Bed Methane (CBM).[66]

---

[63] Osborn et al. 8172; Robert B Jackson et al. 'Increased stray gas abundance in a subset of drinking water wells near Marcellus shale gas extraction' (2013) Vol 110 No 28 *Proceedings of the National Academy of Science of the United States of America* (PNAS) 11250 and 11254; Agency for Toxic Substances and Disease Registry (ATSDR) of the U.S. Department of Health and Human Services '*Health Consultation- Chesapeake ATGAS 2H Well Site Leroy Hill Road, Leroy, Leroy Township, Bradford County, PA November 4 2011*' (2011) Conclusion 1 at page iii, available at: http://www.atsdr.cdc.gov/hac/pha/ChesapeakeATGASWellSite/ChesapeakeATGASWellSiteHC110411Final.pdf [accessed 14 March 2013]; EPA Pavillion 37–9 and xiii; Robeck/Bennett 58.

[64] Donald I Siegel et al. 'Methane Concentrations in Water Wells Unrelated to Proximity to Existing Oil and Gas Wells in Northeastern Pennsylvania' (2015) *Environmental Science & Technology* pages A, F and G available at http://pubs.acs.org/doi/pdf/10.1021/es505775c [accessed 1 April 2015]. Elizabeth W. Boyer et al. '*The Impact of Marcellus Gas Drilling on Rural Drinking Water Supplies*' (Centre for Rural Pennsylvania, Harrisburg 2012) available at: http://www.rural.palegislature.us/documents/reports/Marcellus_and_drinking_water_2012.pdf [accessed 5 July 2013] 4; Energy Institute of the University of Texas '*Fact-Based Regulation for Environmental Protection in Shale Gas Development*' (2012) '*Environmental Impacts of Shale Gas Development*' part 2 'Summary of Findings' 30/31 available at: http://barnettprogress.com/media/ei_shale_gas_regulation120215.pdf [accessed 14 June 2012] (hereinafter: University of Texas).

[65] The study commenced in 1999 and monitored several wells in the US over a time period of 5 years, see EPA 'Hydraulic Fracturing Background Information' http://water.epa.gov/type/groundwater/uic/class2/hydraulicfracturing/wells_hydrowhat.cfm [accessed 28 March 2013].

[66] EPA '*Evaluation of Impacts to Underground Sources of Drinking Water by Hydraulic Fracturing of Coalbed Methane Reservoirs Study (2004)*' available at: http://water.epa.gov/type/groundwater/uic/class2/hydraulicfracturing/wells_coalbedmethanestudy.cfm [accessed 28 March 2013].

It concluded that hydraulic fracturing poses little or no threat to the groundwater.[67]

More recently, EPA has been conducting an even bigger study into shale gas extraction, the results of which are much anticipated by the scientific community.[68] The study was expected to conclude by 2014, but the reporting date has been rolled back.[69] By the time of completion of this manuscript the study was not yet published. However, a preliminary draft of the study has been available for review by the scientific community in June 2015. Although the draft finds that there are mechanisms involved in hydraulic fracturing that could potentially endanger groundwater, it concludes

> We did not find evidence that these mechanisms have led to widespread, systemic impacts on drinking water resources in the United States. Of the potential mechanisms identified in this report, we found specific instances where one or more mechanisms led to impacts on drinking water resources, including contamination of drinking water wells. The number of identified cases, however, was small compared to the number of hydraulically fractured wells.[70]

In line with these results, evidence from small scale monitoring of EU shale gas exploration wells in EU Member States does not suggest the existence of a significant threat to the groundwater.[71]

---

[67] Executive Summary of the study ES-1 and ES-16 available at: http://www.epa.gov/ogwdw/uic/pdfs/cbmstudy_attach_uic_exec_summ.pdf [accessed 28 March 2013]. Although the study was concerned with the hydraulic fracturing of CBM, the findings are transferable as CBM reservoirs are situated much closer to groundwater aquifers than shale plays, see Osborn et al. 8175; Georg Meiners, Michael Denneborg and Frank Müller '*Gutachten für das Umweltbundesamt Umweltauswirkungen von Fracking bei der Aufsuchung und Gewinnung von Erdgas aus unkonventionellen Lagerstätten – Risikobewertung, Handlungsempfehlungen und Evaluierung bestehender rechtlicher Regelungen und Verwaltungsstrukturen*' (Ministry for the Environment, Berlin 2012) B 56 and B57 (hereinafter: Meiners et al. Bund). Moreover, the hydraulic fracturing technique is equally deployed for gas flow stimulation in CBM-reservoirs and in shale plays, since both have very low permeability, see Knut Bjorlykke '*Petroleum Geoscience – From Sedimentary Environments to Rock Physics*' (Springer Verlag, Berlin 2010) 464 (hereinafter: Bjorlykke).

[68] For a detailed account of the study and its importance see Susan L Sakmar '*Energy for the 21st Century*' (Edward Elgar, Cheltenham 2013) 317–19 (hereinafter: Sakmar).

[69] Mint Press News, '*EPA Pushes Back Fracking Impact Study To 2016*' available at: http://www.mintpressnews.com/epa-pushes-back-fracking-impact-study-to-2016/163927/ [accessed 25 February 2015]; Sakmar 319.

[70] US EPA '*Assessment of the Potential Impacts of Hydraulic Fracturing for Oil and Gas on Drinking Water Resources*' (Office of Research and Development, Washington D.C. 2015) at ES-6 (hereinafter: US EPA draft).

[71] Michael Gunzelmann and Mohamed El Hamdaoui 'Grundwassermonitoring im Bereich der Bohrungen Damme 2/3 der Exxon Mobil' available at: http://damme.

To sum up, contradictory and inconclusive scientific findings on groundwater contamination by shale gas extraction exist. Despite the fact that there is some evidence for a looming threat, no conclusive link between the process of shale gas extraction and groundwater contamination has been established. Whether shale gas extraction poses a general environmental threat to the groundwater remains scientifically uncertain for now.

The very same is true for the other three major potential threats of shale gas extraction. With regard to 'flow-back' disposal, the US Environmental Protection Agency (EPA) reported 130 incidents of groundwater and soil contamination through 'flow back' since 2008.[72] In America the majority of 'flow back' is being recycled, requiring on site storage.[73] Leakage from these storage basins has been identified as a major source of contamination.[74] In EU Member States like Germany, however, the most likely option is disposal in 'dead wells' by deep ground injection, being the only effective disposal method currently available.[75] Due to these technological differences between American and European 'flow back' disposal, the US experiences cannot be transferred to Europe. The Rosenwinkel report, a special survey on deep-ground injection (*Verpressen)* of 'flow back' in Germany, concluded that the effects of deep ground injection and possible hazards for water are not adequately investigated or documented.[76] Thus, 'flow back' disposal in Europe is currently surrounded by uncertainty and further research is required.

Discussions about land use and greenhouse gas emissions centre on the questions whether and how coal could be replaced by shale gas as a fuel.[77] There are differing appraisals of the effects that substituting coal with shale gas would have and contradictory studies highlight the need for holistic

---

de/templates/images/news/1272_1.pdf [accessed 5 July 2013] 13; Rosenwinkel 10, 22/23.

72 Maximilian Kuhn and Frank Umbach *'Strategic Perspectives of Unconventional Gas: A Game Changer with Implications for the EU'* (2011) European Centre for Energy and Resource Security (EUCERS) Strategy Paper No 1 https://www.kcl.ac.uk/sspp/departments/warstudies/research/groups/eucers/strategy-paper-1.pdf [accessed 24 April 2012] 22 (hereinafter: EUCERS); UK report I, 44.

73 EUCERS 22.

74 Ibid.; UK report I, 44.

75 Ewen 47; EUCERS 22.

76 Rosenwinkel 64.

77 John Broderick and Kevin Anderson '*Has US Shale Gas Reduced CO_2 Emissions? Examining recent changes in emissions from the US power sector and traded fossil fuels*' (Tyndall Centre at the University of Manchester, Manchester 2012) available at: http://www.tyndall.manchester.ac.uk/public/Broderick_Anderson_2012_Impact_of_Shale_Gas_on_US_Energy_Emissions.pdf [accessed 20 March 2013] 13–15 and 21–4 (hereinafter: Broderick/Anderson).

research.[78] A fierce scientific debate on the climate effects of shale gas extraction and land use in Europe is on its way but scientific uncertainty persists.

To sum up, the existence of the four main potential environmental threats of shale gas extraction has not been scientifically proven. Significant disagreement among scientists on particular findings persist. This, however, is an indicator of scientific uncertainty,[79] the hallmark of the precautionary principle. The apt way to deal with the potential environmental threats of shale gas extraction is hence to apply the precautionary principle.

5.2.3 Precautionary Measures in Shale Gas Regulation

Given the lack of conclusive scientific data on the potential threats of shale gas extraction, the question of how to regulate that method of gas extraction arises. Under the precautionary principle a wide array of precautionary measures is available. They range from banning the activity concerned until conclusive scientific proof for its harmlessness is available[80] to attaching several different precautionary measures to it.[81]

---

78 Broderick/Anderson 13–15 and 21–4; Howarth/Santoro/Ingraffea 684/685; S Pacala and R Socolow '*Stabilization Wedges: Solving the Climate Problem for the Next 50 Years with Current Technologies*' available at: http://www.princeton.edu/mae/people/faculty/socolow/Science-2004-SW-1100103-PAPER-AND-SOM.pdf [accessed 26 February 2014] 17/18 and 25/26 of Supporting On-Line Material; A R Brandt et al. 'Methane Leaks from North American Natural Gas Systems' (2014) Vol. 343 No. 6172 *Science* 733; Daniel Forster and Jonathan Perks '*Climate Impact of potential shale gas production in the EU*' study of 30 July 2012 (European Commission 2012) available at: http://ec.europa.eu/clima/policies/eccp/docs/120815_final_report_en.pdf [accessed 29 October 2012] 67 and v.

79 Stephanie Tai 'Uncertainty about Uncertainty: The Impact of Judicial Decisions on Assessing Scientific Uncertainty' (2009) 11 *University of Pennsylvania Journal of Constitutional Law* 676 and 680; Vern R Walker 'The Myth of Science as a 'Neutral Arbiter' Triggering Precaution' (2006) 26 *Boston College International and Comparative Law Review* 214.

80 Trouwborst 2006, 193 and 200 et sqq; de Sadeleer 131; Joel A Tickner 'A Map Toward Precautionary Decision Making' in Carolyn Raffensperger and Joel A Tickner (eds) '*Protecting Public Health and the Environment: Implementing the Precautionary Principle*' (Island Press Ltd., Washington 1999) 5 (hereinafter: Tickner); Michael D Rogers 'Scientific and technological uncertainty, the precautionary principle, scenarios and risk management' (2001) 4 *Journal of Risk Research* 1; David Dzidzornu 'Four Principles in Marine Environment Protection: A Comparative Analysis' (1998) 29 *Ocean Development and International Law* 100. For German law: BVerwG (1986) *NVwZ* 208; BVerfGE 49, 89 (143) and BVerfGE 53, 30 (58).

81 Sonja Boehmer-Christiansen 'The Precautionary Principle in Germany – enabling government' in Timothy O'Riordan and James Cameron '*Interpreting the*

Essentially every measure can be counted as a precautionary measure, as long as it is a measure that is taken out of precaution.[82] Precautionary measures can broadly include research, the allocation of liability, tax benefits and/or state subsidies, the prescription of environmental impact assessment (EIA).[83] Moreover, the prescription of the use of best available technologies,[84] the prevention of pollution[85] or even prior information and consultation.[86] Precautionary action should, where possible, combine several precautionary measures.[87] With regard to shale gas extraction a number of precautionary measures is conceivable which could feature in shale gas specific regulation.[88]

However, many of these measures may also be brought under other environmental law principles,[89] probably with the exception of two. These are the prescription of the use of 'eco-friendly fracturing fluids' in shale gas operations on the one hand and a ban on shale gas extraction in water protection zones on the other hand. Those two measures are, thus, looked into in the context of the precautionary principle.

Fracturing fluids are the main repository of chemicals in shale gas operations, since chemicals are regularly included in the mixtures that make up the fracturing fluid.[90] Article 10.1 (b) of the EU's 2014 Shale Gas Recommendation[91] asks Member States to ensure that the use of chemical substances in shale gas operations is minimized. Moreover, the same Recommendation asks Member States to encourage

---

Precautionary Principle' (Earthscan Publications Ltd., London 1994) 36 (hereinafter: Boehmer-Christiansen).

[82] Trouwborst 2006 179/180.

[83] To be examined below at 'Sustainable Development'.

[84] Will be examined below at 'Rectification at Source'.

[85] Warwick Gullett 'Environmental Protection and the "Precautionary Principle": A Response to Scientific Uncertainty in Environmental Management' (1997) 14 *EPLJ* 58; Tickner 5.

[86] Jutta Brunnee 'A Conceptual Framework for an International Forests Convention: Customary Law and Emerging Principles' in Canadian Council on International Law (ed.) '*Global Forests and International Environmental Law*' (Kluwer Law International, London 1996) 41–77; Boehmer-Christiansen 34–6.

[87] Trouwborst 2006, 178.

[88] Meiners et al. Bund D 13-D 17; SRU Faulstich 41–5.

[89] See discussions below.

[90] See Chapter 2 above.

[91] Commission Recommendation 2014/70/EU of 22 January 2014 on minimum principles for the exploration and production of hydrocarbons (such as shale gas) using high-volume hydraulic fracturing [2014] OJ L 39/72. Discussed above in Chapter 3.

operators not to use hazardous chemical substances, wherever technically feasible.[92]

This is also supported by alarming evidence from scientific studies in EU Member States on the current status of fracturing fluids. The German environmental ministry recently analysed 25 per cent of all fracturing fluids that had been deployed in Germany between 1983 and 2011 for purposes other than shale gas extraction.[93] Although the composition of fracturing fluids altered from well to well,[94] the additives that were used in these preparations showed stark features. A considerable number had to be characterized as 'water endangering', although only one of the surveyed additives, biocide, had to be characterized as 'strongly water endangering'.[95]

As a consequence of the survey, firms interested in German shale gas projects declared their intention to abandon the use of highly toxic fracturing fluids in their operations.[96] In order to reach the goal of the 2014 Shale Gas Recommendation, shale gas regulation should prescribe the obligatory use of 'eco-friendly fracturing fluids'.[97] A variety of 'fracking' methods that do no not rely on chemical or toxic additives is already available[98] and new developments in this field emerge frequently due to intense research.[99] A stipulation on obligatory use of 'eco-friendly fracturing fluids' could help to fast-track this development.

Such a stipulation would eliminate one of the most important potential environmental perils of shale gas extraction[100] while enabling the extrac-

---

92 Article 10.2 of the EU's 2014 Shale Gas Recommendation.

93 Meiners et al. Bund O 5, A 73 and Annex 2.

94 Meiners NRW; Meiners et al. Bund A 64 and C 21.

95 Meiners et al. Bund C 18 and C 26/27.

96 Meiners NRW 24/25; Meiners et al. B A 74.

97 Meiners et al. Bund D 12; SRU Faulstich 44/45.

98 A good overview of 'eco-friendly fracturing fluids and methods' is provided by Meiners et al. Bund C 56 – C 60. For individual fracturing fluids that are free from chemicals with any toxic potential and, according to the industry, readily available for use in Germany, see Exxon Mobil 'Schiefergas-ohne giftige Stoffe' available at: http://www.erdgassuche-in-deutschland.de/erkundung_foerderung/frac_fluessigkeiten/index.html [accessed 12 March 2015]; Wirtschaftsblatt 'OMV will Mega-Gasvorrat im Weinviertel ab 2020 fördern' available at: http://wirtschaftsblatt.at/home/boerse/wien/1213426/index?_vl_pos=r.1.NT [accessed 22 October 2012]; Gasfrac Energy Systems Inc. 'Realizing the Potential' 11 available at: http://www.gasfrac.com/assets/docs/PDFS/presentations/Investor%20 Presentation%20-%20Realizing%20the%20Potential.pdf [accessed 22 October 2012].

99 Meiners et al. Bund C 56–C 60.

100 Meiners et al. Bund C 60.

tion of domestic gas. It would help to reconcile environmental protection needs with potential energy security benefits and implement the precautionary principle in the case of shale gas extraction.

The second precautionary measure that should be included in shale gas regulation is the banning of shale gas extraction in water protection zones.[101] Many EU Member States currently are only prohibiting shale gas extraction in these zones if it involves the use of water-endangering substances.[102] Cautious, but permissive shale gas regulation should include one particular measure that precludes shale gas extraction in all water protection zones as a precautionary measure.[103]

This general prohibition is justified, because it cannot be ruled out that shale gas extraction, even if done without water-endangering substances ('eco-friendly fracturing fluids', as discussed above) in these particularly vulnerable zones poses a significantly lower threat to water sources.[104] A general prohibition of shale gas extraction in these zones would, in line with precautionary logic, safeguard drinking water supplies whilst not prejudicing extraction in other areas, which could benefit the energy security of EU Member States.

5.3 POLLUTER PAYS

The aim of the polluter pays principle is to internalize otherwise external environmental costs.[105] Internalization means that costs for *remedying* and *preventing* environmental damage of a certain activity must be charged to the operator of the activity.[106] The polluter pays principle, thus, assumes a remediative and a preventive function.[107] On the one hand, it indicates that the costs of remedying pollution should be borne by the person responsible for causing the pollution.[108] On the other hand, the potential polluter

---

[101] SRU Faulstich 42; Meiners et al. Bund C 84.

[102] Meiners et al. Bund B 130/131 and C 84.

[103] For a similar opinion, specific to the situation in Germany, see Meiners et al. Bund C 85/86.

[104] Meiners et al. Bund C 84.

[105] De Sadeleer 25.

[106] Jans/Vedder 43; Krämer Focus 254.

[107] The dual function may be traced in principles 13 and 16 Rio Declaration, see Philippe Sands and Jacqueline Peel '*Principles of International Environmental Law*' 2nd edition (Cambridge University Press, 2003) 232 (hereinafter: Sands/Peel).

[108] Jans/Vedder 43; Sands/Peel 228; Bell/McGillivray 55; Eckard Rehbinder 'Ziele, Grundsätze, Strategien und Instrumente' in Klaus Hansmann and Dieter

should also be responsible for the costs of preventing the occurrence of pollution in the first place.[109]

Although the ECJ has discussed the polluter pays principle in only a couple of cases so far,[110] it endorsed the dual-function of the principle.[111] This duality of remediation and prevention also highlights the strong interconnection between the precautionary principle and the polluter pays principle.[112]

The principle has been spelled out in a 1975 Recommendation of the European Commission on cost allocation and action by public authorities on environmental matters.[113] This Recommendation remains to be the guiding principle for policy on the polluter pays principle, although it is not legally binding.[114] Besides, the polluter pays principle is also included in EU primary law, namely in article 191 (2) TFEU,[115] according to which it shall guide the definition and implementation of EU environmental

---

Sellner '*Grundzüge des Umweltrechts*' 4th edition (Erich Schmidt Verlag, Berlin 2012) paragraph 146 (hereinafter: Rehbinder Ziele).

[109] Bell/McGillivray 244.

[110] Case 172/82 *Syndicat national des fabricants raffineurs d'huile de graissage and others v Groupement d'intérêt économique 'Inter-Huiles' and others* [1983] ECR 555 paragraph 13; Case C-188/07 *Commune de Mesquer v Total France SA and Total International Ltd* [2008] ECR I-4501 paragraphs 71/72, 77, 82 and 89; Case C-378/08 *Raffinerie Mediterranee (ERG) SpA, Polimeri Europa SpA and Syndial SpA v Ministero dello Sviluppo economico and Others* [2010] ECR I-1919 paragraph 57 (hereinafter: *Raffinerie*); Joined Cases 379/08 and 380/08 *Raffinerie Mediterranee (ERG) SpA, Polimeri Europa SpA and Syndial SpA v Ministero dello Sviluppo economico and Others (C-379/08) and ENI SpA v Ministero Ambiente e Tutela del Territorio e del Mare and Others (C-380/08)* [2010] ECR I-02007 paragraph 39. For more, see De Sadeleer 31.

[111] In case C-293/97 *The Queen v Secretary of State for the Environment and Ministry of Agriculture, Fisheries and Food, ex parte H A Standley and Others and D G D Metson and Others* [1999] ECR I-02606 Advocate General Leger argued that the polluter pays principle has a remediatory as well as a preventive dimension, see his Opinion paragraphs 93/94; the court agreed with this, see paragraph 98.

[112] Thorsten Purps '*Umweltpolitik und Verursacherprinzip im Europäischen Gemeinschaftsrecht*' (Carl Heymanns, Köln 1991) 21/22; De Sadeleer 226.

[113] Council Recommendation 75/436/Euratom, ECSC, EEC of 3 March 1975 regarding cost allocation and action by public authorities on environmental matters [1975] OJ L 194/1 (hereinafter: Recommendation 75/436).

[114] Jans/Vedder 43.

[115] Christian Calliess 'EU-Umweltrecht' in Klaus Hansmann and Dieter Sellner '*Grundzüge des Umweltrechts*' 4th edition (Erich Schmidt Verlag, Berlin 2012) paragraph 78 (hereinafter: Calliess EU-Umweltrecht); De Sadeleer 30.

policy.[116] The polluter pays principle also features prominently[117] in secondary EU law, specifically in the Environmental Liability Directive 2004/35/EC.[118]

The polluter, according to the 1975 Recommendation, is someone who, directly or indirectly, damages the environment or who creates conditions leading to such damage.[119] If it is impossible or too difficult to identify the polluter, the costs of pollution can also be charged at the point at which the number of economic operators is least and control is easiest.[120]

With regard to shale gas regulation, the polluter pays principle requires that costs, which arise during the process of shale gas extraction are borne by the industry.[121] Owing to the fact that the shale gas industry creates a potential threat emanating from its sphere of influence, it bears responsibility for containing and controlling that potential threat.[122]

The issue with shale gas extraction is that the sources of pollution are not easily identifiable, as pollution stemming from this activity will mostly be diffuse in character.[123] Groundwater contamination, for instance, is not likely to occur via direct discharge of 'fracking' waste water into the groundwater. Instead, the waste water might mingle with other substances in the ground and this mixture might gradually migrate downwards, which

---

116 De Sadeleer 30.

117 Streinz article 191 AEUV paragraph 100.

118 Council Directive (EC) 2004/35 of 21 April 2004 on environmental liability with regard to the prevention and remedying of environmental damage [2004] OJ L 143/56 (hereinafter: Environmental Liability Directive). With regard to Member State law, the example of Germany might be called upon. In German law, the polluter pays principle is embedded in the German constitution, although different opinions on its exact position exist, see Sebastian Heselhaus 'Verfassungsrechtliche Grundlagen des Umweltschutzes' in Klaus Hansmann and Dieter Sellner '*Grundzüge des Umweltrechts*' 4th edition (Erich Schmidt Verlag, Berlin 2012) paragraphs 49–54; Norbert Bernsdorff 'Positivierung des Umweltschutzes im Grundgesetz 'Positivierung des Umweltschutzes im Grundgesetz (Art. 20a GG)' (1997) 7 *Natur und Recht* 333 (hereinafter: Bernsdorff); Alexander Schink 'Umweltschutz als Staatsziel' (1997) 6 *Die Öffentliche Verwaltung* 226 (hereinafter: Schink). The majority of German scholars, however, accept that the principle is indirectly entailed in the environmental protection stipulation of article 20a German constitution, see: Michael Kloepfer *'Umweltrecht'* 3rd edition (C H Beck, Berlin 2004) § 3 paragraph 25 (hereinafter: Kloepfer); Jarass/Pieroth article 20a paragraph 9; V Mangoldt article 20 a paragraph 74; Dreier article 20a paragraph 69.

119 Paragraph 3 Recommendation 75/436.

120 Ibid.

121 SRU Faulstich 45

122 SRU Faulstich 40.

123 SRU Faulstich 39.

can result in groundwater pollution.[124] This so called diffuse pollution where multiple causes produce single effects and single causes multiple effects,[125] has recently been tackled to some extent by European and Member State courts.

The ECJ discussed the issue of liability and diffuse pollution under the polluter pays principle in a case in 2008.[126] The court ruled that diffuse pollution could be subject to the polluter pays principle and the associated[127] regime of the Environmental Liability Directive if a causal link between a polluter and the pollution could be established.[128] The ECJ even considered it in principle sufficient that a government merely presumes such a causal link, if this presumption is based on plausible evidence.[129]

The court ruled that plausible evidence consists of at least two features: first, an installation must be located in close proximity to the pollution.[130] Second, a correlation between the identified pollutants and the substances used by the operator must exist.[131] If these two requirements are fulfilled, the Environmental Liability Directive applies to the respective pollution and the operator is liable under a strict liability regime, regardless of fault or negligence in his conduct.[132]

Case law from the courts of Member States points in a similar direction. In a case concerned with two agricultural plants that discharged a dangerous substance, the German Federal Civil Court (*Bundesgerichtshof BGH*) ruled that both plant operators are responsible for the resulting damages.[133] This, the court assessed, follows from polluter pays logic, according to which all potential polluters are liable, if the definite causer of the damage cannot be reliably identified.[134] The court decided that such a result is just and fair because no damage would have occurred had the two plants not carried out the discharges.[135] The victim of pollution in that case

---

124 See Chapter 2 above.

125 For that issue in general, see De Sadeleer 41.

126 Case C-378/08 *Raffinerie Mediterranee (ERG) SpA, Polimeri Europa SpA and Syndial SpA v Ministero dello Sviluppo economico and Others* [2010] ECR I-1919 paragraph 57 (hereinafter: *Raffinerie*).

127 See the references to the polluter pays principle in preamble 2 and 18 as well as in article 1 of the Environmental Liability Directive for that connection.

128 Raffinerie paragraphs 52 and 54.

129 Raffinerie paragraphs 56/57.

130 Raffinerie paragraph 57.

131 Ibid.

132 Raffinerie paragraphs 62 and 70.

133 BGHZ 52, 257.

134 Ibid.

135 Ibid.

merely needed to prove that the substances which provoked the damage were discharged by any of the installations but not that a specific emission was the cause of the damage.[136]

These judgements reversed the burden of proof in the case of the polluter pays principle. The ECJ ruled that the reversal of the burden of proof represents a legitimate implementation of the polluter pays principle.[137] An implementation of the polluter pays principle can thus include a strict civil liability regime for environmental damages.[138]

Strict civil liability could and should be extended to all shale gas operations. The legal lever to do so is article 8 No 4 Environmental Liability Directive. It provides Member States with the opportunity to decide whether they want to establish a strict civil liability system in environmental matters or one that is merely based on fault and negligence. Thus, implementation of such a regime is left to the Member States.

There are a number of positive effects for shale gas regulation that would be the result of such action. Shale gas operators would be unable to exclude themselves from responsibility for damages when a causal link between their activity and the pollution can be presumed. As compensation claims could involve considerable sums of money, shale gas firms would have a strong incentive to implement suitable precautionary measures to ensure nothing happens in the first place. It could also mean that operators adopt a particularly strict precautionary approach to drilling in environmentally sensitive areas, as the costs for 'cleaning up the mess' could be considerable. This measure would hence bring together the remediation and the preventive aspect of the polluter pays principle. Such an extension of the strict civil liability regime would enable shale gas extraction, while making sure that active steps are taken to minimize the risk of environmental pollution and effective remedies are provided should harm occur.

---

136 Ibid.; De Sadeleer 56.

137 Case C-254/08 *Futura Immobiliare srl Hotel Futura and Others v Comune di Casoria* [2009] ECR I-06995 paragraph 43 et sqq.; Case C-188/07 *Commune de Mesquer v Total France SA and Total International Ltd* [2008] ECR I-4501 paragraphs 71/72. More on this issue may be found at De Sadeleer 336.

138 Rehbinder Ziele paragraph 152.

5.4 SUSTAINABLE DEVELOPMENT

Although sustainable development as a principle is enshrined in several articles of the European Treaties,[139] it is notably absent from article 191 (2) TFEU, the article that otherwise includes all relevant European environmental law principles. However, article 191 (1) TFEU says that EU policy should contribute to the prudent and rational utilization of natural resources. The 'prudent use of natural resources', according to the European Council, is a cipher for sustainable development.[140] Thus, the concept of sustainable development features, albeit indirectly, in article 191 (1) TFEU.[141]

A clear definition, however, is not included in European law and the concept means different things to different people.[142] As a fixed definition may also not be deduced from the jurisprudence of the ECJ,[143] recourse must be had to international law.[144] Sustainable development has been defined in broad terms by the Brundtlandt report, a UN report on the state of the globe,[145] but received its first genuinely workable definition in the Rio Declaration.[146]

The Rio Declaration took a dual approach to sustainable development.[147] It prescribes a right to development that must be fulfilled by equitably meeting the *developmental* and *environmental* needs of present and future generations.[148] In line with this widely accepted[149] definition, the International Court of Justice (hereinafter: ICJ) considered sustainable development in the *Gabcikovo-Nagymaros* case.[150] This is probably

---

139 Article 3 (3) and (5) Treaty on the European Union (hereinafter: TEU); article 21 (2) (d) and (f) TEU; Article 3 (3) and (5) TFEU; article 21 (2) (d) and (f) TFEU; article 11 TFEU; article 140 (1) TFEU.

140 Council Decision 2002/1600/EC of 22 July 2002 laying down the Sixth Community Environment Action Programme [2002] OJ L 242/1.

141 Jans/Vedder 30.

142 Bell/McGillivray 55.

143 Bell/McGillivray 58.

144 Bell/McGillivray 56/57.

145 World Commission on Environment and Development '*Our Common Future*' (Oxford University Press, 1987) ix and 8 (hereinafter: Common Future); Cordonier Segger/Khalfan 19; Bell/McGillivray 57.

146 Davies 28. For criticism of the definition included in the 'Brundtlandt Report', see Jarass/Pieroth article 20a paragraph 10 et sqq.

147 Cordonier Segger/Khalfan 105.

148 Principles 3 and 4 Rio Declaration.

149 Davies 28; Cordonier Segger/Khalfan 15/16.

150 *Gabcikovo-Nagymaros Project (Hungary v Slovakia), Judgment, ICJ Reports 1997,* 7 available at: http://www.icj-cij.org/docket/files/92/7375.pdf [accessed 9 February 2015] (hereinafter: Danube Dam).

the best known judgement on sustainable development to date.[151] In this case, the ICJ ruled that the principle of sustainable development necessitates policies to reconcile economic development with environmental protection.[152]

To comply with this definition, shale gas regulation should thus aim to reconcile the interests of economic development with environmental protection needs. One key pre-condition of economic development is energy security, since modern economies are based on a secure and reliable supply of energy.[153] The impact of sustainable development means that energy projects must operate within ecological limits.[154] They must protect critical natural capital and exercise prudence where impacts are uncertain or unknown.[155]

Although all sorts of measures could be included in shale gas regulation to achieve reconciliation of these interests,[156] the conduct of Environmental Impact Assessments (EIAs) for shale gas projects has been particularly emphasized by legal scholars.[157] According to them, sustainable development does not mean halting necessary developments or rushing them through, but making informed, economically, environmentally and socially sound development choices.[158] EIAs are designed to deliver just that.[159]

As discussed before,[160] the carrying out of EIAs is not compulsory for shale gas projects at EU level, because these projects are not likely to reach

---

151 Bell/McGillivray 58.

152 Danube Dam paragraph 140.

153 Susan Y Noe and George Rock Pring 'The "Fear Factor": Why We Should Not Allow Energy Security Rhetoric to Trump Sustainable Development' in Barry Barton et al. (eds) *Energy Security: managing risk in a dynamic legal and regulatory environment* (Oxford University Press, Oxford 2004) 432 (hereinafter: Noe/Pring). Also discussed above in Chapter 1.

154 Noe/Pring 435.

155 Ibid.

156 Davies 31. Past measures to implement sustainable development ranged from safeguarding the right to environmental information to the long-term advancement of sustainability by empowering the individual citizen to take steps to enforce obligations of an environmental nature, see Cordonier Segger/Khalfan 100.

157 Nay Htun 'EIA and Sustainable Development' (1990) Vol 8 No 1–2 *Impact Assessment* 15 et sqq.; Hugh Wilkins 'The need for subjectivity in EIA: discourse as a tool for sustainable development' (2003) Vol 23 No 4 *Environmental Impact Assessment Review* 401.

158 Noe/Pring 434.

159 Ibid.

160 See Chapter 3 above.

the prescribed threshold of 500,000 cubic metres daily gas production. Only for projects with an anticipated production beyond that threshold obligatory EIA has to be conducted.[161]

However, given the severity of potential environmental impacts of shale gas extraction and little experience with these projects in Europe,[162] It makes sense to have obligatory EIAs for all shale gas extraction activities in Europe, regardless of their scale.[163] An exception should only apply to very early stages of the shale gas extraction process, when the drilling pad is being prepared.[164] This preparation and the prior carrying out of seismic surveys cannot reasonably be expected to have significant environmental impacts since no substantial interference with the environment is required for such activities.[165]

The outcome of the respective EIA would then determine whether or not shale gas extraction is allowed at a certain location.[166] This case-by-case approach would ensure that shale gas extraction is only carried out where it is environmentally maintainable.[167] This tailor-made approach to shale gas regulation would make it possible to take individual circumstances into account and ensure an early balancing of economic development interests with environmental protection needs.

5.5 PUBLIC PARTICIPATION

Public participation is not listed as an environmental law principle in article 191 (2) TFEU. However, contemporary legal literature argues that public participation is so intricately linked to the other principles and so widely used[168] that it has reached the status of a customary principle of

[161] Philippe & Partners Law Firm 'Final Report on Unconventional Gas in Europe' (2011) paragraph 145 available at: http://ec.europa.eu/energy/studies/doc/2012_unconventional_gas_in_europe.pdf [accessed 12 March 2012] (herein after: Philippe & Partners); Meiners et al. Bund B 30.

[162] Meiners et al. Bund B 76/77.

[163] Meiners et al. Bund C 75.

[164] Meiners et al. Bund C 78.

[165] Ibid.

[166] SRU Faulstich 42/43.

[167] SRU Faulstich 43.

[168] Common Future paragraph 81; Cordonier Segger/Khalfan 157; Benjamin J Richardson and Jona Razzaque 'Public Participation in Environmental Decision-Making' in Benjamin J Richardson and Stepan Wood, '*Environmental Law for Sustainability: A Reader*' (Hart Publishing, Portland 2006) 166 (hereinafter: Richardson/Razzaque); Bell/McGillivray 54.

environmental law.[169] To give an example: paragraph 23.2 of Agenda 21[170] states: 'one of the fundamental prerequisites for the achievement of sustainable development is broad public participation in decision-making'. This linkage has also been stressed by the legal literature.[171] In particular, Kenneth T Kristl demonstrated how the environmental law principles of sustainable development and public participation may be translated into concrete rules on shale gas extraction.[172]

Public participation has been a main pillar of EU environmental policy since the conclusion of the Aarhus Convention and its implementation in 2005.[173] The main aim of the Aarhus convention is to assist decision-makers in understanding and identifying public interests while they are formulating environmental policies.[174] In the EU, the Aarhus Convention has been implemented by the Public Participation Directive[175] and the Environmental Information Directive.[176]

The concept of public participation under the Aarhus Convention

---

169 De Sadeleer 285.

170 United Nations Conference on Environment and Development Agenda 21 U.N. Doc. A/CONF.151.26 (1992).

171 For instance: John C Dernbach 'Sustainable Development: Now More than Ever' (2002) 32 *Envtl. Law Institute* 10003–15; Morgan 145 and 147; Diana Stares, James McElfish and John Ubinger Jr 'Sustainability and community responses to local impacts' in John C Dernbach and James R May (eds) '*Shale Gas and the Future of Energy Law and Policy for Sustainability*' (Edward Elgar Publishing, Cheltenham, 2016) 114–21 (hereinafter: Stares/McElfish/Ubinger);

172 Kenneth T Kristl 'Public participation and sustainability: how Pennsylvania's shale gas program thwarts sustainable outcomes' in John C Dernbach and James R May (eds) '*Shale Gas and the Future of Energy Law and Policy for Sustainability*' (Edward Elgar Publishing, Cheltenham, 2016) 126/127 (hereinafter: Kristl).

173 Council Decision 2005/370/EC of 17 February 2005 on the conclusion, on behalf of the European Community, of the Convention on access to information, public participation in decision-making and access to justice in environmental matters [2005] OJ L 124/1; Title III of Regulation (EC) 1367/2006 of 6 September 2006 on the application of the provisions of the Aarhus Convention on Access to Information, Public Participation in Decision-making and Access to Justice in Environmental Matters to Community Institutions and Bodies [2006] OJ L 264/13; Jans/Vedder 332.

174 Richardson/Razzaque 165.

175 Council Directive (EC) 2003/35 of 26 May 2003 providing for public participation in respect of the drawing up of certain plans and programmes relating to the environment and amending with regard to public participation and access to justice Council Directives 85/337/EEC and 96/61/EC [2003] OJ L 156/17.

176 Council Directive (EC) 2003/4 of 28 January 2003 on public access to environmental information and repealing Council Directive 90/313/EEC [2003] OJ L 41/26.

comprises, inter alia, transparency and the dissemination of information.[177] Interestingly, these two components mirror two main demands that the European public has with regard to shale gas extraction. In 2013 the European Commission released the results of a Europe-wide public consultation on shale gas extraction[178] which was conducted prior to the establishment of the 2014 Shale Gas Recommendation[179] and Communication.[180] Public consultations are generally considered to constitute the best means for realizing the right to public participation[181] as they go beyond the dissemination of information and invite interested parties to set out their points of view.[182]

The shale gas consultation resulted in more than 22,000 responses, broadly falling into three categories: those in favour and against the development of shale gas resources and those asking for strict environmental and health safeguards to be put in place.[183] Despite this divergence of personal opinions, a broad consensus among all groups exists on two demands. First, the regulator should implement measures to address the potential challenges of shale gas extraction and second, transparency and improved dissemination of information are deemed essential at all stages.[184] While the first point should be addressed under all other applicable principles of environmental law, the second particularly pertains to the issue of public participation.

In order to implement these demands in shale gas regulation under the principle of public participation, two measures should be taken. First, a comprehensive public participation process should be set up[185] and second, EIAs should be made compulsory for shale gas projects.[186]

---

177 Cordonier Segger/Khalfan 158/159; Richardson/Razzaque 165.

178 European Commission '*Analysis and presentation of the results of the public consultation "Unconventional fossil fuels (e.g. shale gas) in Europe" Final report*' (Bio Intelligence Service, Paris 2013) (hereinafter: Shale Gas Consultation).

179 Commission Recommendation 2014/70/EU of 22 January 2014 on minimum principles for the exploration and production of hydrocarbons (such as shale gas) using high-volume hydraulic fracturing [2014] OJ L 39/72 (hereinafter: 2014 Shale Gas Recommendation).

180 European Commission 'Communication on the exploration and production of hydrocarbons (such as shale gas) using high volume hydraulic fracturing in the EU' (Communication) *COM (2014) 23 final/2*. See Chapter 3 above.

181 De Sadeleer 285.

182 Ibid.

183 Shale Gas Consultation 13.

184 Shale Gas Consultation 14.

185 For details on that, see Kristl 140–42; Stares/McElfish/Ubinger 114–21; Morgan 145–7.

186 Meiners et al. Bund B 105 and preceding section.

Kenneth Kristl pointed out, with regards to the first point, that a well-designed public participation process in shale gas cases could lead to more knowledgeable, more credible and more transparent decisions, for a variety of reasons.[187] Linking in with the existing models for structuring public participation, he identifies five, particularly important pillars of public participation processes in the case of shale gas extraction.

First, the reasonable opportunity to participate should be given to the general public, not just to a limited circle of people.[188]

Second, public access to relevant information and the opportunity for the public to comment on proposals in writing or during public hearings is paramount.[189] Shale gas regulation should prescribe the instalment of a unified registry that contains all current and proposed shale gas operations.[190] The creation of such a registry would also implement paragraph 15 of the 2014 Shale Gas Recommendation and the pertaining Communication. This paragraph stipulates that Member States shall ensure the effective dissemination of information on chemical substances and volumes of water used in shale gas projects.[191] More specifically, information should be provided on the number of wells completed and on planned projects, the number of permits granted, the names of operators involved and the permit conditions, monitoring results, incidents and accidents, as well as the results of inspections, non-compliance and sanctions.[192] This could be done by means of a public registry. Such a registry would have to be elaborated by the Member States. It would make crucial environmental information about shale gas activities in the respective regions easily accessible to the public.

Third, adequate time for public input to the decision-making process should be provided. In the case of shale gas extraction in the US state of Pennsylvania, for instance, the possibility of the general public to provide input and comments was limited to a handful of days only. This was not deemed sufficient by the local population.[193]

Fourth, instead of strict timetables, structuring the process as a series of sequential steps with each step only triggered after the previous step is completed would increase the adaptability of the process.[194]

---

187 More details at Kristl 129/130.
188 Kristl 140.
189 Ibid.
190 Dannwolf et al. AP8-1.
191 Paragraph 15 (a) 2014 Shale Gas Recommendation and paragraph 15 (a) Commission Shale Gas Communication.
192 Paragraph 15 (b) and (c) Commission Shale Gas Communication.
193 Kristl 134/135.
194 Kristl 141.

Fifth and finally, decision makers should be obliged to be open to public comments and to consider public input during the decision-making process as well as to reply to comments that the decision-makers received from the public.[195]

A second measure, besides the improvement and adaptation of the outline of public participation procedures concerning shale gas, could be taken. It relates to Environmental Impact Assessments and the Public Participation Directive. Public participation is well developed in EIA-regulation[196] and this mechanism could be used for the scrutiny of individual shale gas projects.[197] As public participation already forms a part of the process of elaborating an EIA, the obligation to carry out EIAs for every shale gas project, discussed above, would ensure public scrutiny of individual shale gas projects.[198]

By implementing these two measures, legislators would help to translate the principle of public participation into concrete regulation and allow for a procedure that is working with the biggest amount of information available, as well as with the full spectrum of possible views. By taking all different, and maybe divergent, views into account this process in itself would help to ensure (albeit not guarantee) that shale gas regulation caters for both environmental protection interests as well as energy security needs.

5.6 RECTIFICATION AT SOURCE

The principle of rectification at source is included in article 191 (2) TFEU[199] and in the constitutions of EU Member States.[200] It implies that damages to the environment should be prevented and, if necessary, combated at their source as early as possible.[201] Therefore, emissions

---

195 Ibid.

196 Richardson/Razzaque 179; more on EIAs may be found above in Chapter 3.

197 Lechtenböhmer et al. 61; Philippe & Partners Law Firm '*Final Report on Unconventional Gas in Europe*' (2011) 49 paragraph 145 available at: http://ec.europa.eu/energy/studies/doc/2012_unconventional_gas_in_europe.pdf [accessed 12 March 2012].

198 Meiners et al. Bund C 79.

199 Calliess/Ruffert article 191 paragraph 32 et sqq.

200 For instance, article 20a German constitution, see Jarass/Pieroth article 20a paragraph 10 et sqq.

201 Epiney Umweltrecht chapter 5 paragraph 23; Calliess/Ruffert article 191 paragraph 34; Kloepfer § 9 paragraph 103.

should be prevented as a priority, rather than regulated by environmental quality standards.[202] The ECJ applied this principle in particular in cases concerned with waste disposal.[203] The principle of rectification at source is closely related to the precautionary and the preventive principle in the sense that all three principles aim at combating environmental nuisances as early as possible.[204] However, rectification at source applies further down the line than the precautionary and the preventive principles.[205] After an environmental nuisance has come into existence, the principle of rectification at source demands that it must be combated right at its source and not by 'end-of-the pipe' technologies.[206]

With regard to shale gas extraction, this principle requires the taking of certain measures in order to prevent, or at least abate, the occurrence of environmental damages.[207] To give a concrete example: a scientific study on European shale gas extraction suggested that national shale gas regulations should include the use of best available techniques to prevent the release of greenhouse gases from shale gas wells.[208] The use of the best available techniques – standard in shale gas regulation – has already been discussed above in Chapter 2 in the context of the Mining Waste Directive.

The EU should summarize these techniques in Best Available Technique References Documents (so called EC-BREFs),[209] which guide Member States when drawing up their own regulations.[210] This would not only implement the principle of rectification at source but also contribute to the balancing of environmental protection with energy security in shale gas cases. Ultimately, the prescription of the use of best available technologies could help to create environmentally sound shale gas extraction.

---

[202] Jans/Vedder 42.

[203] Case C-2/90 *Commission of the European Communities v Kingdom of Belgium* [1992] ECR I-4431 paragraph 34 et sqq.; Case C- 209/98 *Entreprenørforeningens Affalds/Miljøsektion (FFAD) v Københavns Kommune* [2000] ECR I-3743 paragraph 51 (hereinafter: Kobenhavns Kommune); Davies 51.

[204] Epiney Umweltrecht chapter 5 paragraph 24.

[205] Ibid.

[206] Jans/Vedder 42; Calliess/Ruffert article 191 paragraphs 33/34.

[207] SRU Faulstich 31/35; Meiners et al. Bund C 86/C 87.

[208] Meiners et al. Bund C 86/C 87.

[209] European Commission 'IED and BREF Revision' page 13 and 14 available at: http://www.cepi.org/system/files/public/epw-presentations/2012/BREFseminar/Paper%20week.pdf [accessed 9 March 2013] (hereinafter: Commission BREF); Lechtenböhmer et al. 62.

[210] Lechtenböhmer et al. 62.

5.7 CONCLUSION

Environmental law principles are not just hollow declarations of political intent. As demonstrated in this chapter, they can be translated into concrete measures for shale gas regulation. They have a concrete influence on the balancing of environmental protection interests with energy security demands.

However, one environmental law principle alone is unable to provide comprehensive directions for the regulation of a complex issue like shale gas extraction. As different environmental law principles may be translated into concrete, sometimes overlapping, measures, they all contribute to the evolution of an ever-closer net of shale gas regulations.

Crucially, all of the five principles that have been discussed in this chapter can assume a precautionary dimension, in the sense that they may be used to implement measures aiming to minimize the potential environmental threats of shale gas extraction. The precautionary principle thus provides paramount directions for the elaboration of shale gas regulation.

Measures that could be deduced from the five scrutinized environmental law principles tackle parts of the most salient potential environmental threats of shale gas extraction. Measures include the requirement to exclusively deploy 'eco-friendly fracturing fluids' in shale gas operations and to prohibit shale gas extraction in water protection zones. Moreover, reversing the burden of proof and implementing a strict civil liability regime for shale gas operations as well as the obligatory prescription of EIAs for all shale gas projects. This should be supplemented by a tailor-made public participation process which requires, as a first step, the creation of a unified registry that includes all current and planned shale gas operations. Finally, in order to prevent the release of greenhouse gases from shale gas wells, the use of best available techniques should be regulated.

Shale gas regulation that includes these measures would be cautious but permissive in character and capable of striking a balance between environmental protection interests and energy security needs. Although all of the named measures have environmental protection as their main aim, they do not disregard the potential energy security benefits that shale gas extraction could bring about.

6. Rules and shale gas

6.1 INTRODUCTION

The third and final category of norms that are part of the trias[1] from which cornerstones of shale gas specific regulation could be derived is the category of concrete rules.

The application of (quasi-) constitutional objectives and environmental law principles to shale gas extraction in the previous chapters has helped to derive a set of concrete measures. However, these measures do not address all of the four most salient potential threats of shale gas extraction.[2] In particular, there are two potential threats that have not yet received sufficient attention in previous chapters.[3] First, shale gas extraction is an additional use of land in a densely populated area like Europe which could increase spatial pressure. Second, current monitoring requirements which do not cover all potential threats of shale gas extraction.[4]

These two issues are not unique to shale gas extraction, but have been discussed in the context of another 'new' energy technology,[5] Carbon Capture and Storage (hereinafter: CCS).[6] CCS is regulated by legally

1 Described in Chapter 4 above.

2 Identified in Chapters 1 and 2 above.

3 For such measures, see in particular Chapter 5 above.

4 Compare with the discussion on potential environmental threats of shale gas extraction in Chapter 1 above. While land use has been addressed there as a stand-alone potential threat, insufficient monitoring has been discussed in the context of climate change impacts of shale gas extraction. Both issues have, to some extent, been addressed by the 2014 Shale Gas Recommendation and the pertaining Communication, which were discussed in Chapter 2 above. However, these EU norms are not legally binding and must be implemented by Member States to attain legal force.

5 For the ambiguity of this terminology see Chapter 1 above.

6 Mark G Little and Robert B Jackson 'Potential Impacts of Leakage from Deep CO_2 Geosequestration on Overlying Freshwater Aquifers' (2010) 44 *Environmental Science & Technology* 9225 and 9230; IPPC Special Report 12/13; Deutscher Bundestag 'Bericht des Ausschusses für Bildung, Forschung und Technikfolgenabschätzung (18. Ausschuss) gemäß § 56a der Geschäftsordnung Technikfolgenabschätzung (TA) CO_2-Abscheidung und -Lagerung bei Kraftwerken' of 1 July 2008 Bundestagsdrucksache 16/9896 available at: http://

binding EU norms and these regulations include solutions to tackle the two named potential threats in a way that could be adopted for the regulation of shale gas extraction, as this chapter shows.

CCS is an appropriate analogy because certain technological aspects, potential environmental issues and possible energy security benefits of CCS resemble those of shale gas extraction.[7] As the technology might allow for the continued use of (domestically recoverable) fossil fuels, a careful balance has to be struck between potential energy security benefits and environmental protection considerations[8] – just as in shale gas regulation. Lessons from the regulation of CCS are hence used in this chapter to elaborate further cornerstones of shale gas regulation.

CCS is a (recently evolving) technology to store carbon dioxide deep beneath the surface.[9] In essence, the CCS technology involves capturing CO_2 from sources of emission (for instance coal fired power plants or other power generation facilities but also other industrial processes) and injecting it into geological formations ('stores') deep beneath the surface.[10] This procedure aims to permanently sequester the CO_2 underground.[11]

Once again,[12] it would be over-ambitious for this book to scrutinize the

---

dip21.bundestag.de/dip21/btd/16/098/1609896.pdf [accessed 16 December 2013] 22 et sqq. (hereinafter: Bundestag Committe Report CCS); Ralf E Krupp 'Gutachten zur geplanten Kohlendioxid-Einlagerung (CCS) in der Antiklinal-Struktur Neutrebbin, Ostbrandenburg' (2011) available at: http://www.co2bombe.de/joomla/images/stories/co2/krupp_gutachten_1_neutrebbin_final.pdf [accessed 5 March 2013] 21, 40 and 48.

7 Discussed in Chapter 5 above.

8 Tim Dixon et al. 'Legal and Regulatory Developments on CCS' (2015) Vol 40 *International Journal of Greenhouse Gas Control* 445 (hereinafter: Dixon et al.).

9 The technical aspects of CCS have been explained at M Granger Morgan (et al.) '*Carbon Capture and Sequestration Removing the Legal and Regulatory Barriers*' (RFF Press Routledge, New York 2012) 12–46; Stuart Haszeldine 'Geological Factors in Framing Legislation to Enable and Regulate Storage of Carbon Dioxide Deep in the Ground' in Ian Havercroft, Robert Macrory and Richard Stewart (eds) '*Carbon Capture and Storage: Emerging Legal and Regulatory Issues*' (Hart Publishing, Oxford and Portland 2011) 8 et sqq. (hereinafter: Haszeldine); Christian Fouillac 'CO_2 Capture, Transport and Storage, a Promising Technology for Limiting Climate Change' in Jean Bernard Saulnier and Marcello Varella (eds) '*Global Change, Energy Issues and Regulation Policies*' (Springer International, Cham 2013) 121–41. Specifically for Germany, see Federal Economic Ministry (*Bundesministerium für Wirtschaft und Energie*) 'Die weitere Entwicklung von CCS-Technologien' available at: http://www.bmwi.de/DE/Themen/Industrie/Industrie-und-Umwelt/ccs.html [accessed 18 March 2015].

10 Ibid.

11 Ibid.

12 This issue has already been covered in Chapter 3 above.

regulatory framework of 28 different EU Member States on CCS. Instead, the chapter focusses on the three jurisdictions that have already been consistently referred to in this book[13] to discuss shale gas regulation: the UK, Germany and France. These examples reveal three, rather distinct, approaches to the transposition of the CCS Directive.

The UK was the first European country to put CCS-specific legislation into place.[14] Germany is an example of a major European country that transposed the Directive in a cautious and restrictive CCS legislation. France rolled out a supportive and permissive regime very quickly.[15] Lessons for shale gas regulation can be learnt from all three approaches, from the first attempt on regulation, to prohibitive, to permissive laws.[16]

The chapter starts by outlining some of the features of the CCS Directive[17] that the EU put into place, because this Directive predetermines national legislation on CCS. Subsequently the chapter focusses on the three named jurisdictions. However, it does not deliver a comprehensive assessment of the legal framework applicable to CCS in these countries, but rather discusses particular features of the respective laws that lend themselves to becoming, *mutatis mutandis,* role models for shale gas regulation. Therefore, each of the discussed features is immediately matched by a possible equivalent in shale gas regulation. The resulting measures tackle the remaining potential threats and contribute to a framework for shale gas regulation that balances environmental protection interests with energy security needs.

---

[13] Chapter 3 above.

[14] Alla Shogenova et al. 'CCS Directive transposition into national laws in Europe: progress and problems by the end of 2011' (2013) Vol 37 *Energy Procedia* 7725 (hereinafter: Shogenova et al. 2011).

[15] Alla Shogenova et al. 'Implementation of the EU CCS Directive in Europe: results and development in 2013' (2014) Vol 63 *Energy Procedia* Fig. 1 on page 6664 (hereinafter: Shogenova et al. 2013).

[16] In particular, given that these countries are amongst the highest emitters of CO_2 in the European Union. At the same time, all three countries have sufficient theoretical geological capacity to store parts of their own CO_2 emissions and are amongst those with the most advanced level of CCS research and technology, see Shogenova et al. 2011 at 7727 and 7730; Shogenova et al. 2013 at 6666.

[17] Council Directive (EC) 2009/31 of 23 April 2009 on the geological storage of carbon dioxide and amending Council Directive 85/337/EEC, European Parliament and Council Directives 2000/60/EC, 2001/80/EC, 2004/35/EC, 2006/12/EC, 2008/1/EC and Regulation (EC) No 1013/2006 [2009] OJ L 140/63 (hereinafter: CCS Directive).

6.2 EU CCS DIRECTIVE AND SHALE GAS

CCS is a controversial topic in Europe and development and uptake of the technology have been slower than expected in the EU.[18] By 2015 the ROAD project in the Netherlands was the first and only commercial CCS project that had been awarded all necessary permits under the new legislative regime of the CCS Directive.[19] However, the project has not proceeded yet, due to financial factors.[20]

Early policy documents on CCS, like the IPCC Special Report on Carbon Dioxide Capture and Storage 2005,[21] noted a lack of dedicated regulations on geological storage of CO_2 and uncertainty in relation to subsurface property rights and liability for stored CO_2.[22] These issues have since been addressed to varying degrees in different jurisdictions, through the adoption of over 50 legal instruments relating to the permanent storage of carbon dioxide.[23]

In the EU, the main legal instrument on CCS is the CCS Directive, which aims to ensure safe management of all environmental and health risks related to CCS. Risks include the acidification of the soil and/or the surrounding ecosystems.[24] The CCS Directive was adopted in 2009 and by 2014 all Member States had transposed it into national legislation.[25]

The CCS Directive amended a number of other Directives, such as the IPPC Directive, EIA Directive, Environmental Liability Directive, ETS Directive, Water Framework Directive and Waste Framework Directive.[26] As all of them have been discussed earlier in this book, they

---

18 Dixon et al. 431–3; See also further below in this chapter.

19 Dixon et al. 445.

20 Ibid.

21 Intergovernmental Panel on Climate Change '*IPCC Special Report on Carbon Dioxide Capture and Storage*' (Cambridge University Press 2005) (IPCC Special Report).

22 Dixon et al. 432.

23 Dixon et al. 431 and 432/433.

24 Martina Doppelhammer 'The CCS Directive, its Implementation and the Co-financing of CCS and RES Demonstration Projects under the Emissions Trading System (NER 300 Process)' in Ian Havercroft and Robert Macrory and Richard Stewart (eds) '*Carbon Capture and Storage: Emerging Legal and Regulatory Issues*' (Hart Publishing, Oxford and Portland 2011) 94 (hereinafter: Doppelhammer); Dixon et al. 432.

25 Dixon et al. 432.

26 Doppelhammer 94; Dixon et al. 437; Shogenova et al. 2013 at 6665.

are left outside the scope of this section, which focuses only on the main features of the CCS Directive.[27]

The CCS Directive regulates three relevant phases of an underground storage project: site-selection, injection and storage of CO_2 and abandonment/closure of the site.[28] The current assessment focuses on the injection and storage of CO_2, as the CCS Directive itself similarly focuses on storage.[29] Moreover, the regulations on selection and abandonment of a site do not address potential threats that are similar to those involved in shale gas extraction.[30]

The CCS Directive provides a framework for national CCS storage licensing regimes. However, this framework is, by and large, comparable to the framework currently in place for hydrocarbon-licensing under the Hydrocarbons Licensing Directive, albeit it is possible to observe some differences between the two regimes.[31]

There are three measures in the CCS Directive[32] that put into place a stringent monitoring regime, which could be of help to tackle, by analogy, the potential threat of insufficient monitoring requirements in shale gas extraction. These three measures pertain to monitoring, reporting and inspection obligations which follow the overall aim to ensure safe and secure injection and storage of CO_2.

First, the CCS Directive asks Member States with regards to monitoring to keep tight control over injection and storage activities to avoid damage to the environment, because there is little experience with geological storage of CO_2.[33] Such tight control will be guaranteed, inter alia, by

---

[27] For a discussion of these Directives within the shale gas context, See Chapter 2 above and for a discussion in the context of the CCS Directive, see Doppelhammer 94; Dixon et al. 437; Shogenova et al. 2013 at 6665.

[28] Martha M Roggenkamp 'Regulating Underground Storage of CO_2' in Martha M Roggenkamp and Edwin Woerdman '*Legal Design of Carbon Capture and Storage Developments in the Netherlands from an International and EU Perspective*' (Intersentia, Antwerp 2009) 207/208 (hereinafter: Roggenkamp CCS).

[29] Dixon et al. 437; Aileen McHarg and Mark Poustie 'Risk, Regulation, and Carbon Capture and Storage: The United Kingdom Experience' in Donald N Zillman et al. (eds) '*The Law of Energy Underground Understanding New Developments in Subsurface Production, Transmission and Storage*' (Oxford University Press, 2014) 266 (hereinafter: McHarg/Poustie).

[30] An, unfortunately superficial, comparison between CCS and shale gas extraction can be found at Haszeldine 9 et sqq.

[31] A comparison of both regimes is available at Roggenkamp CCS 213–16.

[32] A good overview is provided by Marijn Holwerda '*EU Regulation of Cross-Border Carbon Capture and Storage*' (Intersentia Publishing, Cambridge 2014) 33–48 (hereinafter: Holwerda); Roggenkamp CCS 207–9.

[33] Roggenkamp CCS 210.

strict monitoring obligations. Article 13 CCS Directives requires carbon storage facility operators to monitor the injection facilities, the storage complex and the surrounding environment.[34] In order to do so operators are obliged to elaborate and submit a monitoring plan, which should ensure that monitoring of underground containment of CO_2 over long periods is possible.[35] For the eventuality that leakage occurs, the monitoring plan must include mechanisms to detect any detrimental environmental impacts of such leaks as well as monitor the effectiveness of resolution mechanisms.[36]

Second, reporting obligations are an important feature of the CCS Directive which accompany and supplement monitoring requirements.[37] Every year monitoring results have to be submitted to the respective competent national authorities, together with a lot of other data, according to article 14 CCS Directive.

Third, article 15 CCS Directive provides for the obligation to perform inspections. Member States are obliged to organize a system of routine inspections of all storage sites on their territory.[38] These checks are not limited to the underground storage facility itself, but must also include surface and injection facilities, as well as the inspection of records.[39] According to article 15 (3) CCS Directive such inspections have to be carried out at least once a year and non-routine inspections will be carried out in particular circumstances.[40] Inspection results will be communicated to the operator and made available to the public.[41] In case of significant irregularities or leakage, the operator needs to notify the competent national authority immediately and is obliged to implement corrective measures to resolve the issue.[42]

It must be emphasized that the subject of these provisions is the respective national authority of an EU Member State. Under the Directive, the European Commission, however, has a role to play in reviewing and providing comments on any draft storage permit which is not binding upon

---

34 Holwerda 42.
35 Doppelhammer 96; Roggenkamp CCS 210.
36 Ibid.
37 Holwerda 43.
38 Ibid.
39 Article 15 (2) CCS Directive.
40 Article 15 (4) CCS Directive.
41 Article 15 (5) CCS Directive.
42 Article 16 CCS Directive.

Member States.[43] However, the Member States are required to provide reasons if they decide against the Commission's opinion.

Overall, Martha Roggenkamp in her assessment of the CCS Directive concluded that the Directive is 'leading the way in which the EU Member States (. . .) design their legal regimes facilitating CCS'.[44] This statement showcases the important role that the respective national regime for CCS regulation has and emphasizes that the national regulator is in charge.

This is most clearly expressed in the provisions of article 4 (1) and preamble 19 CCS Directive. According to these stipulations it is up to the Member States to decide which areas are made available as potential storage sites.[45] Conversely, Member States also have the right not to allow any storage in parts or all of their territory.[46]

Article 39 (1) CCS Directive obliges Member States to bring into force the laws, regulations and administrative provisions necessary to comply with the CCS Directive by 25 June 2011. They are also obliged to report to the Commission every three years on the status of the implementation of the CCS Directive.[47]

In May 2014 the European Commission started a review process of the CCS Directive, in line with article 38 CCS Directive.[48] As part of this process a report was published in January 2015, which concluded that

> The overall need for CCS (. . .) remains high. The Directive has a useful and important part to play in this. Progress in the uptake of CCS has been slower than predicted (. . .) making detailed evaluation very difficult. There are some concerns with specific aspects of the CCS Directive, but there is not yet enough experience with it to justify (. . .) changes. Revising the Directive (. . .) will create increased regulatory risk and thus cause additional delays (. . .).[49]

This statement shows that CCS is progressing slower than anticipated by the Directive and that, due to a substantial lack of experience, the actual

---

43 According to article 10 CCS Directive.

44 Roggenkamp CCS 206.

45 Article 4 (1) and preamble 19 CCS Directive.

46 This is according to the clear wording of article 4 (1) CCS Directive and one of the main persons involved in drafting the CCS Directive, see Doppelhammer 95. However, this does not seem to be unchallenged as Roggenkamp, for instance, observes that the CCS Directive requires Member States to develop a specific activity – subsoil storage of CO_2, see Roggenkamp CCS 207.

47 Article 27 (1) CCS Directive.

48 Dixon et al. 437.

49 Triple E, Ricardo AEA, TNO '*Support to the review of Directive 2009/31/EC on the geological storage of carbon dioxide (CCS Directive)*' (Publications Office of the European Union, Luxembourg 2014) xvii (hereinafter: Triple E).

impacts of the legal framework upon CCS projects cannot yet be fully assessed. Despite this, some preliminary lessons might be learnt from the transposition procedure itself as well as from those Member States that are more advanced in CCS activities than others.

6.3 MEMBER STATES' TRANSPOSITION OF THE CCS DIRECTIVE – LESSONS FOR SHALE GAS

6.3.1 United Kingdom

The UK is primarily reliant (90 per cent) on fossil fuels for its energy supplies, with the need to cover 28 per cent of total demand by imports.[50] CCS could provide a way to maintain fossil fuels within the energy mix, while ensuring the reduction of GHG emissions in the UK.[51] Accordingly, three strong drivers for CCS in the UK are identifiable: energy security concerns; international and national climate change policies and the business opportunities presented by meeting both challenges.[52]

The pioneering work in CCS legislation amongst EU Member States was undertaken by the UK, where CCS policy has acknowledged the opportunity to be a world leader and first-mover in the technology.[53] In fact, as in other jurisdictions, the process of putting into place a regulatory framework is, in the UK, well ahead of commercial needs. This has been explained with the perception that CCS seems unlikely to be deployed at commercial scale unless a suitable legal and regulatory framework is

---

[50] Chiara Armeni '*Case studies on the implementation of Directive 2009/31/EC on the geological storage of carbon dioxide. United Kingdom.*' (2011) available at: http://blogs.ucl.ac.uk/law-environment/files/2012/11/Chiara-Armeni-CCLP-EU-Case-Studies-UK-2011.pdf [accessed 25 November 2016] 8 (hereinafter: Armeni).

[51] House of Commons Select Committee on Energy and Climate Change's Inquiry 'The UK's Energy Supply, Security or Independence?' (Crown, 2011) Question 213 available at: http://www.publications.parliament.uk/pa/cm201012/cmselect/cmenergy/1065/1065.pdf [accessed 20 November 2016].

[52] McHarg/Poustie 251; Armeni 8.

[53] HM Government '*Clean Coal: An Industrial Strategy for the Development of Carbon Capture and Storage Across the UK*' (Crown 2011) 11 available at: https://ukccsrc.ac.uk/system/files/publications/ccs-reports/DECC_Coal_154.pdf [accessed 30 November 2016] (hereinafter: HM Government Clean Coal); Department of Energy and Climate Change (DECC) '*CCS Roadmap: Supporting Deployment of Carbon Capture and Storage in the UK*' (Crown 2012) 15 (hereinafter: DECC CCS Roadmap); Armeni 12.

already in place which enables the management or mitigation of associated risks.[54]

The UK put into place a regulatory framework that rests on two pillars: first, the UK Energy Act 2008 including the first comprehensive licensing framework for CCS in Europe and one of the first in the world.[55] Second, the Planning Act 2008, which includes a so called 'development consent order' for 'nationally significant infrastructure projects', and which applies to CCS projects.[56]

Remarkably, this framework was issued prior to the CCS Directive, which went into force in 2009.[57] The UK Energy Act 2008 focussed on a regulatory regime for offshore CO_2 storage in the UK but, as it turned out later, also provided sufficient flexibility to transpose the CCS Directive, which applies onshore as well as offshore.[58]

Accordingly, the UK finalized its national transposition process in early 2012.[59] The UK's implementation of the CCS Directive is one of 'infusion' rather than of 'transplant', as Armeni noted.[60] That is to say the CCS Directive has been integrated in the existing and well-established oil and gas legislation, rather than creating a stand-alone regime for the activity.[61] The provisions of the Directive have been transposed by the Energy Act 2008 and by dedicated regulations and amendments to existing laws.[62] Additional provisions to incentivize CCS projects, which go

---

54 McHarg/Poustie 249.

55 Armeni 14 and 38; Global CCS Institute 'United Kingdom CCS legislation' available at: https://hub.globalccsinstitute.com/publications/dedicated-ccs-legislation-current-and-proposed/united-kingdom-ccs-legislation [accessed 30 November 2016] (hereinafter Global CCS Institute UK); McHarg/Poustie 250.

56 Meyric Lewis and Ned Westaway 'Public Participation in UK CCS Planning and Consent Procedures' in Ian Havercroft and Robert Macrory and Richard Stewart (eds) '*Carbon Capture and Storage Emerging: Legal and Regulatory Issues*' (Hart Publishing, Oxford and Portland) 277 (hereinafter: Lewis/Westaway); Armeni 14/15.

57 Shogenova et al. 2011 at 7725.

58 It was amended for that purpose and now includes the new section 17 (3A) and (4) Energy Act 2008, see McHarg/Poustie 266 and 269; Shogenova et al. 2011 at 7726.

59 Ironically, this means that the UK did not meet the deadline for transposition, the year 2011, despite already having a dedicated CCS regime in place prior to the coming into force of the CCS Directive, as noted by Armeni 16; Shogenova et al. 2011 at 7726. The transposition procedure turned out to be more complicated than initially thought, see Armeni 16/17 and 38.

60 Armeni 16.

61 Armeni 4.

62 Part 1 chapter 3 Energy Act 2008; Energy Act 2008 (Consequential Modifications) (Offshore Environmental Protection) Order 2010; The

beyond the CCS Directive, are included in the Energy Acts of 2010, 2011 and 2013.[63]

Overall, the UK has one of the most proactive CCS policies in Europe and, indeed, the world.[64] Political support for CCS began in 2002 and at present there is shared political agreement on CCS deployment in the UK.[65] The government has put into place a number of incentives to get to the stage of commercial deployment of CCS. Lessons from that regime for shale gas regulation can be learnt in two ways, which are discussed now. The first lesson pertains to the powers of the state to adopt a permissive or prohibitive approach to the technology in line with local needs. The second lesson relates to the ways in which decision-makers can incentivize the move offshore of controversial energy technologies like CCS and shale gas to ease spatial pressure onshore.

6.3.1.1 Devolved powers

CCS regulation lies at the intersection between a number of generally reserved matters and areas that are devolved in the UK system of governance. The question of whether or not Scotland, Northern Ireland and Wales could influence CCS regulation or even ban CCS in their respective territories has been posed by the legal literature.[66] As Armeni concluded, the answer depends on the grounds for the ban and the extent of the powers that have been devolved, but it is possible.[67] The opening up of the option for regions to implement similar bans on shale gas extraction could help to alleviate the potential threat of overbearing land-use in regions that are not suitable for shale gas extraction.

Planning is a devolved matter and in Scotland and Northern Ireland

Energy Act 2008 (Storage of Carbon Dioxide) (Scotland) Regulations 2011; The Storage of Carbon Dioxide (Licensing etc.) Regulations 2010; Storage of Carbon Dioxide (Licensing etc.) (Scotland) Regulations 2011; Storage of Carbon Dioxide (Termination of Licences) Regulations 2011; Storage of Carbon Dioxide (Amendment of the Energy Act 2008 etc.) Regulations 2011; Storage of Carbon Dioxide (Access to Infrastructure) Regulations 2011; Storage of Carbon Dioxide (Inspections etc.) Regulations 2012.

[63] In particular, Part 1 Energy Act 2010; sections 107 and 108 Energy Act 2011; section 58 Energy Act 2013; in the Energy Act of 2016, CCS is not mentioned specifically.

[64] McHarg/Poustie 266; Shogenova et al. 2011 at 7728.

[65] This does not include the Green Party, which, however, is not as influential as it is in other Member States like Germany, see Armeni 10.

[66] Armeni 4.

[67] Ibid.

the respective authorities are in charge of planning laws.[68] Scotland and Northern Ireland could decide on a policy against granting planning permission to CCS plans, which would be permitted under current legislation.[69] The result would be a factual ban on CCS, rather than a formal one. A precedent is available with respect to nuclear power in Scotland, where the Scottish parliament effectively banned new nuclear power plants, and Wales now has powers on environmental protection matters, which means it could ban CCS on environmental grounds.[70]

A similar possibility could be envisaged for shale gas regulation. In fact, Scotland and Northern Ireland have, in the past, imposed moratoria on shale gas extraction.[71] Such a possibility for individual regions of a Member State to prohibit shale gas extraction could ease the spatial pressure that might potentially be created by land-use for shale gas extraction in Europe, by confining shale gas to those regions that are suitable for and receptive to shale gas extraction. The UK legislation highlights that such regional 'opt-out' provisions for shale gas extraction are possible in Member States that operate under a similar mode of governance (devolution).

6.3.1.2 Offshore

The UK is keen on offshore carbon dioxide storage for three reasons: first, suitable storage is abundantly available offshore in the UK;[72] second, no complex property rights issues are expected offshore; and third, offshore storage is less likely to attract public opposition.[73] These aspects are, obviously to different degrees, also applicable to shale gas extraction. Wherever the factual circumstances allow, shale gas regulation might similarly provide incentives for going offshore. Such a move could help to mitigate the potential threat of increased competition for land-use in an already densely populated area like Europe.

The UK policy-preference for CO_2 storage offshore is implemented through licensing provisions and financial incentives.[74] First, the licensing

---

68 This point has already been discussed above in Chapter 3. For the particular CCS context, see Armeni 15 and 17.

69 Armeni 17.

70 Armeni 17/18. However, according to Armeni, this is unlikely to happen for the practical reason that opposition to CCS in these states is not as high as it is in other EU Member States.

71 See Chapter 3 above.

72 Namely the possibility to convert abandoned offshore oil and gas infrastructure to CCS demonstration projects.

73 McHarg/Poustie 268/269; Armeni 9.

74 Shogenova et al. 2013 at 6668.

regime itself is twofold: alongside the actual storage licence,[75] a lease of land from the Crown Estate is required, according to section 1 Energy Act 2008.[76] The Crown Estate is a large property portfolio that is owned by the reigning monarch 'in right of the Crown' and managed by the Crown Estate Commissioners.[77] It has a significant offshore component, including almost all of the seabed within the UK territorial sea and the continental shelf, including a so called Gas Importation and Storage Zone.[78] Thus, a licence from the government and a lease from the Crown Estate is required to conduct CCS offshore in the UK.

Initially, it was a prerequisite for obtaining a CCS licence that the carbon dioxide be stored offshore.[79] This limitation faced substantial criticism for failing to accurately transpose the CCS Directive, which applies both onshore and offshore.[80] As a result, the Storage of Carbon Dioxide (Amendment to the Energy Act 2008 etc.) Regulations 2011 came into being, which inserted sections 17 (3A) and (4) into the Energy Act 2008.[81] There is, however, no current UK government policy to deploy CCS onshore.[82]

Vice versa, the Energy Act 2011 facilitates offshore storage of carbon dioxide further by allowing the conversion of abandoned offshore installations and submarine pipelines for CCS demonstration projects[83] and the compulsory acquisition of rights over land for CCS pipeline-conversion.[84]

Financial incentives from the UK government for CCS favour offshore projects and EU funding is also available for offshore CCS projects.[85] The UK government also dedicated money to studies that demonstrate full

---

75 Section 17 (1) Energy Act 2008.

76 McHarg/Poustie 269.

77 Section 1 Crown Estate Act 1961;

78 Ben Milligan 'Planning for offshore CO_2 storage: Law and policy in the United Kingdom' (2014) 48 *Marine Policy* 167 (hereinafter: Milligan); McHarg/Poustie 269.

79 Sections 17 (2) and (3) Energy Act 2008; Armeni 14 and 20; Global CCS Institute UK.

80 Armeni 20.

81 For Scotland, the same has been achieved by the Energy Act 2008 (Storage of Carbon Dioxide) (Scotland) Regulations 2011.

82 Armeni 21.

83 Section 107 Energy Act 2011, with the exception of Scotland, see Section 107 (2) 30A (2) and (3) Energy Act 2011.

84 Section 108 Energy Act 2011.

85 A good example is the White Rose Carbon Capture and Storage Project. This project will be located on land adjacent to the exiting Drax Power Station, near Selby in North Yorkshire and it is the only UK project, so far, to have secured funding from the European Commission. Ninety per cent of the CO_2 produced

chain offshore storage of CO_2 on the UK continental shelf.[86] Section 1 of the Energy Act 2010 enables the Secretary of State to provide financial assistance in respect of CCS demonstration projects or additional CCS use at demonstration sites.[87]

These measures could, *mutatis mutandis*, be applied to shale gas regulation and help to mitigate the potential threat of additional spatial pressure and the possible result of overbearing use of land. Although our oceans are also becoming more and more crowded, many European countries now have marine spatial planning systems in place.[88] Offshore shale gas extraction could be integrated into these systems.

The re-use of abandoned oil and gas platforms for shale gas extraction is arguably even less complicated than the conversions required for CCS. Technical changes would be confined to some technical changes and these platforms could be switched from conventional to unconventional gas extraction, provided they are located at or transferred to suitable locations.

Admittedly, however, there are limitations to what can be delivered by such measures. Above all, viable shale gas reservoirs that are located offshore are required. In order to re-use immovable oil and gas installations it would in some cases even be necessary to find locations where shale gas occurs in combination with conventional reservoirs – these spots might be punishingly hard to find.

However, offshore shale gas extraction in Europe is already being explored and might become a viable option in the long-term.[89] In 2014, the Department of Energy and Climate Change (DECC) awarded three licences for shale gas exploration in the Irish Sea, where the world's first offshore fracking well is planned.[90] The British Geological Survey

---

by the plant will be captured and transported by pipeline for permanent offshore storage beneath the North Sea seabed, see Shogenova et al. 2013 at 6668.

[86] McHarg/Poustie 260–62; Shogenova et al. 2013 at 6669.

[87] This has been made use of, see Armeni 11; McHarg/Poustie 260.

[88] These systems might prioritize certain uses, see for the UK example Milligan 166/167 and 169.

[89] According to Dieter Helm of Oxford University, interviewed by Ben King for *BBC News* 'Shale gas pioneer plans world's first offshore wells in Irish Sea' available at: http://www.bbc.com/news/business-26157228 [accessed 9 December 2016] (hereinafter: *BBC* Offshore Fracking).

[90] House of Lords Economic Affairs Committee 'The Economic Impact on UK Energy Policy of Shale Gas and Oil' (The Stationery Office Limited, London 2014) 32/33 also available at: http://www.publications.parliament.uk/pa/ld201314/ldselect/ldeconaf/172/172.pdf [accessed 9 December 2015] (hereinafter: House of Lords Shale).

has estimated that the UK's total offshore shale gas resources could be between five and 10 times the size of the resources available onshore.[91] In any case, the House of Lords has been told that there is a great potential for offshore shale gas extraction that has not really been investigated yet.[92]

This is not UK-specific but applies to many estimates of European shale gas reserves, which are currently very immature, due to the lack of research.[93] New offshore shale plays that are viable for exploitation might be found in the future. If so, shale gas regulation in these countries could prioritize these offshore reserves for exploration and production.[94]

By putting into place planning restrictions and financial incentives that are similar to those encountered in UK CCS legislation, the chances of finding suitable shale gas reservoirs might increase. Offshore storage of CCS is more publicly acceptable than onshore storage.[95] The same effect is to be expected with regard to shale gas. By concentrating on offshore shale gas extraction, wherever this is possible, the potential threat of additional spatial pressure onshore in a densely populated region like Europe could be mitigated. Although our oceans are also becoming increasingly crowded, spatial pressures are still not at the same level like those that may be encountered onshore.

6.3.2 Germany

Local public protests against CCS at several proposed storage sites in Germany have had a significant impact on legislative and factual developments in Germany.[96] The German CCS demonstration project[97] in Jänschwalde and the exploration of a potential storage site there were

---

[91] *BBC* Offshore Fracking.

[92] House of Lords Shale 32.

[93] As discussed above in Chapter 1.

[94] Commercial viability is another issue. Shale gas extraction in Europe is already rather expensive compared to the US, as described in Chapter 1 above. A move offshore would surely top-up that bill. See also House of Lords Shale 32/33 on this.

[95] Armeni 20.

[96] Shogenova et al. 2013 at 6667; Krämer CCS 4.

[97] There is, however, another project nearby, which is still operative, the CO_2 storage facility at Ketzin, see: GFZ 'Pilot Site Ketzin' available at: http://www.co2ketzin.de/en/pilot-site-ketzin/summary.html [accessed 28 November 2016]. In fact, this is the longest running European test project on onshore CO_2 storage. Having started in 2008, CO_2 injection was complete by 2013, see Global CCS Institute Germany.

abandoned in December 2011 by the operator, Vattenfall.[98] Vattenfall cited two reasons for the decision to withdraw: first, the project was no longer economically viable and second, the long-standing uncertainty about the legal framework for CCS in Germany became intolerable to the company.[99]

Germany, indeed, failed to transpose the CCS Directive before 25 June 2011.[100] This was partly due to public opposition which, by that time, had led to prolonged internal discussions within the German government and the parties supporting it.[101] The Directive was ultimately transposed on 17 August 2012 and arguably because of pressure from the EU Commission, which initiated an infringement procedure against Germany.[102]

The EU CCS Directive has been transposed into two different laws in Germany which were resolved on the same day by the German parliament.[103] However, they have now been brought together into one law, the German Carbon Capture and Storage Law (hereinafter: CCS law).[104] Article 1 of the CCS Law contains the new provisions on CCS (46 Articles), while the rest of its articles amend existing German legislation.[105]

---

[98] Vattenfall Press Release of 5 December 2011 available at www.Vattenfall.de [accessed 25 November 2016]; EurActiv 'Verzicht auf EU-Millionen: Vattenfall stoppt CCS in Brandenburg' of 6 December 2011 available at: http://www.euractiv.de/energie-und-klimaschutz/artikel/verzicht-auf-eu-millionen-vattenfall-stoppt-ccs-in-brandenburg-005708 [accessed 24 October 2013]; Fischer 275.

[99] Ibid.

[100] Wolfgang Fischer 'No CCS in Germany Despite the CCS Act?' in Wilhelm Kuckshinrichs and Jürgen-Friedrich Hake (eds), '*Carbon Capture, Storage and Use Technical, Economic, Environmental and Societal Perspectives*' (Springer International, Cham 2015) 275 (hereinafter: Fischer).

[101] For more on that see Fischer 259–79.

[102] Fischer 275. The infringement procedure has since been terminated, see Fischer 277.

[103] Besides the CCS law there is also the Law on the demonstration of permanent storage of carbon dioxide (*Gesetz zur Demonstration der dauerhaften Speicherung von Kohlendioxid (Kohlendioxid-Speicherungsgesetz–- KSpG).*

[104] Federal Act Concerning the Demonstration and Application of Technologies for the Capture, Transport and Permanent Storage of Carbon Dioxide (*Gesetz zur Demonstration und Anwendung von Technologien zur Abscheidung, zum Transport und zur dauerhaften Speicherung von Kohlendioxid)* (hereinafter: CCS law). The law is based on Council Directive (EC) 2009/31 of 23 April 2009 on the geological storage of carbon dioxide and amending Council Directive 85/337/EEC, European Parliament and Council Directives 2000/60/EC, 2001/80/EC, 2004/35/EC, 2006/12/EC, 2008/1/EC and Regulation (EC) No 1013/2006 [2009] OJ L 140/63.

[105] For a more detailed analysis see Ludwig Krämer, 'Case studies on the implementation of Directive 2009/31/EC on the geological storage of carbon dioxide. Germany.' (2011) available at: http://blogs.ucl.ac.uk/law-environment/

Contentwise, the German CCS law represents a compromise between advocates and opponents of CCS, allowing the research and development of CCS technologies,[106] but not their commercial-scale application in Germany.[107] This is mirrored in article 1 German CCS law, which reads:

> this law serves to safeguard the permanent storage of CO2 in underground geological formations for the protection of humans and the environment, but also with responsibility for future generations. It is, tentatively, regulating the research, piloting and demonstration of technologies for the permanent storage of CO_2 in underground geological formations.[108]

This is not an accurate transposition of article 1 CCS Directive, which says that the CCS Directive establishes a legal framework for the environmentally safe geological storage of carbon dioxide to contribute to the fight against climate change. This includes all stages of CCS and all scales, including commercial. The German CCS law, however, emphasizes its tentative character and does not apply to commercial CCS activities, but only to research, piloting and demonstration of CCS. The European Commission has the discretion to initiate proceedings against Germany for not having transposed the full scope of the CCS Directive and the matter may eventually have to be decided by the ECJ.[109]

A more in-depth assessment of the German CCS law is omitted here.[110] Instead, subsequent explanations focus on three concrete measures of the CCS law that could be transplanted to shale gas regulation to tackle

---

files/2012/11/Ludwig-Kraemer-CCLP-EU-Case-Studies-Germany-2011.pdf [accessed 25 November 2016] 13 (hereinafter: Krämer CCS). Global CCS Institute '6. German CCS legislation' available at http://hub.globalccsinstitute.com/publications/dedicated-ccs-legislation-current-and-proposed/german-ccs-legislation [accessed 28 November 2016] (hereinafter: Global CCS Institute Germany).

[106] Fischer 255.

[107] Krämer CCS 12/13; Global CCS Institute Germany.

[108] Article 1 German CCS law.

[109] For this position, see Global CCS Institute Germany. A more cautious position has been taken by Krämer, who thinks the scope of the law is compatible with the CCS Directive, see Krämer CCS 13.

[110] Bundestag 'Bill of the German Federal Government law on the demonstration and application of technologies concerned with Carbon Capture, Transport and Storage of 9 May 2011' 'Gesetzentwurf der Bundesregierung Entwurf eines Gesetzes zur Demonstration und Anwendung von Technologien zur Abscheidung, zum Transport und zur dauerhaften Speicherung von Kohlendioxid' Bundestagsdrucksache 17/5750 page 62 available at: http://dip21.bundestag.de/dip21/btd/17/057/1705750.pdf [accessed 24 October 2016] (hereinafter: CCS law Draft and Explanatory Memorandum).

the two remaining potential threats of increased competition for land and insufficient monitoring.

6.3.2.1 Spatial planning restrictions

Under the CCS Directive, Member States have the right 'not to allow for any storage in parts or in the whole of their territory.'[111] Germany made use of this possibility to 'opt-out'. According to § 2 (5) German CCS Law, all German states (*Länder*) are entitled to determine independently whether or not they wish to embark on CCS trials under the so called 'state clause'.[112] The power of the *Länder* to impose spatial planning restrictions under the German CCS law is similar to the powers that have been transferred to the individual countries of the United Kingdom, discussed above.[113]

In addition to this 'state clause', which was only brought about by the German CCS law, German planning law includes an interesting possibility for individual municipalities to have a say in energy projects. The German Federal Spatial Planning Law (*Raumordnungsgesetz*) provides municipalities with the option to work together and elaborate a regional plan for their respective territories.[114] In this regional plan, municipalities may dedicate areas for certain mining activities, but may also abstain from such a dedication.[115] There are areas in Germany where shale gas extraction would compete with other land uses like agriculture, water management, forestry, human settlements and recreation.[116] The overlap of shale gas extraction with these other uses and users is called 'spatial resistance' (*Raumwiderstand*).[117] A scientific study showed that those German areas with 'very high' and 'high' spatial resistance are generally unsuitable for shale gas extraction.[118] The opportunity for municipalities to influence the siting of shale gas projects via planning law could help to ease the land-use pressure.

The opportunity for local councils to have a say on whether they would

---

[111] Article 4 (1) and preamble 19 CCS Directive.

[112] Fischer 272.

[113] Although it has to be emphasized that this approach is very new in Germany. Until now, technologies were applied within the whole German territory and no *Land* had the power to refuse or veto, for example, the construction of a nuclear power plant or another infrastructure project, see Krämer CCS 16.

[114] According to § 8 (5) No 2 (b) in conjunction with § 8 (4) German Federal Spatial Planning Law (*Raumordnungsgesetz*).

[115] § 8 (4) and (5) Federal Spatial Planning Law.

[116] SRU Faulstich 33; Meiners NRW 3. See also Chapter 1 above.

[117] Meiners NRW 9.

[118] Meiners NRW 10.

like to see CCS and/or shale gas extraction on their respective territories would enable tailor-made local solutions. Such tailor-made solutions have been advocated by the European Commission for the appraisal of new technologies and their regional application.[119] An adequate regional balance of environmental protection interests with energy security demands in shale gas cases could, thus, be struck.

6.3.2.2 Quantitative restrictions

The issue of, potentially overbearing, land use could also be tackled by quantitative restrictions on the amount of shale gas that may be extracted. Such a cap would restrict the number of shale gas installations and, therefore, also limit interference with the landscape, as the example of CCS highlights. The German CCS law imposes a cap on the amount of carbon dioxide that may be disposed of in a single storage reservoir. This cap currently lies at a maximum of 1.3 million tons of carbon dioxide per year.[120] According to Fischer, this is equivalent to the annual CO_2 emissions of a lignite-fired power plant, producing some 186 megawatts of electricity in base-load mode (8 250 hours/year).[121] In addition, new CCS plants may only be permitted if the annual amount of CO_2 stored in the whole of Germany does not exceed 4 million tons.[122] The interplay of both stipulations ensures that only a very limited number of CCS plants can operate in Germany and this is considered to be an important precautionary measure in its own right.[123]

Quantitative restrictions allow for a thorough assessment of the, yet uncertain, environmental repercussions of CCS in practice.[124] The stipulation is flexible, as the amount of CO_2 to be stored can be altered in line with increases in knowledge and demand.[125] A similar stipulation in shale gas regulations would cap the total amount of shale gas that can be extracted each year and/or curtail the volume of shale gas to be produced at an individual installation.

---

[119] European Commission 'Communication on the Precautionary Principle' (Communication) *COM (2000) 1* paragraphs 6.3.4. and 4 (hereinafter: Communication on precaution).

[120] § 2 (2) No 2 CCS law.

[121] Fischer 276 at footnote 44.

[122] § 2 (2) No 3 CCS law.

[123] CCS law Draft and Explanatory Memorandum page 62 available at: http://dip21.bundestag.de/dip21/btd/17/057/1705750.pdf [accessed 24 October 2013].

[124] CCS law Draft and Explanatory Memorandum 62.

[125] Ibid.

6.3.2.3 Registry

The German CCS law obliges the Federal German Institute for Geology to maintain a registry of all CCS projects, proposals and associated infrastructure in Germany, as well as of sites that were closed or where responsibility was transferred to a public authority.[126] The registry should also include information on whether or not the CO_2 storage site may be used for other purposes, in particular for geo-thermal purposes, as § 6 (2) No 6 German CCS law prescribes. The registry must contain maps that show the geographic extension of the sites[127] and the permanent environmental impacts.[128]

This new, nation-wide registry is a unique development for mining activities in Germany, which hitherto have not been listed in a single, unified and nation-wide registry.[129] Law requires that the registry is kept up-to-date.[130] The German Environmental Information Act (*Umweltinformationsgesetz*) (*UIG*) applies to the registry and guarantees that information on shale gas sites is made available to the public.[131]

The installation of a registry for shale gas projects has already been discussed in Chapter 5 above in the context of public access to shale gas related information under the principle of public participation. In addition to the discussion there, a nation- or Europe-wide registry, containing all current and proposed shale gas operations would also improve the, currently insufficient, monitoring arrangements for shale gas extraction. It would facilitate ongoing regulatory oversight and systematic monitoring.[132] The registry would make it very easy for the regulator to ensure that inspections are performed within the required time-cycles, that production caps are adhered to, etc.

As new procedures to extract shale gas are evolving quickly, constant monitoring of all shale gas activities is paramount to gain systematic insights into potential threats[133] and potential long-term environmental changes.[134] The inclusion of a requirement to install a unified shale gas

---

126 § 6 CCS law; Krämer 18.
127 § 6 (2) No 1 German CCS law.
128 § 6 (2) No 5 German CCS law.
129 Meiners et al. Bund B 105; Krämer 29.
130 § 6 (3) CCS law.
131 § 3 (1) German Environmental Information Act (*Umweltinformationsgesetz*).
132 SRU Faulstich 45.
133 Meiners et al. Bund A 59 and C 87; Meiners NRW 53/54 and 64.
134 SRU Faulstich 27, 31 and 38; Meiners NRW 64.

registry would help to tackle the last identified issue, the hitherto insufficient monitoring arrangements.[135]

6.3.3 France

As opposed to Germany, France swiftly moved to align its legal framework with the CCS Directive. As a result, the country finished national transposition of the CCS Directive in 2011, within the time-frame provided for by the legislator.[136] The transposition was accepted by the European Commission in January 2012.[137]

Similar to the UK and Germany, the French chose not to transpose the CCS Directive into one, stand-alone CCS law. Instead, the Directive was implemented by amending a multitude of pre-existing national laws. The two laws that have been most affected by CCS-related changes are the Environmental Code and the Mining Code.[138]

The new legal framework is twofold: it addresses the identification of adequate storage sites and the operation of such sites separately.[139] The identification of adequate storage sites, is mainly governed by articles L229-27 et sqq. Environmental Code.[140] It is subject to the same rules that apply to mining sites pursuant to the French Mining Code, which means that the search for a storage site requires an exploration permit.[141]

The actual operation of carbon dioxide storage sites is subjected to an authorization procedure and specific conditions under articles L512-1 and

---

[135] Although the comprehensive monitoring requirements of the CCS Directive triggered criticism, the first fully permitted CCS plant in Europe, the ROAD project, showed that monitoring requirements were initially considered rather stringent, but when the monitoring programme was planned it was found that they were easier to meet than initially thought, see Dixon et al. 445.

[136] Shogenova et al. 2011 at 7729.

[137] Ibid. Despite this activity, CCS is, as of 2016, not yet deployed in France at a commercial scale. There are plans for a commercial scale demonstration project in northern France, which is expected to store carbon dioxide in onshore saline formations, but this plant is not yet operative, see Thierry Lauriol 'Energy Law in France' in Martha M Roggenkamp et al. (eds) '*Energy Law in Europe*' 3rd edition (Oxford University Press, Oxford 2016) paragraph 7.303 (hereinafter: Lauriol).

[138] Lauriol paragraph 7.302; Global CCS Institute '*5. French CCS legislation*' available at http://hub.globalccsinstitute.com/publications/dedicated-ccs-legislation-current-and-proposed/french-ccs-legislation [accessed 28November 2016] (hereinafter:Global CCS Institute France). Currently a major revision of the French Mining Code is on the way.

[139] Lauriol paragraph 7.304.

[140] Ibid.; Global CCS Institute France.

[141] Lauriol paragraph 7.304.

L229-37 Environmental Code. Article 229-37 Environmental Code, in particular, regulates particular issues like the composition of the carbon dioxide stream.[142] Article L229-38 Environmental Code requires the applicant to demonstrate the financial capability to shoulder all obligations following on from the storage permit.[143]

Overall, the French transposition does not substantially deviate from the licensing provisions of the CCS Directive.[144] According to the new articles L229-27 and L229-32 Environmental Code, the CCS regulations apply both onshore and offshore and the explicit purpose of the application of CCS in France is to combat global warming.[145] Nevertheless, there are some French peculiarities that arose during the transposition process which might be of interest when considering analogies for shale gas regulation.

6.3.3.1 Priorities in land use

France operates a system under which CCS use has equal priority compared to other land-uses.[146] CCS should not add additional pressure to the system, but may not be dismissed as inferior when potential conflicts with other uses arise.[147] This system is based on the CCS Directive and in some respect goes beyond its prerequisites.[148]

Regarding the identification of a suitable underground formation, article L229-30 Environmental Code clarifies that such works require a CCS licence which is exclusive in nature. Should the identified underground formations already be covered by mining authorizations for other purposes, CCS exploration can only be undertaken with the consent of the holder of these authorizations. Where the existing authorization-holder

---

142 This provision is discussed in more depth immediately below.

143 Lauriol paragraph 7.305.

144 Similar assessment made by Global CCS Institute France.

145 According to articles L229-28 and L229-32 Environmental Code. The CCS Directive talks about climate change, see article 1 (1) CCS Directive.

146 Shogenova et al. 2011 at 7728.

147 Global CCS Institute France; Roggenkamp CCS 207.

148 Competition for the use of land onshore is not an issue confined to the surface area; the allocation of the subsoil for certain uses also becomes challenging due to the multitude of possible uses of particular spots, see Haszeldine 11. The European Commission, when drafting the CCS Directive, appears to have been aware of the fact that CCS would be a new use of the underground that might compete with other user(s) of the subsoil. Articles 5 (4) and 6 (1) CCS Directive state that Member States should ensure, for the period of validity of the exploration and storage permits, that no other uses are allowed at a similar location under- and overground, see Preamble 23, articles 5 (4) and 6 (1) and (3) CCS Directive.

does not agree to CCS exploration, the dispute has to be submitted for arbitration to the ministry that is responsible for mines.[149]

Article L229-37 Environmental Code covers land-use conflicts during or prior to the actual commencement of storage of carbon dioxide. According to article L229-37 Environmental Code, this activity requires a permit and such a storage permit can only be issued if the applicant proves, inter alia, that the intended site is permanently unsuitable for other uses.[150]

The most common conflicts of interests are with hydrocarbon exploration and production, natural gas storage, geothermal resources and drinking water production.[151] Article L229-37 focusses in particular on the production of drinking water and stipulates that carbon dioxide storage cannot take place in areas where water is being produced for agricultural as well as domestic uses.[152]

Shale gas extraction could be regulated in a similar vein to ensure that there is no, or at least mitigated, competition between shale gas extraction and other land uses. This could enable a balance between environmental protection demands and energy security requirements. Applicable national regulations could include a provision that introduces a 'barrier' to shale gas projects: when there are competing, established land uses these take priority over shale gas extraction.

This should, for instance, be the case for drinking water production.[153] This is already stipulated in the Water Acts of some Member States. A clear regulation of priorities is particularly paramount for shale gas exploration. Here, considerably bigger parts of the ground have to be worked, as compared to conventional hydrocarbon exploration[154] and existing land-users are more likely to be affected.

6.3.3.2 No exemptions from authorization procedure

In France an authorization for CCS projects has to be obtained, regardless of the size of the project.[155] The European CCS Directive is more lenient in that regard and does not apply to projects with a total intended storage amount of less than 100 kilotons carbon dioxide storage.[156] This stringent measure was borne out of the French feeling that CCS could be a technol-

149 Ibid.
150 Lauriol paragraph 7.305.
151 Haszeldine 11; Shogenova et al. 2011 at 7728.
152 Lauriol paragraph 7.305.
153 For the example of Germany, see Chapter 3 above.
154 See Chapter 1 above.
155 Article L229-37 Environmental Code.
156 Article 2 (2) and preamble 18 CCS Directive.

ogy beset with potential threats that should be watched closely.[157] Similar concerns have been put forward with regard to shale gas extraction, which makes an analogy with this French implementation of the CCS Directive interesting to balance environmental protection concerns with energy security interests in shale gas cases.

The drafters of the CCS Directive felt that storages with a total intended storage amount of less than 100 kilotons carbon dioxide were simply too small to cause significant environmental concern.[158] This sentiment was supported by the industry which suggested that such an exemption could stimulate CCS projects, albeit failing to back this claim by evidence.[159]

In contrast to the CCS Directive, article L229-37 Environmental Code is very specific in prescribing that

> the exploitation of sites for geological storage of carbon dioxide, *including those with a total storage capacity envisaged at less than 100 kilotons* (. . .) is subject to the requirement of an authorization (. . .).[160]

This applies to the operation of all carbon dioxide storage facilities, regardless of whether they are utilized for research or for commercial purposes.[161] This stipulation is a clear diversion from the text of the EU Directive. Article 193 TFEU establishes explicitly that Member States have the right to adopt *more* stringent environmental measures than those adopted under EU legislation.[162] The extension of the authorization requirement to minor storage facilities is a more stringent measure than the one envisaged by the CCS Directive.

National shale gas regulation could prescribe a regime that is equally strict by demanding a full authorization even for sites where only research into shale gas extraction shall be conducted. Such a measure would help to alleviate the concerns about insufficient monitoring, since national authorizations of EU Member States have to include provisions on

---

157 Global CCS Institute France.

158 Triple E 116–118.

159 Triple E 118.

160 Emphasis added. The original French text has been translated by the author and is: '*L'exploitation de sites de stockage géologique de dioxyde de carbone, y compris ceux d'une capacité de stockage totale envisagée inférieure à 100 kilotonnes entrepris à des fins de recherche et développement ou d'expérimentation de nouveaux produits et procédés, est soumise à l'obtention d'une autorisation délivrée en application de l'article* L. 512-1 *et des dispositions particulières prévues par la présente section.*'

161 Ibid.

162 Jans/Vedder 69. See the discussion of that point above in Chapter 2.

monitoring.[163] In this way it would ensure that all shale gas projects, big or small, are subject to regulatory oversight.

6.4 CONCLUSION

Having tackled the last remaining issues, it is time to reflect upon the insights gained. Viewed together, the last three chapters demonstrated how the trias of (quasi-) constitutional objectives, environmental law principles and analogies from concrete rules (CCS) can be used as a tool to bring about concrete shale gas regulation.

Such regulation has to *strive for* an 'optimal' reconciliation of the two objectives at play, environmental protection and energy security. A balance between the two can only be struck if four (five together with earth tremors) most salient potential threats of European shale gas extraction are being addressed.

The three potential threats of groundwater contamination/issues with well integrity, irresponsible disposal of 'flow-back' and greenhouse gas emissions can be abated by measures discussed in Chapter 5. However, these measures fail to address two remaining potential threats: the potential increase in competition for land-use in densely populated European countries,[164] and insufficient monitoring requirements.

These issues are not unique to shale gas extraction. In fact, they have recently had to be addressed in the context of a new, emerging energy technology, CCS. CCS legislation in the UK, Germany and France aims to strike a balance between environmental protection needs and energy security interests by implementing a set of precautionary measures. Albeit having a different emphasis in their respective national CCS legislation, all assessed legislative frameworks address the two potential threats and tackle them by innovative solutions like incentives to move offshore, unified registries, quantitative restrictions as well as allowing states or municipalities to establish spatial priorities. As the current chapter has shown, these measures can be applied, *mutatis mutandis,* to the regulation

---

[163] This is due to articles 1 (1) and 6 (4) Council Directive (EC) 94/22 of 30 May 1994 on the conditions for granting and using authorizations for the prospection, exploration and production of hydrocarbons OJ L 164/3 (Hydrocarbons Licensing Directive). In its article 1 (1) it clarifies that authorization and monitoring go hand in hand and article 6 (4) asks Member States to ensure sufficient monitoring of entities under an authorization, stressing however that such monitoring shall not go beyond what is strictly necessary.

[164] Compared to the USA.

of shale gas extraction and provide tailor-made solutions for individual regions.

By supplementing the measures that have been deduced from the application of law principles to shale gas extraction with the measures described in the current chapter, a comprehensive set of means to regulate shale gas extraction evolves. This comprehensive list of measures allows conciliation of environmental protection interests with energy security needs in the case of shale gas regulation. This conciliation is the result of the application of the methodological trias, put forward by this book.

PART III

Conclusions

7. Conclusions

7.1 MAIN RESULT

This book highlighted two things. First, it showed why cautious, but permissive regulation is the legally soundest way to administer shale gas extraction in the European Union. Second, it used shale gas as a case-study to demonstrate how a new methodology for energy regulation, the trias of (quasi-) constitutional objectives, law principles and rules, is leading to a concrete regulatory framework.

The lessons to be learnt are twofold and transcend the context of shale gas regulation. The first lesson concerns the role of environmental protection and energy security for energy regulation in the European Union. In cases where both ideas are competing with each other, the interests need to be reconciled by practical concordance to create a legally sound framework. This interplay is a crucial, pre-empting factor for the shape of regulatory frameworks on emerging energy technologies.

The second lesson is the trias of (quasi-) constitutional objectives, law principles and rules that is thought of as a methodological tool for decision-makers to provide orientation and facilitate the design of a legal framework for new energy technologies. Accordingly, the conclusions of this book stand on two legs: the first concerns the interplay of environmental protection and energy security in shale gas regulation and the second leg consists of the new trias-methodology.

7.2 FIRST LEG OF CONCLUSIONS

The point of departure for this book was the fact that shale gas extraction is not a 'new' technology in its own right because it is brought about by a combination of two pre-existing technologies, directional/horizontal drilling and hydraulic fracturing. As this combination of technologies opens up previously inaccessible areas for gas extraction, the number and scope of potential environmental threats also increases, compared to conventional gas extraction. However, at the same time there are potential benefits, primarily the possibility of boosting Europe's energy security with the help of domestically produced shale gas.

Four (to five) most salient potential environmental threats of shale gas extraction may be contrasted with four main potential benefits for Europe's energy security. It is crucial to note that neither the four potential energy security benefits nor the four potential environmental threats have yet been established by scientists beyond reasonable doubt. All threats and benefits are still shrouded, to a greater or lesser extent, in scientific uncertainty concerning their scale, severity and even their very existence. The scientific argument is vivid and numerous studies with inconclusive results have been conducted and basic research is on-going. That is why it is apt to point out that these are *potential* threats and benefits.

The grouping of potential threats and benefits into those related to environmental protection and those related to energy security requires a conceptual look at the interplay of these two concepts. Although both concepts interact strongly in shale gas regulation, each concept exists in its own right. Despite a recent trend in legal literature to view environmental protection as part of the definition of energy security, this book takes the position that the two concepts should be distinguished. An inclusion of environmental considerations in the very definition of energy security would complicate the legal conversation and is not necessary. As this work showed, each concept anyway constrains the scope of the other, as they represent different poles in the energy 'trilemma'.

The concepts of environmental protection and energy security are not simply matters of societal or political interest, as frequently discussed in the literature.[1] In the past it has been overlooked that both concepts have actual legal leverage. Environmental protection and energy security are enshrined in European primary law and in Member State constitutions in the form of (quasi-) constitutional objectives. Constitutional objectives are very abstract norms and their purpose is to guide the creation of specific, in the current case shale gas specific, regulation. Thus, environmental protection and energy security are steering political *and* legal decision making on shale gas extraction at EU and Member State level.

Member States have to pay attention to the existing secondary EU law framework for the environmental aspects of shale gas extraction. But the regulation of the process as such is largely left to the Member States. This

---

1 See for example Raphael J Heffron, Darren McCauley and Benjamin K Sovacool 'Resolving society's energy trilemma through the Energy Justice Metric' (2015) 87 *Energy Policy* 168/169 with further references. These authors put energy security firmly into the group of political considerations.

led to the emergence of two groups of EU Member States, those in favour of and those opposed to shale gas extraction. It seems as if the latter position has been adopted by a majority of Member States. Member States that are opposed to shale gas extraction did not, however, adopt uniform prohibitions. Measures range from 'bans by law' to 'moratoria by law' to 'political moratoria'.

The central question is how much leverage a Member State wishes to apportion to the constitutional objectives of environmental protection and energy security. Theoretically, both constitutional objectives are equal in their abstract rank, but in practice they compete with each other in the case of shale gas extraction.

The strongest environmental protection can be achieved by a moratorium or a ban on shale gas extraction, but this solution would disregard the energy security objective. Under a moratorium or a ban, the extraction of shale gas would be illegal and domestic shale gas reserves could not be developed to boost energy security. Vice versa, the best result with a view to energy security may be achieved if shale gas extraction can proceed without any environmental safeguard measures attached. However, this solution would disregard the environmental protection objective.

As a solution, this book has put forward the 'meta'-principles of unity of the constitution and practical concordance. Reconciliation of the competing objectives of environmental protection and energy security can be achieved in shale gas cases by the implementation of both objectives into concrete shale gas regulation with the help of these 'meta'-principles.

However, the obligation to strive for 'optimization' of the interplay between objectives under the principles of unity of the constitution and practical concordance finds its ultimate limitation in the principle of proportionality. The legislator has discretion and is free to adopt whichever shale gas regulation he prefers. The approach to shale gas regulation adopted in any given jurisdiction is, ultimately, a political decision. 'Optimization' might only occur to the extent that this discretion is not curtailed in a disproportionate manner. Allowing the legislator to reconcile constitutional objectives only at the highest point of realization of both competing objectives would, in effect, remove that discretion. This would infringe the separation of powers under the rule of law.

Meaningful shale gas regulation should *strive for* an 'optimal solution', wherever this is practical, as the term *practical* concordance suggests. However, the legislator is not obliged to adopt that particular solution. The prohibitive as well as the cautious, but permissive approach to regulation are both open to the legislator. As this is a discretionary decision, the adoption of either solution is theoretically possible under current law.

The final conclusion of this book with respect to the first leg is that a cautious, but permissive approach to shale gas regulation is legally sounder than prohibitive regulation. It would translate the competing objectives environmental protection and energy security into one regulatory framework for shale gas extraction that could cater for both interests at the same time. Shale gas could be extracted, but only at an environmentally sustainable scale.

Cautious, but permissive shale gas regulation would even constitute the best option for a shale gas-sceptic legislator. It may encompass such a broad variety of measures that its effective impact can be almost similar to that of a moratorium, while the danger that it would be struck down by a court for non-compliance with relevant (quasi-) constitutional objectives is minimal.

7.3 SECOND LEG OF CONCLUSIONS

The second leg of conclusions pertains to the methodological trias put forward by this book. Constitutional objectives like environmental protection and energy security should be translated into legal principles in decision-making processes and these principles, in turn, should be translated into concrete rules. All three categories of norms, objectives, principles and rules, represent gradations of one legal thought and are, hence, interlinked with each other.

It should be stressed that the trias may be of general use as a methodology for the regulation of new energy technologies. Since energy law is a very young discipline that has to keep up with rapid technological developments,[2] it is important to have a regulatory template for the elaboration of new rules for new energy technologies. The methodology put forward here could be applied *mutatis mutandis* to the development of regulatory regimes on energy technologies that are today unheard of, but which might be developed in the future.

The obvious field of application is that of innovations in the extractive industry. This book has been written against the backdrop of depleting conventional hydrocarbon reservoirs. Judging by the remarkable proliferation of innovative technologies to extract hydrocarbons that the world witnessed during the last 20 years, it is likely that even more of these technologies are going to emerge in the near future. Many of those future

---

[2] Adrian J Bradbrook 'Energy Law as an Academic Discipline' (1996) Vol 14 No 2 *Journal of Energy & Natural Resources Law* 193–217.

technologies will bring about uncertainties, concerning potential threats and benefits that are similar to those observed in the context of shale gas extraction. Results from this study might hence help to guide regulatory decision making on future energy regulation.

However, the potential use of the trias is not confined to the extractive industry or renewable energy production, but it might also be put to good use in the regulation of non-producing branches of the energy sector, such as smart grids,[3] an offshore grid in the North Sea,[4] electricity storage[5] and energy interconnectors,[6] to name but a few.

Besides multiple advantages, the trias also comes with some disadvantages attached to it. Its main setback is the fact that it will be used for the development of new regulation. It applies *prior* to the adoption of particular regulations. In doing so it focuses on the particular perspective of constitutional theory on state objectives.[7] A comprehensive assessment, however, also needs to encompass a check of the proposed framework for compliance with fundamental rights and that is not delivered by the trias.

This omission, however, is justified by the point in time when the trias applies and the state of research. Checking compliance of new regulations with fundamental rights has to be done *post* adoption of the particular regulation. Inter alia, the exact wording of individual paragraphs as well as their position within a legal framework, but also the guise and content of fundamental rights in a given jurisdiction, are decisive for such an assessment.

Sticking with the example of shale gas extraction, the compatibility of

---

[3] More on the legal dimension of smart grids can be found at Anita Ronne 'Smart Grids and Intelligent Energy Systems: A European Perspective' in Martha M Roggenkamp et al. (eds) '*Energy Networks and the Law*' (Oxford University Press, Oxford 2012) 141–4 and 156–9; Lea Diestelmeier and Dirk Kuiken 'Sustaining Universal Service Conditions in Smart Electricity Systems' (2016) Vol 18 No 3 *Network Industries Quarterly* 7/8.

[4] Hannah Katharina Müller '*A Legal Framework for a Transnational Offshore Grid in the North Sea*' (Intersentia, Cambridge 2015).

[5] Manuel Goetz et al. 'Renewable Power-to-Gas: A technological and economic review' (2016) Vol 85 *Renewable Energy* 1371–90.

[6] A good introduction on the legal aspects of electricity and gas interconnectors may be found at Hans Vedder et al. 'EU Energy Law' in Martha M Roggenkamp et al. (eds) '*Energy Law in Europe*' 3rd edition (Oxford University Press, Oxford 2016) paragraphs 4.263–4.292.

[7] For an instrumental introduction to the topic of constitutional rights theory in respect of constitutional objectives: Ronald Dworkin '*Taking Rights Seriously*' (Harvard University Press, Cambridge 1977) 24 (hereinafter: Dworkin); Robert Alexy '*A Theory of Constitutional Rights*' (Oxford University Press, 2004) 44–7 (hereinafter: Alexy); Hesse paragraph 72.

shale gas regulation with fundamental rights has been discussed in various other places, inter alia, by scientific studies,[8] scholarly writings[9] and a recent French court ruling.[10] The author feels that the importance of state objectives for energy regulation, by contrast, received too little attention in the past.

Despite these challenges there are some further upsides and advantages of the trias. To start with, it is a rather clear and stringent approach to law-making which, nonetheless, encompasses a considerable amount of flexibility. As such it might be used by various groups and people. The most obvious addressees are decision-makers involved in the regulation of new energy technologies. However, the trias methodology can also be used by lawyers, NGOs and the industry to check whether or not the legislator put into place a legal framework that is coherent and takes all relevant interests into account.

Furthermore, the proposed trias sits well with at least two parts of the energy 'trilemma'.[11] By way of a reminder: energy policy-makers in all countries aim to achieve the same three objectives which are called the energy 'trilemma':[12] maximizing supply security, minimizing environmental impacts and providing equitable access to energy.[13]

---

[8] For instance SRU Faulstich 39/40.

[9] Edyta Materka 'End of Transition? Expropriation, Resource Nationalism, Fuzzy Research, and Corruption of Environmental Institutions in the Making of the Shale Gas Revolution in Northern Poland' (2011) Vol 19 Issue 3 *Journal of Contemporary Central and Eastern Europe* 599 et sqq.

[10] French Constitutional Court 'Decision no. 2013-346 QPC of 11 October 2013' (n° 2013-346 QPC *Société Schuepbach Energy LLC [Interdiction de la fracturation hydraulique pour l'exploration et l'exploitation des hydrocarbures – Abrogation des permis de recherches]*) available at: http://www.conseil-constitutionnel.fr/conseil-constitutionnel/english/priority-preliminary-rulings-on-the-issue-of-constitutionality-qpc-/sample-of-decisions-qpc/2013/decision-no-2013-346-qpc-of-11-october-2013.138596.html [accessed 8 Nov 2013] (hereinafter: French Constitutional Court).

[11] World Energy Council 'World Energy Trilemma' available at: https://www.worldenergy.org/work-programme/strategic-insight/assessment-of-energy-climate-change-policy/ [accessed 22 September 2016] (hereinafter: World Energy Council); Cristelle Maurin and Vlado Vivoda 'Shale Gas and the Energy Policy "Trilemma"' in Tina Hunter (ed.) *'Handbook of Shale Gas Law and Policy'* (Intersentia, Cambridge 2016) 369–81 (hereinafter: Maurin/Vivoda); Raphael J Heffron, '*Energy Law: an Introduction*' (Springer International, Cham 2015) 3–5; Raphael J Heffron, Darren McCauley and Benjamin K Sovacool 'Resolving society's energy trilemma through the Energy Justice Metric' (2015) 87 *Energy Policy* 168.

[12] Maurin/Vivoda 369.

[13] World Energy Council.

The relationship between energy security and environmental protection in particular is being cared for under the proposed trias. At the first level of the trias, both energy and environmental interests can be represented in an even manner. Via their translation into law principles at the second level of the trias (in the shale gas context predominantly environmental law principles), energy and environmental interests can both impact upon emerging legislation.

With regard to the third aspect of the 'trilemma', equitable access to energy, this relationship is not so clear. However, it could be argued that the competition between energy security and environmental protection will often be a predominant aspect of the regulation of all new energy technologies. As the trias is aiming at reconciliation of both interests it will frequently lead to cautious, but permissive regulation. By allowing a certain new energy technology to advance, synergies that drive down costs will appear. Lower costs, in turn, can result in more equitable access to energy.

To sum up, the application of the trias to the regulation of new energy technologies delivers a set of concrete measures for energy regulation. Every single one of these concrete measures should reflect the balance between the overarching constitutional objectives at play. Ideally, they result in a coherent regulatory framework, when put together. A German proverb, '*Das Große spiegelt sich im Kleinen*' (the whole is a reflection of its parts), best sums up this magic of the trias.

Bibliography

OFFICIAL DOCUMENTS

Agency for Toxic Substances and Disease Registry, 'Health Consultation Chesapeake ATGAS 2HWell Site Leroy Hill Road, Leroy Township, Bradford County' Study on behest the US Environment Protection Agency available at: http://www.atsdr.cdc.gov/hac/pha/ChesapeakeATGASWell Site ChesapeakeATGASWellSiteHC110411Final.pdf [accessed 26 April 2012]

AMEC Ltd., 'Technical Support for Assessing the Need for a Risk Management Framework for Unconventional Gas Extraction' (2014) available at: http://ec.europa.eu/environment/integration/energy/uff_studies_en.htm [accessed 4 September 2014]

Andruleit H et al., 'Kurzstudie- Reserven, Ressourcen und Verfügbarkeit von Energierohstoffen 2011' (Deutsche Rohstoffagentur (DERA) in der Bundesanstalt für Geowissenschaften und Rohstoffe (BGR), Hannover November 2011) available at: http://www.bgr.bund.de/DE/Themen/Energie/Downloads/Energiestudie-Kurzf-2011.pdf?__blob=publicationFile&v=3 [accessed 31 October 2012]

Asia Pacific Energy Research Centre, 'A Quest for Energy Security in the 21st Century' (Asia Pacific Energy Research Centre, Tokyo 2007)

Bartel H, 'Wasserwirtschaft in Deutschland Teil 1: Grundlagen' (Umweltbundesamt, Dessau-Roßlau 2010)

Böhm M, 'Rechtsgutachten Voraussetzungen einer bergrechtlichen Erlaubnis nach § 7 BBergG unter besonderer Berücksichtigung der Versagungsgründe des § 11 Nr. 10 BbergG' available at: http://frackingfrei.files.wordpress.com/2013/06/2013-03-20-rechtsgutachten-bc3b6hm-zu-fracking.pdf [accessed 9 January 2015]

Boxtel Council, 'Motie De raad van de gemeente Boxtel in openbare vergadering bijeen op 5 maart 2012' at http://www.breda.nl/system/files/120305motieschaliegasvrijegemeenteboxtel.pdf [accessed 15 June 2012]

Boyer E W et al., 'The Impact of Marcellus Gas Drilling on Rural Drinking Water Supplies' (Center for Rural Pennsylvania, Harrisburg 2012) available at: http://www.rural.palegislature.us/documents/reports/Marcellus_and_drinking_water_2012.pdf [accessed 5 July 2013]

Broderick J and Anderson K, 'Has US Shale Gas Reduced CO_2 Emissions? Examining recent changes in emissions from the US power sector and traded fossil fuels' (Tyndall Centre at the University of Manchester, October 2012) available at: http://www.tyndall.manchester.ac.uk/public/Broderick_Anderson_2012_Impact_of_Shale_Gas_on_US_Energy_Emissions.pdf [accessed 20 October 2012]

Broomfield M, 'Support to the identification of potential risks for the environment and human health arising from hydrocarbons operations involving hydraulic fracturing in Europe' (AEA Technology, Didcot 2012) available at: http://ec.europa.eu/environment/integration/energy/uff_studies_en.htm [accessed 4 September 2014]

Bulgarian National Assembly, 'Ad-hoc Committee to study and analyse and discuss best practices and legislative decisions concerning the regulation of activities in exploration and mining and effects on the environment 'MINUTES № 1' of 11/April/2012' page 3 (Временна комисия за проучване, анализ и обсъждане на добри практики и законодателни решения във връзка с регулирането на дейности по проучване и добив на подземни богатства при опазване на околната среда 'ПРОТОКОЛ № 1' of 11 April 2012, 3 available at: http://www.parliament.bg/bg/parliamentarycommittees/members/1464/steno/ID/2430 [accessed 29 April 2014]

Bulgarian National Assembly, 'Assessment and Discussion Committee for the Environment and Water, of the Legislative Proposal' (КОМИСИЯ ПО ОКОЛНАТА СРЕДА И ВОДИТЕ Разглеждане и обсъждане на законопроект) of 1 November 2011 available at: http://www.parliament.bg/bg/parliamentarycommittees/members/234/steno/ID/2267 [accessed 20 April 2014]

Bulgarian National Assembly, 'Information on the legal project to outlaw exploration for and production of shale gas' (Информация за законопроектИме на законопроектаЗаконопроект за забрана на проучване и добив на шистов газ) available at: http://www.parliament.bg/bg/bills/ID/1756 [accessed 30 April 2014]

Bulgarian National Assembly, 'Legal project to outlaw exploration and production of shale gas' (Проект ЗАКОН ЗА ЗАБРАНА НА ПРОУЧВАНЕ И ДОБИВ НА ШИСТОВ ГАЗ) available at: http://parliament.bg/bills/41/154-01106.pdf [accessed 30 April 2014]

Bulgarian National Assembly, 'Stenographic report of 304th Plenary Sittings, Sofia, Wednesday, 18 January, 2012' (Стенограми от пленарни заседания ТРИСТА 7 ЧЕТВЪРТО ЗАСЕДАНИЕ София, сряда, 18 януари 2012 г.) available at: http://www.parliament.bg/bg/plenaryst/ns/7/ID/2688 [accessed 29 April 2014]

Bulgarian National Assembly Committee for Economic Policy, Energy and Tourism, 'Report on the legislative proposal to outlaw exploration for and production of shale gas № 154-01-106' (КОМИСИЯ ПО ИКОНОМИЧЕСКАТА ПОЛИТИКА, ЕНЕРГЕТИКА И ТУРИЗЪМ Д О К Л А Д по законопроект за забрана на проучване и добив на шистов газ № 154-01-106) of 16 January 2012 available at: http://www.parliament.bg/bg/parliamentarycommittees/members/224/reports/ID/3181 [accessed 30 April 2014]

Bulgarian National Assembly Committee for Economic Policy, Energy and Tourism, 'Report on a draft resolution on a moratorium on exploration and production of shale gas in Bulgaria, № 154-02-93 and a draft decision to ban mining shale gas and the method of hydraulic fracturing and similar methods at its exploration and production of oil and natural gas on the territory of the Republic of Bulgaria, № 154-02-101' (КОМИСИЯ ПО ИКОНОМИЧЕСКАТА ПОЛИТИКА, ЕНЕРГЕТИКА И ТУРИЗЪМ Д О К Л А Д относно проект за решение за налагане на мораториум върху проучванията и добива на шистов газ на територията на Република България, № 154-02-93 и проект за решение за забрана на добива на шистов газ и прилагането на метода на хидравлично разбиване и сходни 7 него методи при проучване и добив на нефт и природен газ на територията на Република България, № 154-02-101) of 16 January 2012 available at: http://parliament.bg/bg/parliamentarycommittees/members/224/reports/ID/3180 [accessed 1 May 2014]

Bundesanstalt für Geowissenschaften und Rohstoffe (BGR), 'Abschätzung des Erdgaspotenzials aus dichten Tongesteinen (Schiefergas) in Deutschland' (Bundesanstalt für Geowissenschaften und Rohstoffe, Hannover 2012)

Bundesanstalt für Geowissenschaften und Rohstoffe (BGR), 'Schieferöl und Schiefergas in Deutschland Potenziale und Umweltaspekte' (Bundesanstalt für Geowissenschaften und Rohstoffe, Hannover 2016) 13/14

Bundesanstalt für Geowissenschaften und Rohstoffe (BGR) 'Wissenswertes über Schieferöl und Schiefergas' available at: https://www.bgr.bund.de/DE/Themen/Energie/Projekte/laufend/NIKO/FAQ/faq_inhalt.html [accessed 30 June 2016]

Bundesrat, 'Draft Ordinance on alteration of the ordinance on Environmental Impact Assessment and mining activities' (Entwurf einer Verordnung zur Änderung der Verordnung über die Umweltverträglichkeitsprüfung bergbaulicher Vorhaben) available at: http://www.bundesrat.de/SharedDocs/drucksachen/2012/0701-0800/747-12(B).pdf?__blob=publicationFile&v=1 [accessed 16 April 2014]

Bundesrat, 'Drucksache 144/15 German Ministry of the Economy and Energy Ordinance on the introduction of Environmental Impact Assessments and on mining requirements in deployment of the fracking technology and deep drills' (*Verordnung zur Einführung von UmweltverträglichkeitsprüfungenundüberbergbaulicheAnforderungenbeim Einsatz der Fracking-Technologie und Tiefbohrungen*) available at: http://www.bundesrat.de/SharedDocs/beratungsvorgaenge/2015/0101-0200/0144-15.html;jsessionid=86AAD3036B5C2CDA8A8A2B4B565717AF.2_cid349?cms_templateQueryString=Suchbegriff&cms_fromSearch=true [accessed 7 April 2016]

Bundesrat, 'Drucksache 353/1/16 (Empfehlungen der Ausschuesse 30.06.2016)' available at: http://www.bundesrat.de/SharedDocs/drucksachen/2016/0301-0400/353-1-16.pdf?__blob=publicationFile&v=1 [accessed 13 July 2016]

Bundesrat, 'Drucksache 353/16 (Beschluss des Bundesrates)' available at: http://www.bundesrat.de/SharedDocs/drucksachen/2016/0301-0400/353-16(B).pdf?__blob=publicationFile&v=1 [accessed 13 July 2016]

Bundesrat, 'Drucksache 358/16 (Beschluss des Bundesrates)' available at: http://www.bundesrat.de/SharedDocs/drucksachen/2016/0301-0400/358-16(B).pdf?__blob=publicationFile&v=1 [accessed 13 July 2016]

Bundesrat, 'Drucksache 358/16 (Empfehlungen der Ausschüße 05.07.2016)' available at: http://www.bundesrat.de/SharedDocs/drucksachen/2016/0301-0400/358-16.pdf?__blob=publicationFile&v=1 [accessed 13 July 2016]

Bundesrat, 'Drucksache 754/1/12 Committee Recommendations on the resolution of the Bundesrat on the handling of the application of fracking-technologies of 21 January 2013' (Empfehlungen der Ausschüsse zur Entschließung des Bundesrates zum Umgang mit dem Einsatz von Fracking-Technologien) Bundesratsdrucksache 754/1/12 available at: http://www.bundesrat.de/SharedDocs/beratungsvorgaenge/2012/0701-0800/0754-12.html [accessed 16 April 2014]

Bundesrat, 'Drucksache 754/12 Resolution of the Bundesrat on the handling of the application of fracking-technologies with environmentally-toxic chemicals during the exporation and production of unconventional deposits of 3rd December 2012' (Entschließung des Bundesrates zum Umgang mit dem Einsatz von Fracking-Technologien mit umwelttoxischen Chemikalien bei der Aufsuchung und Gewinnung von Erdgas aus unkonventionellen Lagerstätten) Bundesratsdrucksache 754/12 available at: http://www.bundesrat.de/SharedDocs/beratungsvorgaenge/2012/0701-0800/0754-12.html [accessed 15 April 2014]

Bundesrat, 'Plenarprotokoll 947. Sitzung 8.7.2016' available at: http://www.bundesrat.de/DE/dokumente/plenarprotokolle/plenarprotokolle-node.html [accessed 15 July 2016]

Bundesrat, 'Stenographic Report 904th session of 14th December 2013' (Stenografischer Bericht 904. Sitzung) Plenarprotokoll 904 available at: http://dipbt.bundestag.de/dip21/brp/904.pdf [accessed 16 April 2014]

Bundesrat, 'Stenographic Report 906th session of 1st February 2013' (Stenografischer Bericht 906. Sitzung) Plenarprotokoll 906 available at: http://www.bundesrat.de/SharedDocs/downloads/DE/plenarprotokolle/2013/Plenarprotokoll-906.pdf?__blob=publicationFile&v=3 [accessed 16 April 2014]

Bundestag, 'A new Mining Act for the 21st Century of 14 December 2011' (Ein neues Bergrecht für das 21. Jahrhundert) Bundestagsdrucksache 17/8133

Bundestag, 'Bill of the German government draft on a Mining Act' (Gesetzentwurf der Bundesregierung Entwurf eines Bundesberggesetzes) Bundestagsdrucksache 8/1315 available at: http://suche.bundestag.de/search_bt.do?actualPage=165&resultsSubCategoryFilter=&sort=da&queryAll=Energie&tab=all&datumBis=%20&oneCategoryOnlySearch=false&drucksachennr=false&schlagwort=false&titel=false&resultsPerPage=10&resultsCategoryFilter=&displayCategories=true&language=de&datumVon= [accessed 8 January 2015]

Bundestag, 'Drucksache 18/4713 Bill of the German Federal Government on the alteration of water and environmental protection norms with the aim of prohibiting and minimizing risks associated with the procedures of the fracking technology' (*Gesetzentwurf der Bundesregierung 'Entwurf eines Gesetzes zur Änderung wasser- und naturschutzrechtlicher Vorschriften zur Untersagung und zur Risikominimierung bei den Verfahren der Fracking-Technologie'*) available at: http://dipbt.bundestag.de/doc/btd/18/047/1804713.pdf [accessed 7 April 2015]

Bundestag, 'Drucksache 18/4714 German Government Draft Bill to expand mining liability to wellbore mining and caverns' (*Bundesregierung Entwurf eines Gesetzes zur Ausdehnung der Bergschadenshaftung auf den Bohrlochbergbau und Kavernen*) available at: http://dipbt.bundestag.de/doc/btd/18/047/1804714.pdf [accessed 7 April 2015]

Bundestag, 'Drucksache 18/4949 Report by the government according Drucksache 18/4713 Position of the Bundesrat and rebuttal of the government' available at http://dipbt.bundestag.de/doc/btd/18/049/1804949.pdf [accessed 5 July 2016]

Bundestag, 'Drucksache 18/4952 Report by the government according Drucksache 18/4714 Position of the Bundesrat and rebuttal of the government' available at http://dipbt.bundestag.de/doc/btd/18/049/1804952.pdf [accessed 5 July 2016]

Bundestag, 'Drucksache 18/8907 Recommendation and Report of the Committee for Economic Affairs and Energy concerning Drucksache

18/4714 (*Beschlussempfehlung und Bericht des Ausschusses für Wirtschaft und Energie zu Drucksache 18/4714*)' available at: http://dipbt.bundestag.de/doc/btd/18/089/1808907.pdf [accessed 5 July 2016]

Bundestag, 'Drucksache 18/8916 Recommendation and Report of the Committee for Environment, Nature Protection, Building and Reactor-safety concerning Drucksache 18/4713 (*Beschlussempfehlung und Bericht des Ausschusses für Umwelt, Naturschutz, Bau und Reaktorsicherheit zu Drucksache 18/4713*)' available at: http://dipbt.bundestag.de/doc/btd/18/089/1808916.pdf [accessed 5 July 2016]

Bundestag, 'Guidelines of the German government on environmental precaution by avoidance and reduction of pollutants of 19 September 1986' (Leitlinien der Bundesregierung zur Umweltvorsorge durch Vermeidung und stufenweise Verminderung von Schadstoffen) Bundestagsdrucksache 10/6028

Bundestag, 'Motion guidance notes on transparency and environmental soundness during unconventional gas production of 8 November 2011' (Antrag Leitlinien für Transparenz und Umweltverträglichkeit bei der Förderung von unkonventionellem Erdgas) Bundestagsdrucksache 17/7612 available at: http://dipbt.bundestag.de/dip21/btd/17/076/1707612.pdf [accessed 16 April 2014]

Bundestag, 'Motion no gas extraction at the expense of drinking water – banning fracking for gas extraction purposes of 8 June 2011' (Antrag Keine Erdgasförderung auf Kosten des Trinkwassers – Fracking bei der Erdgasförderung verbieten) Bundestagsdrucksache 17/6097 available at: http://dip21.bundestag.de/dip21/btd/17/060/1706097.pdf [accessed 17 April 2014]

Bundestag, 'Motion transparency and control of unconventional gas production in Germany of 13 April 2011' (Antrag Transparenz und Kontrolle bei der Förderung von unkonventionellem Erdgas in Deutschland) Bundestagsdrucksache 17/5573 available at: http://dip21.bundestag.de/dip21/btd/17/055/1705573.pdf [accessed 16 April 2014]

Bundestag, 'Report of the Committee for Education, Research and Technology Assessment CCS at power plants of 1 July 2008' (Bericht des Ausschusses für Bildung, Forschung und Technikfolgenabschätzung (18. Ausschuss) gemäß § 56a der Geschäftsordnung Technikfolgenabschätzung (TA) CO_2-Abscheidung und -Lagerung bei Kraftwerken) Bundestagsdrucksache 16/9896, available at: http://dip21.bundestag.de/dip21/btd/16/098/1609896.pdf [accessed 16 December 2013]

Bundestag, 'Resolution recommendation and report of the committee on the environment, environmental protection and reactor safety on motions 17/5750, 17/6264 of the German government of 6 July 2011' (Beschlussempfehlung und Bericht des Ausschusses für

Umwelt, Naturschutz und Reaktorsicherheit (16. Ausschuss) zu dem Gesetzentwurf der Bundesregierung – Drucksachen 17/5750, 17/6264) Bundestagsdrucksache 17/6507 available at: http://dip21.bundestag.de/dip21/btd/17/065/1706507.pdf [accessed 16 April 2014]

Bundestag, 'Resolution recommendation and report of the committee on the environment, environmental protection and reactor safety on a) motion of MPs Frank Schwabe, Ingrid Arndt-Brauer, Dirk Becker and others as well as the SPD- parliamentary group b) motion of MPs Oliver Krischer, Hans-Josef Fell, Bärbel Höhn and others as well as the BÜNDNIS 90/DIE GRÜNEN parliamentary group of 26 April 2012' (Beschlussempfehlung und Bericht des Ausschusses für Umwelt, Naturschutz und Reaktorsicherheit (16. Ausschuss) a) zu dem Antrag der Abgeordneten Frank Schwabe, Ingrid Arndt-Brauer, Dirk Becker, weiterer Abgeordneter und der Fraktion der SPD – Drucksache 17/7612 – b) zu dem Antrag der Abgeordneten Oliver Krischer, Hans-Josef Fell, Bärbel Höhn, weiterer Abgeordneter und der Fraktion BÜNDNIS 90/DIE GRÜNEN – Drucksache 17/5573 –) Bundestagsdrucksache 17/9450 available at: http://dipbt.bundestag.de/dip21/btd/17/094/1709450.pdf [accessed 16 April 2014]

Bundestag, 'Resolution Recommendation and Report of the committee on consumer protection, nutrition and agriculture' (Beschlussempfehlung und Bericht des Ausschusses für Verbraucherschutz, Ernährung und Landwirtschaft (10. Ausschuss)' Bundestagsdrucksache 15/3344

Bundestag, 'Resolution Recommendation and Report of the committee on the economy and technologies on the motion of MPs Johanna Voß, Dr. Barbara Höll, Eva Bulling-Schröter and others as well as the LINKE-parliamentary group of 28 March 2012' (Beschlussempfehlung und Bericht des Ausschusses für Wirtschaft und Technologie (9. Ausschuss) zu dem Antrag der Abgeordneten Johanna Voß, Dr. Barbara Höll, Eva Bulling-Schröter, weiterer Abgeordneter und der Fraktion DIE LINKE. – Drucksache 17/6097) Bundestagsdrucksache 17/9196 available at: http://dipbt.bundestag.de/dip21/btd/17/091/1709196.pdf [accessed 17 April 2014]

Bundestag, 'Speech of federal agriculture minister Ilse Aigner on the 23/April/2009' reproduced in Plenarprotokoll 16/217, 23540 available at: http://dipbt.bundestag.de/dip21/btp/16/16217.pdf [accessed 16 April 2014]

Bundestag, 'Stenographic Report 178th session of 10th May 2012' (Stenografischer Bericht 178. Sitzung) Plenarprotokoll 17/178 available at: http://dipbt.bundestag.de/dip21/btp/17/17178.pdf [accessed 17 April 2014]

Bundestag , 'Stenografischer Bericht 180. Sitzung Plenarprotokoll 18/180' at 17795 available at: http://dipbt.bundestag.de/dip21/btp/18/18180.pdf#P.17790 [accessed 30 June 2016]

Bundestag, 'Third report on emission protection' (Dritter Imissionsschutzbericht) Bundestagsdrucksache 10/1345 (Bonn 1984)

Burgoyne J H, 'Offshore Safety: Report of the Committee' (Cmnd. 7866, 1980)

Cantabrian Parliament, 'Minutes of Parliamentary Meeting on 8 April 2013' (Parlamento Cantabria 'Diario de Sediones 8 de abril de 2013') available at: http://www.parlamento-cantabria.es/sites/default/files/diarios/Punto%201%20Pleno%28%2066-A%29.pdf [accessed 28 April 2014]

Cantabrian Parliament, 'Session of the environmental committee of the Parliament of Cantabria of 21 March 2013' ('Estudio, debate y votación del Informe de la Ponencia y de las enmiendas presentadas al Proyecto de Ley de Cantabria por la que se regula la prohibición en el territorio de la Comunidad Autónoma de Cantabria de la técnica de fractura hidráulica. [8L/1000-0011]') available at: http://www.parlamento-cantabria.es/sites/default/files/diarios/Punto%20%C3%BAnico%20C%20Media%20Ambiente%20%28122-B%29.pdf [accessed 28 April 2014]

Carbon Capture and Storage Association, 'Submission to the UNFCCC: Carbon dioxide capture and storage in geological formations as clean development project activities' available at: http://unfccc.int/resource/docs/2012/smsn/ngo/154.pdf [accessed 16 February 2013]

Clingendael International Energy Programme, 'Study on Energy Supply Security and Geopolitics Final Report' (2004) available at: http://www.clingendael.nl/publications/004/200401000_ciep_study.pdf [accessed 25 April 2013]

Committee for Energy, 'Infrastructure and Rural Development of Mecklenburg-Hither Pomerania "Recommendation and Report" Drucksache 6/750 of 16/May/2012' available at: http://www.landtag-mv.de/fileadmin/media/Dokumente/Ausschuesse/Energieausschuss/Drs06-0750.pdf [accessed 16 October 2012]

Constitutional Council France, Decision 79-105 DC, Right to Strike on Radio and Television, 25 July 1979, Rec. 33 Constitutional Council France, Decision 2008-564 DC, Genetically Modified Organisms, of 19 June 2008, Rec. 313

Council of Europe, 'Pan-European Biological and Landscape Diversity Strategy of 25 October 1995' (Council of Europe Publishing, Strasbourg 1998)

Council of the Organization for Economic Cooperation and Development, 'Recommendation of the Council on Integrated Cooperation and Development, Pollution Prevention and Control' of 31 January 1991, C(90)164

Cuadrilla Resources Ltd, 'Geomechanical Study of Bowland Shale Seismicity Synthesis Report' (2011) available at: http://www.cuadrilaresources.com/wp-content/uploads/2012/02/Geomechanical-Study-of-Bowland-Shale-Seismicity_02-11-11.pdf [accessed 24 April 2012]
Cullen D, 'The Public Inquiry into the Piper Alpha Disaster' (Cm 1310, 1990)
Czech House of Representatives, 'Sněmovní tisk 902 Novela z. v souvislosti se zákazem hydraul.štěpení hornin' available at: http://www.psp.cz/sqw/historie.sqw?o=6&t=902 [accessed 23 May 2014]
Czech Senate, 'Minutes of the 23rd Senate meeting (2nd day of meeting – 14.06.2012)' (Těsnopisecká Zpráva z 23. schůze Senátu (2. den schůze – 14.06.2012)) available at: http://www.senat.cz/xqw/xervlet/pssenat/hlasovani?action=steno&O=8&IS=4817&D=14.06.2012#b12986 [accessed 23 May 2014]
Czech Senate, 'Minutes of the 25th meeting of the Senate (2nd day of meeting – 16 August 2012)' (Těsnopisecká zpráva z 25. schůze Senátu (2. den schůze – 16.08.2012)) available at: http://www.senat.cz/xqw/xervlet/pssenat/hlasovani?action=steno&O=8&IS=4860&D=16.08.2012#b13127 [accessed 23 May 2014]
Czech Senate, 'Minutes of the fourth Senate Meeting (1st day of meeting – 30 January 2013)' (Těsnopisecká zpráva z 4. schůze Senátu (1. den schůze – 30.01.2013) available at: http://www.senat.cz/xqw/xervlet/pssenat/htmlhled?action=doc&value=67073 [accessed 27 March 2014]
Czech Senate, 'Minutes of the fourth Senate meeting (2nd day of meeting – 31 January 2013)' (9. funkční období Těsnopisecká zpráva z 4. schůze Senátu (2. den schůze – 31.01.2013)) available at: http://www.senat.cz/xqw/webdav/pssenat/original/67114/56512 [accessed 27 March 2014]
Czech Senate, 'Proposal for a Senate bill by Senators Pakosty Peter, George Oberfalzer and Paul Trpák' (Návrh senátního návrhu zákona, senátorů Petra Pakosty, Jiřího Oberfalzera a Pavla Trpáka) available at: http://www.senat.cz/xqw/webdav/pssenat/original/64897/54755 [accessed 23 May 2014]
Czech Senate Committee on National Economy, Agriculture and Transport, '292nd RESOLUTION of the 29th meeting held on July 12, 2012' (292. USNESENÍ z 29. schůze konané dne 12. července 2012) available at: http://www.senat.cz/xqw/webdav/pssenat/original/65340/55088 [accessed 23 May 2014]
Czech Senate Committee on National Economy, Agriculture and Transport, '307th resolution of 31st meeting held on August 14, 2012' (307. USNESENÍ z 31. schůze konané dne 14. srpna 2012) available at: http://www.senat.cz/xqw/webdav/pssenat/original/65586/55273 [accessed 23 May 2014]

Czech Senate Committee on National Economy, Agriculture and Transport, '330th RESOLUTION of the 32nd meeting held on October 23, 2012' (330. USNESENÍ z 32. schůze konané dne 23. října 2012) available at: http://www.senat.cz/xqw/webdav/pssenat/original/66036/55644 [accessed 23 May 2014]
Czech Senate Committee on Regional Development, Public Administration and Environment, '19th RESOLUTION of the 4th meeting held on January 30, 2013' (19. USNESENÍ z 4. schůze konané dne 30. ledna 2013) available at: http://www.senat.cz/xqw/webdav/pssenat/original/67109/56508 [accessed 23 May 2014]
Czech Senate Committee on Regional Development, Public Administration and Environment, '120th RESOLUTION of the 29th meeting held on August 8, 2012' (120. USNESENÍ z 29. schůze konané dne 8. srpna 2012) available at: http://www.senat.cz/xqw/webdav/pssenat/origi nal/65564/55259 [accessed 23 May 2014]
Czech Senate Committee on Regional Development, Public Administration and Environment, '125th RESOLUTION of the 31st meeting held on October 17, 2012' (125. USNESENÍ z 31. schůze konané dne 17. října 2012) available at: http://www.senat.cz/xqw/webdav/pssenat/origi nal/65991/55607 [accessed 23 May 2014]
Dannwolf U et al., 'Umweltauswirkungen von fracking bei der Aufsuchung und Gewinnung von Erdgas insbesondere aus Schiefergaslagerstätten Teil 2' (Umweltbundesamt, Dessau-Roßlau 2014)
Davey E, 'Written Ministerial Statement by Edward Davey: Exploration for shale gas 13 December 2012' available at: http://www.decc.gov.uk/en/content/cms/news/wms_shale/wms_shale.aspx [accessed 07 March 2013]
Department of Energy and Climate Change, 'CCS Roadmap: Supporting Deployment of Carbon Capture and Storage in the UK' (Crown 2012)
Department of Energy and Climate Change, 'SEA for a 14th and Subsequent Onshore Oil & Gas Licensing Rounds – Environmental Report' (Crown 2011)
Department of Environmental Quality Michigan 'Hydraulic Fracturing in Michigan' available at: http://www.michigan.gov/deq/0,4561,7-135-3311_4231-262172--,00.html [accessed 30 June 2016]
Dutch Government, 'Derde Nota Waddenzee Derde Nota Waddenzee, Deel 1: ontwerp planologische kernbeslissing 2001' (Third Paper on the Waddenzee, Part 1: draft for a crucial planning decision 2001)
Economic Committee of the German Parliament, 'Resolution recommendation and report of the committee on the economy of 30.04.1980' (Beschlußempfehlung und Bericht des Ausschusses für Wirtschaft (9. Ausschuß zu dem von der Bundesregierung eingebrachten Entwurf eines Bundesberggesetzes (BbergG)) Drucksache 8/3965

Electoral Commission '*EU Referendum count processes and results*' available at: http://www.electoralcommission.org.uk/__data/assets/pdf_file/0013/206113/Media-briefing-EU-Referendum-count-processes-and-results.pdf [accessed 29 June 2016]

Element Energy for The North Sea Basin Task Force, 'One North Sea: A Study into North Sea cross-border CO2 transport and storage Executive Summary 18th March 2010' available at: http://www.cgseurope.net/UserFiles/file/News/CGS%20Europe%20report%20_D2_10_State%20of%20play%20on%20CO2%20storage%20in%2028%20European%20countries(1).pdf [accessed: 24 November 2016]

Energy Agency, 'World Energy Outlook 2015' (OECD/International Energy Agency, Paris, 2015)

Energy Institute of the University of Texas, 'Fact-Based Regulation for Environmental Protection in Shale Gas Development' (2012) in Part 2 'Summary of findings' available at: http://barnettprogress.com/media/ei_shale_gas_regulation120215.pdf [accessed 14 June 2012]

Environmental Protection Agency (EPA), 'Evaluation of Impacts to Underground Sources of Drinking Water by Hydraulic Fracturing of Coalbed Methane Reservoirs Study (2004)' available at: http://water.epa.gov/type/groundwater/uic/class2/hydraulicfracturing/wells_coalbedmethanestudy.cfm [accessed 28 March 2013]

European Carbon Capture and Storage Demonstration Project Network, 'Situation Report 2012' available at: http://cdn.globalccsinstitute.com/sites/default/files/publications/115876/network-situation-report-2012.pdf [accessed 25 October 2013]

European Commission, 'Analysis and presentation of the results of the public consultation "Unconventional fossil fuels (e.g. shale gas) in Europe" Final report' (Bio Intelligence Service, Paris 2013)

European Commission, Commission Staff Working Document 'The January 2009 gas supply disruption to the EU: an Assessment' SEC (2009) 977

European Commission 'Energy and Environment Hydrocarbons BREF' available at: http://ec.europa.eu/environment/integration/energy/hc_bref_en.htm [accessed 6 June 2016]

European Commission, 'Executive Summary of the Impact Assessment on Communication COM (2014) 23 final' of 22nd January 2014 available at: http://eur-lex.europa.eu/legal-content/EN/TXT/PDF/?uri=CELEX:52014SC0022&from=EN [accessed 2 April 2014]

European Commission, 'Guidance note on the application of Directive 85/337/EEC to projects related to the exploration and exploitation of unconventional hydrocarbon' of 12 December 2011 available at: http://

ec.europa.eu/environment/integration/energy/pdf/guidance_note.pdf [accessed 28 April 2014]

European Commission, 'IED and BREF Revision' available at: http://www.cepi.org/system/files/public/epw-presentations/2012/BREFseminar/Paper%20week.pdf [accessed 9 March 2013]

European Commission, 'Roadmap Report on the Effectiveness of Commission Recommendation 2014/70/EU' available at: http://ec.europa.eu/smart-regulation/roadmaps/docs/2015_env_021_shale_gas_fracking_en.pdfhttp://ec.europa.eu/smart-regulation/roadmaps/docs/2015_env_021_shale_gas_fracking_en.pdf [accessed 7 June 2016]

European Council, 'Conclusions of 4 February 2011 EUCO 2/1/11 Rev 1' available at: https://www.consilium.europa.eu/uedocs/cms_data/docs/pressdata/en/ec/119175.pdf [accessed 4 February 2015]

European Council, 'Recommendation on the Broad Guidelines of the Economic Policies of the Member States and the Community' [2001] OJ L 179/1

European Council, Submission of the EC to the WTO 'European Council Resolution on the Precautionary Principle' (2001) G/SPS/GEN/225 G/TBT/W/154 WT/CTE/W/181 available at: http://trade.ec.europa.eu/doclib/docs/2006/december/tradoc_117313.pdf [accessed 04 June 2012]

European Court of Justice, 'Opinion of Advocate General Bot in Case C-204/12 through to C 208/12 Essent Belgium NV v Vlaamse Reguleringsinstantie voor de Elektriciteits' available at: EurLex http://curia.europa.eu/juris/celex.jsf?celex=62012CC0204&lang1=en&type=NOT&ancre= [accessed 3 February 2014]

European Court of Justice, 'Opinion of Advocate General Bot in Case C-573/12 Alands Vindkraft v Energimyndigheten' available at: EurLex: http://eur-lex.europa.eu/LexUriServ/LexUriServ.do?uri=CELEX:62012CC0573:EN:NOT [accessed: 3 February 2014]

European Court of Justice, 'Opinion of Advocate General Cosmos in Case C-318/98 Criminal Proceedings against Giancarlo Fornasar, Andrea Strizzolo, Giancarlo Toso, Lucio Mucchino, Enzo Peressutti and Sante Chiarcosso' [2000] ECR I-4785

European Court of Justice, 'Opinion of Advocate General Geelhoed in Case C-320/03 Commission v Austria' [2005] ECR I-9873

European Court of Justice, 'Opinion of Advocate General Lagrange in Case 13/57 Wirtschaftsvereinigung v Hohe Behörde' [1958] ECR 288

European Court of Justice, 'Opinion of Advocate General Leger delivered on 8 October 1998 in case C-293/97 The Queen v Secretary of State for the Environment and Ministry of Agriculture, Fisheries and Food, ex parte H.A. Standley and Others and D.G.D. Metson and Others' [1999] ECR I-02606

European Court of Justice, 'Opinion of Advocate General Saggio in Case C-319/97 Criminal Proceedings against Antoine Kortas' [1999] ECR I-3143 available at: http://eur-lex.europa.eu/LexUriServ/LexUriServ.do?uri=CELEX:61997CC0319:EN:PDF [accessed 7 August 2013]

European Environment Agency, 'Late Lessons from Early Warnings: The Precautionary Principle 1896-2000' (Environmental Issue Report No. 22, Copenhagen 2001)

European Parliament, '2011/2308(INI) – 21/11/2012 Text adopted by Parliament, single reading' available at: http://www.europarl.europa.eu/oeil/popups/summary.do?id=1235116&t=d&l=en [accessed 3 July 2013]

European Parliament, '2012/0297(COD) – 12/03/2014 Text adopted by Parliament, 1st reading/single reading' available at: http://www.europarl.europa.eu/oeil/popups/summary.do?id=1342382&t=d&l=en [accessed 22 April 2014]

European Parliament, 'Amendments adopted by the European Parliament on 9 October 2013 on the proposal for a directive of the European Parliament and of the Council amending Directive 2011/92/EU of the assessment of the effects of certain public and private projects on the environment (COM (2012)0628 – C7-0367/2012 – 2012/0297(COD))' amendments 31, 79, 112 and 126 available at: http://www.europarl.europa.eu/sides/getDoc.do?type=TA&reference=P7-TA-2013-0413&language=EN#def_1_1 [accessed 22 April 2014]

European Parliament 'Debate of 20 November 2012 Environmental Impacts of Shale Gas and Shale Oil Extraction Activities – Industrial, Energy and Other Aspects of Shale Gas and Oil CRE 20/11/2012 – 11' available at: http://www.europarl.europa.eu/sides/getDoc.do?type=CRE&reference=20121120&secondRef=ITEM-011&language=EN [accessed 2 September 2014]

European Parliament, 'Entwurf eines Berichts über Industrie-, Energie und andere Aspekte von Schiefergas und –Öl' 2011/2309 (INI) (European Parliament, Brussels 2012)

European Parliament, 'European Parliament legislative resolution of 12 March 2014 on the proposal for a directive of the European Parliament and of the Council amending Directive 2011/92/EU of the assessment of the effects of certain public and private projects on the environment (COM (2012)0628 – C7-0367/2012 – 2012/0297(COD)) (Ordinary legislative procedure: first reading)' available at: http://www.europarl.europa.eu/sides/getDoc.do?type=TA&language=EN&reference=P7-TA-2014-0225 [accessed 22 April 2014]

European Parliament, 'European Parliament resolution of 21 November 2012 on the environmental impacts of shale gas and shale oil extraction activities (2011/2308(INI))' available at: http://www.europarl.europa.

eu/sides/getDoc.do?pubRef=-//EP//NONSGML+TA+P7-TA-2012-0443+0+DOC+PDF+V0//EN [accessed 23 April 2014]

European Parliament, 'Procedure 2012/0297(COD) 'Proposal for a directive Recital 23 a (new)' available at: http://www.europarl.europa.eu/sides/getDoc.do?type=TA&language=EN&reference=P7-TA-2013-413 [accessed 12 December 2013]

European Parliament, 'Shale gas: member states need robust rules on fracking' available at: http://www.europarl.europa.eu/news/en/news-room/content/20120917IPR51525/html/Shale-gas-member-states-need-robust-rules-on-fracking [accessed 23 April 2014]

European Parliament, 'Shale gas: new fracking projects must pass environmental test' available at: http://www.europarl.europa.eu/pdfs/news/expert/infopress/20131004IPR21541/20131004IPR21541_en.pdf [accessed 22 April 2014]

European Policy Centre, 'Occasional Paper April 2001 Towards a Proportionate Implementation of the Precautionary Principle' (The European Policy Centre, Brussels 2001)

European Union Assembly of Regional and Local Representatives, 'Draft Opinion of the Committee of the Regions local and regional authorities perspective on shale/tight gas and oil (unconventional hydrocarbons)' available at: http://cor.europa.eu/en/news/Pages/fracking-environmental impact.aspx [accessed 4 September 2014]

European Union Technical Working Group, 'First Meeting on 27 January 2012, Brussels final on environmental aspects of unconventional fossil fuels, in particular shale gas, Summary Report' available at: http://ec.europa.eu/transparency/regexpert/index.cfm?do=groupDetail.groupDetailDoc&id=8487&no=1.html [accessed 10 December 2014]

Ewen C et al., 'Risikostudie Fracking Übersichtsfassung der Studie' available at: http://dialog-erdgasundfrac.de/risikostudie-fracking [accessed 15 June 2012]

Ewen C et al., 'Hydrofracking Risk Assessment- Executive Summary' available at: http://dialog-erdgasundfrac.de/sites/dialog-erdgasundfrac.de/files/Ex_HydrofrackingRiskAssessment_120611.pdf [accessed 15 June 2012]

Federal Council of Germany, 'Entwurf einer Verordnung zur Änderung der Verordnung über die Umweltverträglichkeitsprüfung bergbaulicher Vorhaben' BR-Drs. 388/11 of 29.06.2011

Forster D and Perks J, 'Climate Impact of potential shale gas production in the EU' (AEA Technology, Didcot 2012) available at: http://ec.europa.eu/clima/policies/eccp/docs/120815_final_report_en.pdf [accessed 29 October 2012]

French Constitutional Council, 'Juin 2014 : La Charte de l'Environnement de 2004' available at http://www.conseil-constitutionnel.fr/conseil-constitutionnel/francais/a-la-une/juin-2014-la-charte-de-l-environnement-de-2004.141685.html [accessed 15 December 2016]

French Joint Committee on the draft law on the exploration and exploitation of hydrocarbons, 'Gathering in the Senate Wednesday, 15 June 2011' (Commission mixte paritaire sur la proposition de loi relative à l'exploration et l'exploitation des mines d'hydrocarbures Réunie au Sénat le mercredi 15 juin 2011) available at: http://www.senat.fr/compte-rendu-commissions/20110614/cmp.html#toc4 [accessed 7 May 2014]

French National Assembly, 'Information report on the implementation of the conclusions of the information report (No. 2719) 8 July 2010 on the evaluation of the implementation of Article 5 of the Charter of the Environment on the application the precautionary principle' (17/11/2011 RAPPORT D'INFORMATION déposé en application de l'article 146-3, alinéa 8, du Règlement par le comité d'évaluation et de contrôle des politiques publiques sur la mise en œuvre des conclusions du rapport d'information (n° 2719) du 8 juillet 2010 sur l'évaluation de la mise en œuvre de l'article 5 de la Charte de l'Environnement relatif à l'application du principe de précaution) available at: http://www.assemblee-nationale.fr/13/rap-info/i3970.asp [accessed 12 June 2012]

French National Assembly, 'Law proposal to prohibit the exploration and exploitation of unconventional hydrocarbons, to repeal the exclusive licences for mines of unconventional hydrocarbons and ensure greater transparency in the mining code' (PROPOSITION DE LOI visant à interdire l'exploration et l'exploitation des hydrocarbures non conventionnels, à abroger les permis exclusifs de recherche de mines d'hydrocarbures non conventionnels et a assurer plus de transparence dans le code minier) available at: http://www.assemblee-nationale.fr/13/propositions/pion3690.asp [accessed 12 June 2012]

French National Assembly, 'Minutes of Meeting of Thursday, 6th October 2011 Electronic Summary Unconventional Hydrocarbon Discussion of a proposed law' (Assemblée nationale XIIIe legislature Session ordinaire de 2011–2012 Compte rendu Première séance du jeudi 6 octobre 2011 SOMMAIRE ÉLECTRONIQUE Hydrocarbures non conventionnels Discussion d'une proposition de loi) available at: http://www.assemblee-nationale.fr/13/cri/2011-2012/20120007.asp [accessed: 12 June 2012]

French National Assembly, 'PROHIBITION OF HYDRAULIC FRACTURE Discussion of a bill Official Journal of the French Republic Regular Session 2010–2011 173rd meeting Minutes integral 1st meeting of Tuesday, May 10, 2011' (INTERDICTION DE LA FRACTURATION HYDRAULIQUE Discussion d'une proposition de loi JOURNAL OFFICIEL DE LA RÉPUBLIQUE FRANÇAISE

SESSION ORDINAIRE DE 2010-2011 173e séance Compte rendu integral 1re séance du mardi 10 mai 2011) available at: http://www.assemblee-nationale.fr/13/pdf/cri/2010-2011/20110173.pdf [accessed 7 May 2014]

French National Assembly, 'PROHIBITION OF HYDRAULIC FRACTURE Discussion of a bill Official Journal of the French Republic Regular Session 2010-2011 174th meeting Minutes integral 2nd meeting of Tuesday, May 10, 2011' (INTERDICTION DE LA FRACTURATION HYDRAULIQUE Discussion d'une proposition de loi JOURNAL OFFICIEL DE LA RÉPUBLIQUE FRANÇAISE SESSION ORDINAIRE DE 2010–2011 174e séance Compte rendu integral 2e séance du mardi 10 mai 2011) available at: http://www.assemblee-nationale.fr/13/pdf/cri/2010-2011/20110174.pdf [accessed 7 May 2014]

French National Assembly, 'PROHIBITION OF HYDRAULIC FRACTURE Solemn Vote Official Journal of the French Republic Regular Session 2010–2011 175th meeting Minutes integral 1st meeting of Wednesday, May 11, 2011' (INTERDICTION DE LA FRACTURATION HYDRAULIQUE Vote solennel JOURNAL OFFICIEL DE LA RÉPUBLIQUE FRANÇAISE SESSION ORDINAIRE DE 2010–2011 175e séance Compte rendu integral 1re séance du mercredi 11 mai 2011) available at: http://www.assemblee-nationale.fr/13/pdf/cri/2010–2011/20110175.pdf [accessed 7 May 2014]

French National Assembly, 'Thirteenth legislature Ordinary Session 2010–2011 Report First meeting of Tuesday, June 21, 2011 ELECTRONIC SUMMARY' (XIIIe législature Session ordinaire de 2010–2011 Compte rendu intégral Première séance du mardi 21 juin 2011 SOMMAIRE ÉLECTRONIQUE) available at: http://www.assemblee-nationale.fr/13/cri/2010-2011/20110216.asp#P913_214373 [accessed 2 May 2014]

French National Assembly Commission for Sustainable Development and Spatial Planning,'Consideration of proposed legislation to prohibit exploration and exploitation of unconventional hydrocarbons (3690) (Jean-Paul Chanteguet, Rapporteur 28 September 2011' (Commission du développement durable et de l'aménagement du territoire– Examen de la proposition de loi visant à interdire l'exploration et l'exploitation des hydrocarbures non conventionnels (n° 3690) (M. Jean-Paul Chanteguet, rapporteur) Mercredi 28 septembre 2011) available at: http://www.assemblee-nationale.fr/13/cr-dvp/10-11/c1011063.asp [accessed 12 June 2012]

French National Assembly Commission for Sustainable Development and Spatial Planning, 'Report on the proposed law to ban the exploration and exploitation of oil and gas by hydraulic fracturing mines and to repeal the exclusive licences including projects using this technique 4th May 2011' (RAPPORT FAIT AU NOM DE LA COMMISSION DU DÉVELOPPEMENT DURABLE ET DE L'AMÉNAGEMENT DU TERRITOIRE SUR LA PROPOSITION DE LOI, visant à interdire l'exploration et l'exploitation des mines d'hydrocarbures liquides ou gazeux par fracturation hydraulique et à abroger les permis exclusifs de recherches comportant des projets ayant recours à cette technique 4 mai 2011) available at: http://www.assemblee-nationale.fr/13/pdf/rapports/r3392.pdf [accessed 12 June 2012]

French Senate, 'Meeting of 30 June 2011 (Verbatim Report of Debates)' (Séance du 30 juin 2011 (compte rendu intégral des débats)) available at: http://www.senat.fr/seances/s201106/s20110630/st20110630000.html [accessed 2 May 2014]

French Senate, 'Prohibition of exploring for and mining of hydrocarbons by hydraulic fracturing – Discussion in the accelerated procedure on three legislative proposals in the text of the commission OFFICIAL JOURNAL OF THE FRENCH REPUBLIC REGULAR SESSION OF FULL SENATE REPORT 2010–2011 Meeting on Wednesday, June 1, 2011 (112th sitting day of the session)' (Interdiction de l'exploration et de l'exploitation des mines d'hydrocarbures par fracturation hydraulique. – Discussion en procédure accélérée de trois propositions de loi dans le texte de la commission JOURNAL OFFICIEL DE LA RÉPUBLIQUE FRANÇAISESESSION ORDINAIRE DE 2010–2011 COMPTE RENDU INTÉGRAL Séance du mercredi 1er juin 2011 (112e jour de séance de la session)) available at: http://www.senat.fr/seances/s201106/s20110601/s20110601.pdf [accessed 9 May 2014]

French Senate, 'Prohibition of exploring for and mining of hydrocarbons by hydraulic fracturing – Further discussion accelerated procedure three legislative proposals and adoption of the text of the amended commission OFFICIAL JOURNAL OF THE FRENCH REPUBLIC REGULAR SESSION OF FULL SENATE REPORT 2010–2011 Session Thursday, June 9, 2011 (115th sitting day of the session)' (Interdiction de l'exploration et de l'exploitation des mines d'hydrocarbures par fracturation hydraulique. – Suite de la discussion en procédure accélérée de trois propositions de loi et adoption du texte de la commission modifié JOURNAL OFFICIEL DE LA RÉPUBLIQUE FRANÇAISE SESSION ORDINAIRE DE 2010–2011 COMPTE RENDU INTÉGRAL SÉNAT Séance du jeudi 9 juin 2011 (115e jour de séance de la session)) available at: http://www.senat.fr/seances/s201106/s20110609/s20110609.pdf [accessed 7 May 2014]

French Senate Committee on Economy, Sustainable Development and

Planning, 'Report on the proposed law to ban the exploration and exploitation of mines or liquid hydrocarbons gas by hydraulic fracturing and to repeal the exclusive licences including projects using this technique 25 May 2011' (RAPPORT au nom de la commission de l'économie, du développement durable et de l'aménagement du territoire sur la proposition de loi, visant à interdire l'exploration et l'exploitation des mines d'hydrocarbures liquides ou gazeux par fracturation hydraulique et à abroger les permis exclusifs de recherches comportant des projets ayant recours à cette technique 25 mai 2011) available at: http://www.senat.fr/rap/l10-556/l10-5561.pdf [accessed 12 May 2014]

Fritsche U R and Herling J, 'Energie- und Klimabilanz von Erdgas aus unkonventionellen Lagerstätten im Vergleich zu anderen Energiequellen. Endbericht zum Gutachten für Team Ewen im Rahmen des InfoDialog Fracking' (Öko-Institut, Darmstadt 2012) available at: http://dialog-erdgasundfrac.de/sites/dialog-erdgasundfrac.de/files/OEKO_IINAS-Fracking-Energie-Klimabilanz.pdf [accessed 17 July 2014]

Gabriel S and Hendricks B, 'Brief an die Mitglieder der Fraktionen von CDU/CSU und SPD im Deutschen Bundestag 20.11.2014' available at: http://www.bmwi.de/BMWi/Redaktion/PDF/B/brief-gabriel-und-hendricks-zu-fracking,property=pdf,bereich=bmwi2012,sprache=de,rwb=true.pdf [accessed 2 February 2015]

Gandossi L, 'An overview of hydraulic fracturing and other formation stimulation technologies for shale gas production' (Publications Office of the European Union, Luxembourg 2013)

German Federal Environmental Agency, 'Manual for individual screenings to establish whether or not projects are subject to obligatory EIAs' (Leitfaden zur Vorprüfung des Einzelfalls im Rahmen der Feststellung der UVP-Pflicht von Projekten) available at: http://www.bmu.de/files/pdfs/allgemein/application/pdf/vorpr_uvp_pflicht.pdf [accessed 20 October 2012]

German Federal Environmental Agency, 'Statement: Appraisal of shale gas extraction in Germany December 2011' (Stellungnahme: Einschätzung der Schiefergasförderung in Deutschland Dezember 2011) available at: http://www.umweltbundesamt.de/wasser-und-gewaesserschutz/publikationen/stellungnahme_fracking.pdf [accessed 26 October 2012]

German Federal Government, 'Bill of the German Federal Government on the alteration of water- and environmental protection norms with the aim of prohibiting and minimizing risks associated with the procedures of the fracking technology' (Gesetzentwurf der Bundesregierung 'Entwurf eines Gesetzes zur Änderung wasser- und naturschutzrechtlicher Vorschriften zur Untersagung und zur Risikominimierung bei den Verfahren der Fracking-Technologie') available at: http://www.bmub.

bund.de/fileadmin/Daten_BMU/Download_PDF/Binnengewaesser/fracking_g_entwurf_neu_bf.pdf [accessed 7 April 2015]

German Federal Government, 'Bill of the German Federal Government law on the demonstration and application of technologies concerned with Carbon Capture, Transport and Storage of 9 May 2011' (Gesetzentwurf der Bundesregierung Entwurf eines Gesetzes zur Demonstration und Anwendung von Technologien zur Abscheidung, zum Transport und zur dauerhaften Speicherung von Kohlendioxid) Bundestagsdrucksache 17/5750 available at: http://dip21.bundestag.de/dip21/btd/17/057/1705750.pdf [accessed 24 October 2013]

German Federal Government, 'Bill of the German Federal Government law on the regulation of Carbon Capture, Transport and Storage of 9 May 2011' (Gesetzentwurf der Bundesregierung Entwurf eines Gesetzes zur Regelung von Abscheidung, Transport und dauerhafter Speicherung von Kohlendioxid) Bundestagsdrucksache 16/12782 available at: http://dip21.bundestag.de/dip21/btd/16/127/1612782.pdf [accessed 24 October 2013]

German Federal Government, 'Bill of the German Federal Government law on the reorganization of the Energy Industry Act of 19 March 1997' (Gesetzesentwurf der Bundesregierung – Entwurf eines Gesetzes zur Neuregelung des Energiewirtschaftsrechts) Bundestagsdrucksache 13/7274

German Federal Government, 'Bill to expand mining liability to wellbore mining and caverns' (Bundesregierung Entwurf eines Gesetzes zur Ausdehnung der Bergschadenshaftung auf den Bohrlochbergbau und Kavernen) available at: http://www.bmwi.de/BMWi/Redaktion/PDF/E/entwurf-eines-gesetzes-zur-ausdehnung-der-bergschadenshaftung-auf-den-bohrlochbergbau-und-kavernen,property=pdf,bereich=bmwi,sprache=de,rwb=true.pdf [accessed 7 April 2015]

German Federal Government, 'Energy Concept 2050' (Energiekonzept 2050) available at: http://www.bmu.de/files/pdfs/allgemein/application/pdf/energiekonzept_bundesregierung.pdf [accessed 31 May 2012]

German Federal Government, 'Environmental Programme of the German Government of 14 October 1971' (Umweltprogramm der Bundesregierung) Bundestagsdrucksache VI/2710 available at: http://dipbt.bundestag.de/doc/btd/06/027/0602710.pdf [accessed 21 January 2014]

German Federal Government, 'The way towards energy of the future – safe, affordable and environmentally friendly capstone paper of the German Government concerning the energy turnaround' (Der Weg zur Energie der Zukunft – sicher, bezahlbar und umweltfreundlich Eckpunktepapier der Bundesregierung zur Energiewende) available at:

http://www.bmwi.de/DE/Themen/Energie/energiepolitik,did=405004.html [accessed 31 May 2012]

German Federal Ministry of Economy and Technology, 'Energy in Germany trends and background information on energy supply' (Energie in Deutschland Trends und Hintergründe zur Energieversorgung) (PRpetuum GmbH, Munich February 2013)

German Federal Network Agency, 'Report concerning the status of grind-bound energy supply during the winter 2011/2012' (Bericht zum Zustand der leitungsgebundenen Energieversorgung im Winter 2011/2012) available at: http://www.bundesnetzagentur.de/SharedDocs/Downloads/DE/Allgemeines/Bundesnetzagentur/Publikationen/Berichte/2012/NetzBericht_ZustandWinter11_12pdf.pdf?__blob=publicationFile&v=2 [accessed 26 June 2014]

German Government, 'Kabinettbeschluss Fracking: Mehr Schutz durch strenge Regeln' available at: http://www.bundesregierung.de/Content/DE/Artikel/2015/04/2015-04-01-fracking-gesetz-kabinett.html;jsessionid=7198048376FAAF673F3B194C34B7DE85.s2t2 [accessed 7 April 2015]

German Ministry of Economy and Energy, 'Erdgasimporte und Eigen produktion' available at: http://www.bmwi.de/DE/Themen/Energie/Konventionelle-Energietraeger/gas,did=292324.html [accessed 18 June 2014]

German Parties CDU, CSU and SPD, 'Shaping Germany's future Coalition Agreement between CDU, CSU and SPD 18th Session of Parliament' (Deutschlands Zukunft gestalten Koalitionsvertrag zwischen CDU, CSU und SPD 18. Legislaturperiode) available at: https://www.spd-berlin.de/w/files/spd-2013regierungsprogramm_mp3/koalitionsvertrag-2013.pdf [accessed 12 December 2013]

Glassman D et al., 'THE WATER-ENERGY NEXUS – Adding Water to the Energy Agenda' (World Policy Institute, New York 2011) available at: http://www.tailcatering.com/images/THE_WATER-ENERGY_NEXUS_REPORT.pdf [accessed 07 June 2012]

Gottardo S et al., 'Assessment of the use of substances in hydraulic fracturing of shale gas reservoirs under REACH' (Publications Office of the European Union, Luxembourg 2013)

Government of Canada, 'A Canadian Perspective on the Precautionary Approach/Principle: Proposed Guiding Principles' (2001) available at: http://www.ercb.ca/docs/applications/submissions/1458443/Ex%20002-96-02.pdf [accessed 20 March 2012]

Green C A, Styles P and Baptie B J, 'Preese Hall Shale Gas Fracturing Review and Recommendations for Induced Seismic Mitigation' Report on behalf of the UK Department of Energy and Climate Change http://

og.decc.gov.uk/assets/og/ep/onshore/5075-preese-hall-shale-gas-fracturing-review.pdf [accessed 26 April 2012]

Ground Water Protection Council and ALL Consulting, 'Modern Shale Gas Development in the United States: A Primer' (2009) http://www.eogresources.com/responsibility/doeModernShaleGasDevelopment.pdf [accessed 28 April 2012]

Gunzelmann M and El Hamdaoui M, 'Grundwassermonitoring im Bereich der Bohrungen Damme 2/3 der Exxon Mobil' available at: http://damme.de/templates/images/news/1272_1.pdf [accessed 5 July 2013]

Hessisches Landesamt für Umwelt und Geologie, 'Stellungnahme zu vorliegenden: Gutachten zum Fracking in Deutschland im Zusammenhang mit dem Aufsuchungsantrag der BNK Deutschland GmbH auf Kohlenwasserstoffe im Erlaubnisfeld "Adler South"' available at: http://www.hlug.de/fileadmin/dokumente/geologie/rohstoffe/kw/Fracking_HLUG_lang_260313.pdf [accessed 9 January 2015]

HM Government, 'Clean Coal: An Industrial Strategy for the Development of Carbon Capture and Storage Across the UK' (Crown 2011) available at: https://ukccsrc.ac.uk/system/files/publications/ccs-reports/DECC_Coal_154.pdf [accessed 30 November 2016]

HM Treasury, 'Budget 2013' available at: https://www.gov.uk/government/uploads/system/uploads/attachment_data/file/221885/budget2013_complete.pdf [accessed 22 September 2016]

Hofmann F, 'Kurzgutachten zur Abschätzung der Maispollendeposition in Relation zur Entfernung von Maispollenquellen mittels technischem Pollensammler PMF' of 25 Mai 2007 available at: http://www.bfn.de/fileadmin/MDB/documents/themen/agrogentechnik/07-05-31_Gutachten_Pollendeposition_end.pdf [accessed 5 November 2013]

House of Commons Energy and Climate Change Committee, 'Shale Gas' Fifth Report of Session 2010–12, Vol. I (Crown 2011) House of Commons Energy and Climate Change Committee, 'Shale Gas' Fifth Report of Session 2010–12, Vol. II (Crown 2011)

House of Commons Energy and Climate Change Committee, 'UK Energy Supply: Security or Independence?' Vol. I (Crown 2011)

House of Commons Energy and Climate Change Committee, 'Shale Gas: Government Response to the Committee's Fifth Report of Session 2010–12' (Crown 2011)

House of Commons Select Committee on Energy and Climate Change's Inquiry, 'The UK's Energy Supply, Security or Independence?' (Crown, 2011) available at: http://www.publications.parliament.uk/pa/cm201012/cmselect/cmenergy/1065/1065.pdf [accessed 20 November 2016]

House of Lords, 'European Union – Fourteenth Report' available at:

http://www.publications.parliament.uk/pa/ld200405/ldselect/ldeucom/101/10105.htm [accessed 8 September 2014]

House of Lords, 'The Economic Impact on UK Energy Policy of Shale Gas and Oil' (The Stationery Office Limited, London 2014) also available at: http://www.publications.parliament.uk/pa/ld201314/ldselect/ldeconaf/172/172.pdf [accessed 9 December 2015]

Howarth R W, 'Statement for the EPA Hydraulic Fracturing Public Informational Meeting' (2010) submission to EPA, available at: http://cce.cornell.edu/EnergyClimateChange/NaturalGasDev/Documents/PDFs/Howarth%20statement%20to%20EPA%20--%2015%20Sept%20%202010.pdf [accessed 14 June 2012]

ICF International Ltd., 'Mitigation of climate impacts of possible future shale gas extraction in the EU: available technologies, best practices and options for policy makers' (2014) available at: http://ec.europa.eu/environment/integration/energy/uff_studies_en.htm [accessed 4 September 2014]

Independent Expert Group on Mobile Phones, 'Mobile Phones and Health' (IEGMP Secretariat, Oxon 2000)

Independent Expert Scientific Panel, 'Report on Unconventional Oil And Gas' (Crown 2014) available at: http://www.gov.scot/Resource/0045/00456579.pdf [accessed 8 September 2016]

Inter-Departmental Liaison Group on Risk Assessment (ILGRA), 'The Precautionary Principle: Policy and Application' available at: http://www.hse.gov.uk/aboutus/meetings/committees/ilgra/pppa.pdf [accessed 15 April 2013]

Intergovernmental Panel on Climate Change, 'IPCC Special Report on Carbon Dioxide Capture and Storage' (Cambridge University Press 2005)

Intergovernmental Panel on Climate Change, 'Summary for Policymakers' in IPPC Working Group I Climate Change 2007 (Cambridge University Press, Cambridge 2007)

International Energy Agency, 'Energy Security and Climate Policy – Assessing Interactions' (OECD/International Energy Agency, Paris, 2007)

International Energy Agency, 'Golden Rules for a Golden Age of Gas' (International Energy Agency, Paris 2012)

International Energy Agency, 'Towards A Sustainable Energy Future' available at: http://ccs101.ca/assets/Documents/g8_towards_sustainable_future.pdf [accessed 10 February 2015]

International Energy Agency, 'World Energy Outlook 2009' (International Energy Agency, Paris 2009)

International Energy Agency, 'World Energy Outlook 2014' (OECD/International Energy Agency, Paris, 2014)

International Law Association, 'Berlin Conference Water Resources Law' (2004) http://www.internationalwaterlaw.org/documents/intldocs/ILA_Berlin_Rules-2004.pdf [accessed 24 April 2012]
International Law Association, 'Declaration on Principles of International Law in the Field if Sustainable Development' (2002) http://www.ila-hq.org/ [accessed 24 April 2012]
International Law Association, 'Resolution2/2014 on the Legal Principles Relating to Climate Change (2014)' available at: http://www.ila-hq.org/en/committees/index.cfm/cid/1029 [accessed 7 November 2016]
Kuhn M and Umbach F, 'Strategic Perspectives of Unconventional Gas: A Game Changer with Implications for the EU' (2011) European Centre for Energy and Resource Security (EUCERS) Strategy Paper No 1 http://www.eucers.eu/2011/05/06/eucers-strategy-paper-no1/ [accessed 24 April 2012]
Krupp R E, 'Gutachten zur geplanten Kohlendioxid-Einlagerung (CCS) in der Antiklinal-Struktur Neutrebbin, Ostbrandenburg' (2011) available at: http://www.co2bombe.de/joomla/images/stories/co2/krupp_gutachten_1_neutrebbin_final.pdf [accessed 15 October 2012]
Länderausschuss Bergbau, 'Handlungsempfehlung zur Behandlung von Förderbohrungen' available at http://www.tlba.de/docs/Handlungsempfehlung_Foerderbohrung.PDF [accessed 11 March 2015]
Lechtenböhmer S et al., 'Impacts of shale gas and shale oil extraction on the environment and on human health' (European Parliament 2011) Study IP/A/ENVI/ST/2011-07 available at: http://www.europarl.europa.eu/meetdocs/2009_2014/documents/envi/dv/shale_gas_pe464_425_final_/shale_gas_pe464_425_final_en.pdf [accessed 13 June 2012]
Luciani G, 'Security of Supply for Natural Gas Markets: What is it and what is it not' (2204) INDES Working Papers available at: http://papers.ssrn.com/sol3/papers.cfm?abstract_id=600501 [accessed 25 April 2013]
Mathis P et al.,'Macroeconomic impacts of shale gas extraction in the EU' (2014) available at: http://ec.europa.eu/environment/integration/energy/uff_studies_en.htm [accessed 4 September 2014]
Meiners G, Denneborg M and Müller F, 'Gutachten für das Umweltbundesamt Umweltauswirkungen von Fracking bei der Aufsuchung und Gewinnung von Erdgas aus unkonventionellen Lagerstätten– Risikobewertung, Handlungsempfehlungen und Evaluierung bestehender rechtlicher Regelungen und Verwaltungsstrukturen' of August 2012 (Ministry for the Environment 2012)
Meiners G et al., 'Fracking in unkonventionellen Erdgas-Lagerstätten in NRW Kurzfassung zum Gutachten "Gutachten mit Risikostudie zur Exploration und Gewinnung von Erdgas aus unkonventionellen Lagerstätten in Nordrhein-Westfalen (NRW) und deren Auswir-

kungen auf den Naturhaushalt insbesondere die öffentliche Trinkwasserversorgung"' of 7 September 2012 (Ministry for Climate Protection, Environment, Agriculture, Conservation and Consumer Protection of North Rhine-Westphalia 2012)

Milieu Ltd., 'Regulatory provisions governing key aspects of unconventional gas extraction in selected Member States' (2013) available at: http://ec.europa.eu/environment/integration/energy/uff_studies_en.htm [accessed 4 September 2014]

Ministerie van Economische Zaken Directie Energiemarkt, 'Aanvullend onderzoek naar mogelijke risico's en gevolgen van de opsporing en winning van schalie- en steenkoolgas in Nederland Eindrapport onderzoeksvragen' available at: http://www.rijksoverheid.nl/ministeries/ez/documenten-en-publicaties/rapporten/2013/08/26/aanvullend-onderzoek-naar-mogelijke-risico-s-en-gevolgen-van-de-opsporing-en-winning-van-schalie-en-steenkoolgas-in-nederland-eindrapport-onderzoeksvragen-a-en-b.html [accessed 14 April 2014]

Ministerie van Economische Zaken Directie Energiemarkt, 'Brief aan de Tweede Kamer van minister Kamp (Economische Zaken) over vervolgstappen winning schaliegas' available at: http://www.rijksoverheid.nl/documenten-en-publicaties/kamerstukken/2013/09/18/brief-aan-de-tweede-kamer-vervolgstap-schaliegas.html [accessed 14 April 2014]

Mordick B, 'Risks to Drinking Water from Oil and Gas Wellbore Construction and Integrity: Case Studies and Lessons Learned' available at: http://www.epa.gov/hfstudy/riskstodrinkingwaterfromoilandgaswellboreconstructionandintegrity.pdf [accessed 29 March 2013]

National Research Council, 'Science and Judgment in Risk Assessment' (National Academy Press, Washington D.C. 1994)

National Research Council, 'Understanding Risk: Informing Decisions in a Democratic Society' (National Academy Press, Washington D.C. 1996)

Neumann M, Pleines H and Schröder H, 'Russland Analysen Erdgaskonflikt mit der Ukraine' [2009] No 176 Forschungsstelle Osteuropa an der Universität Bremen available at: http://www.laender-analysen.de/russland/pdf/Russlandanalysen176.pdf [accessed 18 July 2013]

Niedersachsen Landesamt für Bergbau, Energie und Geologie, '*Erdöl und Erdgas in der Bundesrepublik Deutschland 2012*' (Hannover, 2013) available at: http://www.lbeg.niedersachsen.de/portal/live.php?navigation_id=655&article_id=936&_psmand=4 [accessed 15 April 2014]

Niedersachsen Landesamt für Bergbau, Energie und Geologie, '*Erdöl und Erdgas in der Bundesrepublik Deutschland 2014*' (Landesamt für Bergbau, Energie und Geologie, Hannover 2015)

Niedersächsisches Ministerium für Umwelt, Energie und Klimaschutz

'Zulassung von Vorhaben zur Aufsuchung und Gewinnung von Erdgas aus konventionellen Lagerstätten mittels hydraulischer Bohrlochbehandlung zur Risserzeugung in einem Verfahren mit Umweltverträglichkeitsprüfung' available at: http://www.umwelt.niedersachsen.de/aktuelles/pressemitteilungen/strenge-auflagen-bei-der-ergasfoerderung-122495.html [accessed 14 April 2014]

Nitsch J and Pregger T et al., 'Langfristszenarien und Strategien für den Ausbau der erneuerbaren Energien in Deutschland bei Berücksichtigung der Entwicklung in Europa und global' (Deutsches Zentrum für Luft- und Raumfahrt, Stuttgart/Kassel 2012) available at: http://www.dlr.de/tt/Portaldata/41/Resources/dokumente/institut/system/publications/leitstudie2011_bf.pdf [accessed 17 July 2014]

OSPAR Commission, 'Report from the Group of Jurists and Linguists on the placement of carbon dioxide in the OSPAR maritime area. Annex 12 to 2004 Summary Record' available at: http://www.co2geonet.com/FileDownload.aspx?IdFile=222&ViewType=Old&IdType=18&From=News [accessed 24 October 2013]

Parliament of Saxon-Anhalt, 'Motion parliamentary group DIE LINKE – supplementing the resource protection clause of the Mining Act with social and ecological criteria of 21.04.2010' (Antrag Die Linke Ergänzung der Rohstoffsicherungsklausel des Bundesberggesetzes durch soziale und ökologische Kriterien) Drucksache 5/2547

Pearson I et al., 'Unconventional Gas: Potential Energy Market Impacts in the European Union' (Joint Research Centre of the European Commission, 2012) 142 available at: http://ec.europa.eu/dgs/jrc/downloads/jrc_report_2012_09_unconventional_gas.pdf [accessed 20 May 2014]

Philippe & Partners Law Firm, 'Final Report on Unconventional Gas in Europe' (2011) http://ec.europa.eu/energy/studies/doc/2012_unconventional_gas_in_europe.pdf [accessed 12 March 2012]

Potočnik J European Commissioner for Environment, 'A European Strategy for Shale SPEECH/13/840' available at: http://europa.eu/rapid/press-release_SPEECH-13-840_en.htm [accessed 24 April 2014]

Potočnik J European Commissioner for Environment, 'Transmission Note on the EU environmental legal framework applicable to shale gas projects' of 22nd December 2011 available at: http://ec.europa.eu/environment/integration/energy/pdf/legal_assessment.pdf [accessed 24 April 2014]

Research and Information Service of the Northern Ireland Assembly, Briefing Paper 'Key Concerns Surrounding Fracking' available at: http://www.niassembly.gov.uk/Documents/RaISe/Publications/2011/Enterprise-Trade-Investment/17111.pdf [accessed 31 March 2014]

Romanian Parliament, '2012 Expunerea de motive la iniţiativa legislative' available at: http://www.senat.ro/Legis/PDF/2012/12L228EM.pdf [accessed 11 June 2012]

Romanian Parliament, Committee for Industry and Services of the Romanian Parliament, 'RAPORT asupra propunerii legislative privind interzicerea explorărilor şi exploatărilor perimetrelor cu zăcăminte de hidrocarburi lichide sau gazoase prin fracturarea (fisurarea) hidraulică şi anularea licenţelor exclusive de explorare a tuturor proiectelor care recurg la această tehnică' available at: http://www.cdep.ro/comisii/industrii/pdf/2014/rp278_12.pdf [accessed 2 April 2014]

Romanian Parliament, 'Parlamentul României Derularea procedurii legislative' available at: http://senat.ro/Legis/Lista.aspx?cod=16591 [accessed 2 April 2014]

Romanian Senate, 'Minutes of Senate meeting on 21st June 2012 Summary' (Stenograma şedinţei Senatului din 21 iunie 2012 Sumar) available at: http://www.senat.ro/pagini/stenograme/Stenograme2012/12.06.21.pdf [accessed 26 May 2014]

Romanian Senate, 'Minutes of Senate meeting on 5th November 2013 Summary' (Stenograma şedinţei Senatului din 5 noiembrie 2013 Sumar) 92 available at: http://www.senat.ro/PAGINI/Stenograme/Stenograme2013/13.11.05.pdf [accessed 26 May 2014]

Romanian Senate, 'Minutes of Senate meeting on 6th November 2013 Summary' (Stenograma şedinţei Senatului din 6 noiembrie 2013 Sumar) 92 available at: http://www.senat.ro/PAGINI/Stenograme/Stenograme2013/13.11.06.pdf [accessed 26 May 2014]

Rosenwinkel K H, et al., 'Gutachten zur Abwasserentsorgung und Stoffstrombilanz ISAH 2012' available at: http://dialog-erdgasundfrac.de/sites/dialog-erdgasundfrac.de/files/Gutachten%20zur%20Abwasserentsorgung%20und%20Stoffstrombilanz%20ISAH%20Mai%20 2012.pdf [accessed 25 June 2012]

Royal Commission on Environmental Pollution, 'Setting Environmental Standards 21st Report' (Crown, London 1998)

Royal Society and the Royal Academy of Engineering, 'Shale gas extraction in the UK: a review of hydraulic fracturing' (London, 2012) available at: https://royalsociety.org/policy/projects/shale-gas-extraction/report/ [accessed 11 April 2014]

RWE Dea AG, 'Völkersen Nord Z1 Sanierung eines Leckageschadens' available at: http://www.rwe.com/web/cms/mediablob/de/1262156/data/1260112/2/rwe-dea-buergerinformation-voelkersen/rwe-dea-ag-buergerinformation-voelkersen/08.-Dezember-2011-Praesentation-Buergerinformation-Voelkersen-Lagerstaettenwasseraustritt-Voelkersen-Nord-Z1-Massnahmen-Sanierung.pdf [accessed 25 June 2012]

Sachverständigenrat für Umweltfragen (SRU) Faulstich M et al., 'Fracking zur Schiefergasgewinnung - Ein Beitrag zur energie- und umweltpolitischen Bewertung Stellungnahme' available at: http://www.umweltrat.de/SharedDocs/Downloads/DE/04_Stellungnahmen/2012_2016/2013_05_AS_18_Fracking.pdf?__blob=publicationFile [accessed 2 July 2013]
Scottish government 'Policy on unconventional gas' 19 October 2013 available at: http://news.scotland.gov.uk/News/Policy-on-unconventional-gas-547.aspx [accessed 10 June 2016]
Shell UK Ltd., 'Brent Spar Dossier' available at: http://www.static.shell.com/static/gbr/downloads/e_and_p/brent_spar_dossier.pdf [accessed 27 June 2012]
Stern J, 'Security of European Natural Gas Supplies: The Impact of Import Dependence and Liberalization' (2002) available at: http://kms1.isn.ethz.ch/serviceengine/Files/ISN/23033/ipublicationdocument_singledocument/BBD578CB-8574-4923-B727-FA9F1FBF17A0/en/110-Security_Natural_Gas.pdf [accessed 25 April 2013]
Terracon, 'Remedial Investigation Work Plan – Final Crosby 25-3 Natural Gas Well Release Road 1AB Clark, Park County, Wyoming' available at: http://deq.state.wy.us/volremedi/downloads/Web%20Notices/Windsor%20Well_Clark/Work%20Plan%20FINAL%20020108.pdf [accessed 29 March 2013]
Triple E, Ricardo AEA, TNO, '*Support to the review of Directive 2009/31/EC on the geological storage of carbon dioxide (CCS Directive)*' (Publications Office of the European Union, Luxembourg 2014)
UK Government, 'Government response to Royal Academy of Engineering and Royal Society report on "Shale Gas Extraction in the UK: a review of hydraulic fracturing"' (10 December 2012) available at: https://www.gov.uk/government/uploads/system/uploads/attachment_data/file/49541/7269-government-response-sg-report-.pdf [accessed 11 April 2014]
UK Secretary of State for Foreign and Commonwealth Affairs, 'A comparative table of the current EC and EU Treaties as amended by the Treaty of Lisbon' available at: http://www.official-documents.gov.uk/document/cm73/7311/7311.pdf [accessed 9 April 2013]
UKOOG, 'UK Onshore Shale Gas Well Guidelines' (Issue 1, February 2013) available at: https://www.gov.uk/government/uploads/system/uploads/attachment_data/file/185935/UKOOGShaleGasWellGuidelines.pdf [accessed 11 April 2014]
United Nations Conference of the parties to the Convention on Biological Diversity, 'Report of the group of legal and technical experts on liability and redress in the context of paragraph 2 of Article 14 of the convention on biological diversity' of 18/October/2005, UNEP/CBD/COP/8/27/

Add.3 http://www.cbd.int/doc/meetings/cop/cop-08/official/cop-08-27-add3-en.pdf (accessed 27 March 2012)

United Nations Conference on Trade and Development (UNCTAD), 'UNCTAD Series on issues in international investment agreements – Taking of Property' UNCTAD/ITE/IIT/15 (United Nations Publication, New York and Geneva 2000)

United Nations Economic and Social Council, 'Implementation Report submitted by Germany' ECE/MP.PP/IR/2008/DEU, submitted to the meeting of the parties to the Convention on access to information, public participation in decision-making and access to justice in environmental matters, available at: http://www.unece.org/fileadmin/DAM/env/documents/2008/pp/mop3/ece_mp_pp_ir_2008_DEU_e.pdf [accessed 15 October 2012]

United Nations Framework Climate Change Secretariat (UNFCCC) (ed.), 'Understanding Climate Change: A Beginner's Guide to the UN Framework Convention and its Kyoto Protocol' (International Environment House, Geneva 2002)

United Nations International Law Commission, 'Draft articles on Prevention of Transboundary Harm from Hazardous Activities with commentaries' (2001) http://untreaty.un.org/ilc/texts/instruments/english/commentaries/9_7_2001.pdf [accessed 19 April 2012]

United Nations World Health Organisation (WHO), 'International Standards for drinking-water' (WHO, Geneva 1958)

United Nations World Health Organisation (WHO), 'Guidelines for drinking water quality Volume 1' 3rd edition (WHO, Geneva 2004)

US Department of Energy, 'Shale Gas: Applying Technology to Solve America's Energy Challenges' (2011) http://www.netl.doe.gov/technologies/oil-gas/publications/brochures/Shale_Gas_March_2011.pdf [accessed 10 March 2013]

US Energy Information Administration, 'International Energy Outlook 2006' (US Department of Energy, Washington D.C., 2006)

US Energy Information Administration, 'World Shale Gas Resources: An Initial Assessment of 14 Regions outside the US' (April 2011) available at: http://www.eia.gov/analysis/studies/worldshalegas [accessed 6 June 2012]

US Energy Information Administration, 'Technically Recoverable Shale Oil and Shale Gas Resources: An Assessment of 137 Shale Formations in 41 Countries Outside the United States' (US Department of Energy, Washington D.C., June 2013)

US Environmental Protection Agency EPA, '*Assessment of the Potential Impacts of Hydraulic Fracturing for Oil and Gas on Drinking Water Resources*' (Office of Research and Development, Washington D.C. 2015) at ES-6

US Environmental Protection Agency (EPA), 'Draft Investigation of Ground Water Contamination near Pavillion, Wyoming' (2011) available at: http://www.epa.gov/region8/superfund/wy/pavillion/EPA_ReportOn Pavillion_Dec-8-2011.pdf [accessed 14 June 2012]

US Environmental Protection Agency (EPA), 'Draft Plan to Study the Potential Impacts of Hydraulic Fracturing on Drinking Water Resources' (2011) http://water.epa.gov/type/groundwater/uic/class2/hydraulicfracturing/upload/hf_study_plan_110211_final_508.pdfhttp://water.epa.gov/type/groundwater/uic/class2/hydraulicfracturing/upload/hf_study_plan_10211_final_508.pdf [accessed 25 April 2012]

US Environmental Protection Agency (EPA), 'Greenhouse gas emissions reporting from the petroleum and natural gas industry. Background Technical Support Document' available at: http://www.epa.gov/climatechange/emissions/downloads10/Subpart-W_TSD.pdf [accessed 4 June 2012]

Wirtschaftsverband Erdöl- und Erdgasgewinnung e.V., 'Hydraulic Fracturing-Prozess und Perspektiven in Deutschland' available at: http://www.erdoel-erdgas.de/Themen/Erdgas-aus-Deutschland/Hydraulic-Fracturing [accessed 1 July 2014]

Wood R et al., 'Shale gas: a provisional assessment of climate change and environmental impacts' (Tyndall Centre at the University of Manchester, January 2011) available at: http://www.mace.manchester.ac.uk/our-research/centres-institutes/tyndall-manchester/publications/index.htm?year=2011 [retrieved in 2012]

World Commission on Environment and Development, '*Our Common Future*' (Oxford University Press, 1987)

Zoback M, Kitasei S and Copithorne B, 'Addressing the Environmental Risks from Shale Gas Development' (Worldwatch Institute 2010) available at: http://efdsystems.org/Portals/25/Hydraulic%20Fracturing%20Paper%20-%20World%20Watch.pdf [accessed 1 November 2012]

BOOKS

Albrecht R K, '*Zumutbarkeit als Verfassungsmassstab*' (Duncker & Humblot, Berlin 1995)

Alexy R, '*Theorie der Grundrechte*' (Suhrkamp, Frankfurt am Main 1985)

Alexy R, '*A Theory of Constitutional Rights*' (Oxford University Press, 2004)

Appel I, '*Staatliche Zukunfts- und Entwicklungsvorsorge*' (Mohr Siebeck, Tübingen 2005)

Arndt B, *'Das Vorsorgeprinzip im EU-Recht'* (Mohr Siebeck, Tübingen 2009)

Atapattu S U, '*Emerging Principles of International Environmental Law*' (Transnational Publishers, Ardsley 2006)

Aven T, *'Quantitative Risk Assessment'* (Cambridge University Press, Cambridge 2011)

Azar J J, '*Drilling Engineering*' (PennWell Corporation, Tulsa 2007)

Bahgat G, '*Energy Security: An Interdisciplinary Approach*' (Wiley, Hoboken 2011)

Bakshi V (ed.), '*Shale Gas: A Practitioner's Guide to Shale Gas and Other Unconventional Resources*' (Globe Law Publishing, London 2012)

Barton B et al. (eds), '*Energy Security: managing risk in a dynamic legal and regulatory environment*' (Oxford University Press, Oxford 2004)

Battis U, Krautzberger M and Löhr R P, '*Baugesetzbuch Kommentar*' (11th edition, C H Beck, München 2009)

Bauerfeind S, '*Rücknahme und Rückgabepflichten im Umweltrecht*' (Duncker & Humblot, Berlin 1999)

Beaucamp G, '*Das Konzept der zukunftsfähigen Entwicklung im Recht*' (Mohr Siebeck, Tübingen 2002)

Beck U, '*Risk Society: Towards a New Modernity*' (Sage, London 1992)

Bell J, '*French Constitutional Law*' (Oxford University Press 1992)

Bell S and McGillivray D, '*Environmental Law*' 7th edition (Oxford University Press, Oxford 2008)

Bender B, Sparwasser R and Engel R, '*Umweltrecht*' 3rd edition (C F Müller, Heidelberg 1995)

Berendes K, '*Wasserhaushaltsgesetz*' (Erich Schmidt, Berlin 2010)

Bergel J-L, '*Méthodologie juridique*' (Presses Universitaires de France, Paris 2001)

Bergel J-L ,'*Théorie générale du droit*' 4th edition (Dalloz, Paris 2003)

Bergmann Àvila H, '*Theorie der Rechtsprinzipien*' (Duncker & Humblot, Berlin 2006)

Best E, '*Understanding EU Decision-Making*' (Springer International, Cham 2016)

Birnie P W, Boyle A and Redgwell C, '*International Law and the Environment*' 3rd edition (Oxford University Press, Oxford 2009)

Bjorlykke K, '*Petroleum Geoscience – From Sedimentary Environments to Rock Physics*' (Springer Verlag, Berlin 2010)

Bjørnebye H, '*Investing in EU energy security: exploring the regulatory approach to tomorrow's electricity production*' (Kluwer Law International, Alphen aan den Rijn 2010)

Blasberg D, '*Inhalts- und Schrankenbestimmungen des Grundeigentums zum Schutz der natürlichen Lebensgrundlagen*' (Springer, Berlin 2008)

Blümel W et al., '*Verfassungsprobleme im vereinten Deutschland*' 2nd edition (Speyerer Forschungsberichte, Speyer 1993)
Bock B, '*Umweltschutz im Spiegel von Verfassungsrecht und Verfassungspolitik*' (Duncker & Humblot, Berlin 1990)
Bodansky D, Brunnée J and Hey E (eds), '*The Oxford Handbook of International Environmental Law*' (Oxford University Press, Oxford 2008)
Bogdal K-M, '*Heinrich von Kleist: Michael Kohlhaas*' (Wilhelm Fink Verlag, München 1981)
Boldt G and Weller H, '*Bundesberggesetz*' (de Gruyter, Berlin 1994)
Boute A, '*Russian Electricity and Energy Investment Law*' (Brill/Nijhoff, Leiden, 2015)
Britz G and Hellermann J and Hermes G, '*Energiewirtschaftsgesetz Kommentar*' 2nd edition (Beck, München 2010)
Brown M A and Sovacool B J, '*Climate Change and Global Energy Security*' (Massachusetts Institute for Technology, Sabon USA 2011)
Brownlie I, '*Principles of Public International Law*' 3rd edition (Oxford University Press 1979)
Buendia Sierra J L, '*Exclusive Rights and State Monopolies under EC Law: Article 86 (Former Article 90) of the EC Treaty*' (Oxford University Press, Oxford 1999)
Büllesbach R, 'Die rechtliche Beurteilung von Abgrabungen nach Bundes- und Landesrecht' (Duncker & Humblot, Berlin 1994)
Bultmann P F, '*Öffentliches Recht: Mit Vertiefung im Gewerbe-, Wettbewerbs-, Subventions- und Vergaberecht*' 2nd edition (Springer Verlag, Berlin 2008)
Calliess C and Ruffert M, '*EUV/EGV Kommentar*' 4th edition (Beck, München 2011)
Cameron P D, '*Competition in Energy Markets – Law and Regulation in the European Union*' (Oxford University Press, Oxford 2007)
Carbonnier J, '*Droit Civil Introduction*' 27th edition (Presses Universitaires de France, Paris 2002)
Carmines E G and Zeller R A, '*Reliability and Validity Assessment*' (Sage, London 1979)
Clinton B, '*Back to Work: Why We Need Smart Government for a Strong Economy*' (Alfred A. Knopf, New York 2011)
Clò S, '*European Emissions Trading in Practice*' (Edward Elgar Publishing, Cheltenham 2011)
Connelly N and Smith G, '*Politics and the Environment*' 2nd edition (Routledge, London 1999)
Cooney R, '*The Precautionary Principle in Biodiversity Conservation and*

Natural Resource Management' (International Union for Conservation of Nature (IUCN), Gland (Switzerland) and Cambridge 2004)
Cordonier Segger M C and Khalfan A, '*Sustainable Development Law*' (Oxford University Press, 2004)
Corino C, '*Energy Law in Germany*' (Beck, München 2003)
Craig P and de Burca G, '*EU Law Texts Cases and Materials*' 4th edition (Oxford University Press, Oxford 2007)
Craig P and de Burca G, '*EU Law Texts Cases and Materials*' 5th edition (Oxford University Press, Oxford 2011)
Craig P and de Burca G, '*EU Law Texts Cases and Materials*' 6th edition (Oxford University Press, Oxford 2015)
Czychowski M and Reinhardt M, '*Wasserhaushaltsgesetz*' 10th edition, (C H Beck, München 2010)
Danner W und Theobald C, '*Energierecht Kommentar Band 1*' (Beck München 2012)
Dauses M A, '*Handbuch des EU-Wirtschaftsrechts*' 31st edition (Beck, München 2012)
Davies G, '*Nationality Discrimination in the European Internal Market*' (Kluwer Law International, The Hague 2003)
Davies P G G, '*European Union Environmental Law*' (Ashgate, Hants 2004)
de Arechaga E J, '*International law in the past third of a century*' (Martinus Nijhoff Publishers, The Hague 1978)
de Mestral A L C (ed.), '*International Law Chiefly as Interpreted and Applied in Canada*' 5th edition (Emond Montgomery Ltd., Toronto 1993)
de Sadeleer N, '*Environmental Principles: From Political Slogans to Legal Rules*' (Oxford University Press, Oxford 2005)
de Sadeleer N, '*Environmental Law and the Internal Market*' (Oxford University Press, Oxford 2014)
de-Shalit A, '*Why Posterity Matters*' (Routledge Ltd, London 1995)
Deacon R '*Devolution in the United Kingdom*' 2nd edition (Edinburgh University Press, Edinburgh 2012)
Dechsling R, '*Das Verhältnismäßigkeitsgebot*' (Vahlen, München 1989)
Dente B, '*Understanding Policy Decisions*' (Springer, Heidelberg 2014)
Detterbeck S, '*Allgemeines Verwaltungsrecht mit Verwaltungsprozessrecht*' 3rd edition (C H Beck, München 2005)
Deville A and Harding R, '*Applying the Precautionary Principle*' (The Federation Press Ltd., Sydney 1997)
Di Fabio U, '*Risikoentscheidungen im Rechtsstaat: zum Wandel der Dogmatik im öffentlichen Recht, insbesondere am Beispiel der Arzneimittelüberwachung*' Jus Publicum 8 (Mohr Siebeck, Tübingen 1994)
Dixon M, McCorquodale R and Williams S, '*Cases and Materials on International Law*' 5th edition (Oxford University Press, Oxford 2011)

Dolzer R, Graßhof K, Kahl W and Waldhoff C (eds), '*Bonner Kommentar zum Grundgesetz*' (C F Müller Verlag, Heidelberg since 1950)
Dratwa F A et al., '*Energiewirtschaft in Europa*' (Springer, Berlin 2010)
Dreier H, '*Grundgesetz Band II*' (Mohr Siebeck, Tübingen 1998)
Dreier H, '*Grundgesetz Band I*' (Mohr Siebeck, Tübingen 1998)
Drohmann D and Townsend M, '*REACH Best Practice Guide to Regulation No 1907/2006*' (Beck, Muenchen 2013)
Dworkin R, '*Taking Rights Seriously*' (Harvard University Press, Cambridge 1977)
Ebbesson J, '*Compatibility of International and National Environmental Law*' (Kluwer Law International, Dordrecht 1996)
Eberbach W, Lange P and Ronellenfitsch M, '*Recht der Gentechnik und Biomedizin*' Volume 1 (C F Müller Verlag, Karlsruhe 2012)
Emiliou N, '*The Principle of Proportionality in European Law: A Comparative Study*' (Kluwer Law International, London 1996)
Emmerich-Fritsche A, '*Der Grundsatz der Verhältnismäßigkeit als Direktive und Schranke der EG-Rechtsetzung*' (Duncker & Humblot, Berlin 2000)
Epiney A, '*Umweltrecht in der Europäischen Union*' (Heymann, Köln 1997)
Epiney A, '*Umweltrecht in der Europäischen Union*' 2nd edition (Heymann, Köln 2005)
Epiney A, '*Umweltrecht in der Europäischen Union*' 3rd edition (Nomos, Baden-Baden 2013)
Epiney A and Scheyli M, '*Strukturprinzipien des Umweltvölkerrechts*' (Nomos Verlag, Baden-Baden 1998)
Esser J, '*Vorverständnis und Methodenwahl in der Rechtsfindung*' (Ätheneum Verlag, München 1972)
Ferguson S and Gilbert M T, '*Hydraulic Fracturing and Shale Gas Production: Issues, Proposals and Recommendations*' (Nova Science Publishers, New York 2013)
Fischer F and Black M, '*Greening Environmental Policy: The Politics of a Sustainable Future*' (Paul Chapman Publishing, London 1995)
Fischerhof H, '*Deutsches Atomgesetz und Strahlenschutzrecht Kommentar*' 2nd edition (Nomos Verlag, Baden-Baden 1978)
Foster C E, '*Science and the Precautionary Principle in International Courts and Tribunals: Expert Evidence, Burden of Proof and Finality*' (Cambridge University Press, Cambridge 2011)
Frenz W, '*Sustainable Development durch Raumplanung*' (Duncker & Humblot, Berlin 2000)
Frenz W, '*Bergrecht und nachhaltige Entwicklung*' (Duncker & Humblot, Berlin 2001)

Frenz W, '*Handbuch Europarecht Band I Europäische Grundfreiheiten*' 2nd edition (Springer, Berlin 2012)

Frenz W and Unnerstall H, '*Nachhaltige Entwicklung im Europarecht*' (Nomos Verlag, Baden-Baden 1999)

Garner B A (ed.), '*Black's Law Dictionary*' 8th edition (Thomson/West, St. Paul 2004)

Geiger R, Khan D E and Kotzur M, '*European Union Treaties*' (Beck, München 2015)

Gellermann M, '*Grundrechte in einfachgesetzlichem Gewande*' (Mohr Siebeck, Tübingen 2000)

Gény F, '*Can Unconventional Gas be a Game Changer in European Markets?*' (Oxford Institute for Energy Studies, 2010)

Ghiselli E E et al., '*Measurement Theory for the Behavioural Science*' (W H Freeman & Co; San Francisco 1980)

Glaser A, '*Nachhaltige Entwicklung und Demokratie*' (Mohr Siebeck, Tübingen 2006)

Goklany I M, '*The Precautionary Principle: A critical Appraisal of Environment Risk Assessment*' (Cato Institute, Washington D.C. 2001)

Goldsmith B J and Bergkamp L, '*The EU Environmental Liability Directive: A Commentary*' (Oxford University Press, 2013)

Grabitz E, Hilf M and Nettesheim M, '*Das Recht der Europäischen Union Loseblattsammlung*' (Beck, München 2012)

Granger Morgan M et al., '*Carbon Capture and Sequestration Removing the Legal and Regulatory Barriers*' (RFF Press Routledge, New York 2012)

Hager G, '*Rechtsmethoden in Europa*' (Mohr Siebeck, Tübingen 2009)

Haghighi S S, '*Energy Security The External Legal Relations of the European Union with Major Oil- and Gas-Supplying Countries*' (Hart Publishing, Oxford and Portland 2007)

Hahn D, '*Staatszielbestimmungen im integrierten Bundesstaat*' (Duncker & Humblot, Berlin 2010)

Hakemann M, '*Die verfassungsrechtliche Dimension des Verhältnismäßigkeitsgebots in Deutschland*' (GRIN Verlag, Norderstedt 2011)

Haltern U, '*Europarecht Dogmatik im Kontext*' (Mohr Siebeck, Tübingen 2005)

Hansmann K and Sellner D, '*Grundzüge des Umweltrechts*' 4th edition (Erich Schmidt Verlag, Berlin 2012)

Hawke N, '*Environmental Policy: Implementation and Enforcement*' (Ashgate, Dartmouth 2002)

Hayward T, '*Ecological Thought*' (Polity Press, Cambridge 1995)

Heffron R J, '*Energy Law: an Introduction*' (Springer International, Cham 2015)

Hesse K, '*Grundzüge des Verfassungsrechts*' 15th edition (C F Müller, Heidelberg 1985)

Hilson C, '*Regulating Pollution: A UK and EC Perspective*' (Hart, Oxford 2000)

Hinteregger M, '*Environmental Liability and Ecological Damage in European Law*' (Cambridge University Press, Cambridge 2008)

Hohmann H, '*Precautionary Legal Duties and Principles of Modem International Environmental Law*' (Kluwer Law International, The Hague 1994)

Holwerda M, '*EU Regulation of Cross-Border Carbon Capture and Storage*' (Intersentia Publishing, Cambridge 2014)

Hoppe W, Beckmann M and Kauch P, '*Umweltrecht*' 2nd edition (Beck, München 2000)

Hotz W F, '*Zur Notwendigkeit und Verhältnismäßigkeit von Grundrechtseingriffen*' (Dissertation, Zürich 1977)

Hunter T (ed.), '*Handbook on Shale Gas Law and Policy*' (Intersentia, Deventer 2016)

Isensee J and Kirchhof P, '*Handbuch des Staatsrechts Band V*' (C F Müller, Heidelberg 1992)

Isensee J and Kirchhof P, '*Handbuch des Staatsrechts Band VII*' (C F Müller, Heidelberg 1992)

Jakobs M C, '*Der Grundsatz der Verhältnismäßigkeit*' (Carl Heymanns, Köln 1985)

Jans J H and Vedder H H B, '*European Environmental Law*' 3rd edition (Europa Law Publishing, Groningen 2008)

Jans J H and Vedder H H B, '*European Environmental Law*' 4th edition (Europa Law Publishing, Groningen 2012)

Jarass H D, '*Bundesimmissionsschutzgesetz Kommentar*' 6th edition (C H Beck Verlag, München 2005)

Jarass H D and Pieroth B, '*Grundgesetz Kommentar*' 11th edition (C H Beck, München 2011)

Jasanoff S, '*The Fifth Branch: Science Advisers as Policymakers*' (Harvard University Press, Cambridge 1990)

Johnson N, '*Government in the Federal Republic of Germany: The executive at work*' 2nd edition (Pergamon Press, Oxford 1983)

Johnston A and Block G, '*EU Energy Law*' (Oxford University Press, 2012)

Jones C et al., '*EU Energy Law-Volume I: The Internal Market – The Third Liberalisation Package*' 3rd edition (Claeys and Casteels, Leuven 2010)

Kahl W, '*Umweltprinzip und Gemeinschaftsrecht*' (C F Müller, Heidelberg 1993)

Kidd S, Plater A and Frid C (eds), '*The ecosystem approach to marine planning and management*' (Earthscan Publications Ltd., London 2011)
King G et al., '*Designing Social Inquiry: Scientific Inference in Qualitative Research*' (Princeton University Press, Princeton 1994)
Kischel U, '*Die Begründung: Zur Erläuterung staatlicher Entscheidungen gegenüber dem Bürger*' (Mohr Siebeck Verlag, Tübingen 2003)
Kiss A C and Shelton D, '*International Environmental Law*' (Graham & Trotman, London 1991)
Kloepfer M, *'Umweltrecht'* 3rd edition (C H Beck, Berlin 2004)
Kloepfer M, *'Verfassungsrecht Band I'* (C H Beck, München 2011)
Kloepfer M, *'Verfassungsrecht Band II'* (C H Beck, München 2010)
Knack H J and Busch J D, '*Verwaltungsverfahrensgesetz Kommentar*' (C Heymann, Köln 1996)
Kopp F O and Ramsauer U, '*Verwaltungsverfahrensgesetz Kommentar*' (C H Beck, München 2000)
Kosso P, '*Reading the book of nature*' (Cambridge University Press, Cambridge 1992)
Kotulla M, *'Wasserhaushaltsgesetz Kommentar'* (2nd edition, Kohlhammer, Stuttgart 2011)
Kramer E A, '*Juristische Methodenlehre*' 3rd edition (Beck, München 2010)
Krämer L, '*Focus on European Environmental Law*' (Sweet & Maxwell, London 1992)
Krämer L, '*EC Environmental Law*' 6th edition (Sweet & Maxwell, London 2006)
Krämer L, *'EU Environmental Law'* 7th edition (Sweet & Maxwell, London 2011)
Kuhn T S, '*The essential tension*' (University of Chicago Press, Chicago 1977)
Lange B, '*Implementing EU Pollution Control*' (Cambridge University Press, Cambridge 2008)
Larenz K, '*Methodenlehre der Rechtswissenschaft*' 2nd edition (Springer, Berlin 1991)
Larenz K and Canaris W, '*Methodenlehre der Rechtswissenschaft*' 3rd edition (Springer, Berlin 1995)
Laudan L, '*Science and values*' (University of California Press, Berkeley 1981)
Lausche B J, '*Weaving a Web of Environmental Law*' (IUCN Environmental Law Programme, Bonn 2008)
Leal-Arcas R, Filis A and Abu Gosh E S '*International Energy Governance Selected Legal issues*' (Edward Elgar, Cheltenham 2014)

Lecheler H, 'Das Subsidiaritätsprinzip – Strukturprinzipien in der EU' (Duncker & Humblot, Berlin 1993)
Lee M, *'EU Environmental Law: Challenges, Change and Decision-Making'* (Hart Publishing, Oxford 2005)
Lefeber R, '*Transboundary Environmental Interference and the Origin of State*' Liability (Martinus Nijhoff Publishers, Dordrecht 1996)
Lenz C-O and Borchardt K D, '*EU Verträge Kommentar*' 5th edition (Bundesanzeiger Verlag, Köln 2010)
Leopold A, '*A Sand County Almanac; With Essays on Conservation from Round River*' (Oxford University Press, New York 1970)
Longino H L, '*Science as social knowledge*' (Princeton University Press, Princeton 1990)
Luce R D and Raiffa H, '*Games and Decisions*' (John Wiley and Sons, New York 1957)
Manssen G, '*Staatsrecht II Grundrechte*' 11th edition (Beck, München 2014)
Marchant G E and Mossmann K L, '*Arbitrary and Capricious: The Precautionary Principle in the EU Courts*' (Washington D.C., AEI Press 2004)
Marr S, '*The Precautionary Principle in the Law of the Sea: Modern Decision Making in International Law*' (Kluwer Law International, The Hague 2003)
Maunz T and Dürig G (eds), '*Grundgesetz Kommentar*' (CH Beck, München since 1958)
Maurmann D, '*Rechtsgrundsätze im Völkerrecht am Beispiel des Vorsorgeprinzips*' (Nomos Verlag, Baden-Baden 2008)
Meßerschmidt K, '*Europäisches Umweltrecht*' (Beck, München 2011)
Mizdalski F, *'Zur restriktiven Auslegung der Urheberschranken vor dem Hintergund von Art. 5 GG'* (Lang Verlag, Frankfurt am Main 2011)
Molenaar E J, '*Coastal State Jurisdiction over Vessel-Source Pollution*' (Kluwer Law International, Dordrecht 1998)
Morris J (ed.), '*Rethinking Risk and the Precautionary Principle*' (Butterworth-Heinemann, Oxford 2000)
Muchlinski P, '*Multinational Enterprises and the Law*' (Blackwell Ltd., Oxford 1995)
Müller H K '*A Legal Framework for a Transnational Offshore Grid in the North Sea*' (Intersentia, Cambridge 2016)
Müller-Kraenner S, '*Energiesicherheit*' (Kunstmann, München 2007)
Müller-Kraenner S, '*Energy Security: Re-measuring the World*' (Earthscan, London 2008)
Mulvaney D and Robbins P (eds), '*Green Politics*' (SAGE Publications, London 2011)

Murswiek D, '*Ausgewâhlte Probleme des Allgemeinen Umweltrechts*' (Duncker & Humblot, Berlin 2005)

Nötzold A, '*Die Energiepolitik der EU und der VR China*' (Springer, Wiesbaden 2011)

Oliver P, '*Free Movement of Goods in the European Community under Article 28 to 30 of the EC Treaty*' 4th edition (Sweet & Maxwell, London 2003)

Oppermann T, Classen D and Nettesheim N, '*Europarecht*' 4th edition (Beck, München 2009)

Organisation for economic co-operation and development (OECD), '*International Investment Law: A changing landscape*' (OECD publishing, Paris 2005)

Ortino F, '*The WTO Dispute Settlement System, 1995-2003*' (Kluwer Law International, The Hague 2004)

Pascual C and Elkind J, '*Energy Security*' (Brookings Institution Press 2010)

Pearce D and Turner R K, '*Economics of Natural Resources and the Environment*' (Johns Hopkins University Press, Baltimore 1990)

Piens R, Schulte H-W and Graf Vitzthum S, '*Bundesberggesetz*' 2nd edition (Kohlhammer, Stuttgart 2013)

Pieroth B and Schlink B, '*Grundrechte Staatsrecht II*' 25th edition (CF Müller Heidelberg 2008)

Purps T, '*Umweltpolitik und Verursacherprinzip im Europäischen Gemeinschaftsrecht*' (Carl Heymanns, Köln 1991

Radbruch G, '*Rechtsphilosophie*' 2nd edition (C F Müller, Heidelberg 2003)

Rehbinder E, '*Politische und rechtliche Probleme des Verursacherprinzips*' (Erich Schmidt Verlag, Berlin 1973)

Reiter B, '*Entschädigungslösungen für durch Luftverunreinigungen verursachte Distanz- und Summationsstörungen*' (Erich Schmidt Verlag, Berlin 1998)

Rengeling H W (ed.), '*Handbuch zum europäischen und deutschen Umweltrecht Band 1 Allgemeines Umweltrecht*' (Carl Heymanns Verlag, Köln 1998)

Resnik M D, '*Choices: An introduction to decision theory*' (University of Minnesota Press, Minneapolis 1987)

Richardson B J and Wood S, '*Environmental Law for Sustainability: A Reader*' (Hart Publishing, Portland 2006)

Ringeling H-W (ed.), '*Handbuch zum europäischen und deutschen Umweltrecht Band I*' (Heymanns, Köln 1998)

Roller G, '*Genehmigungsaufhebung und Entschädigung im Atomrecht*' (Nomos, Baden Baden 1994)

Rubel R, 'Planungsermessen. Norm- und Begründungsstruktur' (Metzner, Frankfurt am Main 1982)

Ruff A, '*Staatszielbestimmung "Umweltschutz" Artikel 20a Grundgesetz*' (Mensch und Buch Verlag, Berlin 2001)

Rusteberg B, '*Der grundrechtliche Gewährleistungsgehalt*' (Mohr Siebeck, Tübingen 2009)

Rüthers B, Fischer C and Birk A, '*Rechtstheorie*' 8th edition (Beck, München 2015)

Sachs M, '*Grundgesetz Kommentar*' 6th edition (Beck, München 2011)

Säcker F J, '*Berliner Kommentar zum Energierecht Band I*' 2nd edition (Verlag Recht und Wirtschaft, Frankfurt am Main 2010)

Sakmar S L, '*Energy for the 21st Century*' (Edward Elgar, Cheltenham 2013)

Sanden J, '*Introduction to International Environmental Law*' (Juura Ltd., Tallinn 2003)

Sands P and Peel J, '*Principles of International Environmental Law, Vol. I*' (Manchester University Press 1995)

Sands P and Peel J, '*Principles of International Environmental Law*' 2nd edition (Cambridge University Press, 2003)

Sands P, '*International Law in the Field of Sustainable Development*' (Kluwer Law International, London 1995)

Scheepers M et al., '*EU Standards for Energy Security of Supply*' (Energy Research Centre of the Netherlands, The Hague 2006)

Scherzberg A, '*Grundrechtsschutz und "Eingriffsintensität"*' (Duncker & Humblot, Berlin 1989)

Schmidt R,'*Grundrechte*' 12th edition (Verlag Rolf Schmidt, Grasberg 2010)

Schmidt-Bleek F B, '*The Fossil Makers*' (Birkhäuser Ltd., Basel 1993)

Schmidt-Bleibtreu B and Klein F, '*Kommentar zum Grundgesetz*' 10th edition (Wolters Kluwer Deutschland GmbH, München 2004)

Schmitz S, '*Die Europäische Union als Umweltunion*' (Rhombos Verlag, Berlin 1996)

Schneider H, '*Die Güterabwägung des Bundesverfassungsgerichts bei Grundrechtskonflikten*' (Nomos, Baden-Baden 1979)

Schrijver N, '*Sovereignty Over Natural Resources: Balancing Rights and Duties*' (Cambridge University Press 1997)

Schwabe J, '*Probleme der Grundrechtsdogmatik*' (Selbstverlag, Darmstadt 1977)

Schwarze J et al., (eds), '*Europäisches Verwaltungsrecht*' 2nd edition (Nomos, Baden-Baden 2005)

Schwarze J et al., (eds), '*EU Kommentar*' 3rd edition (Nomos, Baden-Baden 2012)

Schwier H, '*Der Schutz der "Unternehmerischen Freiheit" nach Artikel 16 der Charta der Grundrechte der Europäischen Union*' (Peter Lang GmbH, Frankfurt am Main 2008)

Shrader-Frechette K, '*Risk and Rationality*' (University of California Press, Berkeley 1991)

Sieder F and Zeitler H et al., '*Wasserhaushaltsgesetz Abwasserabgabengesetz Band* 1' (C H Beck, München 2016)

Sjafjell B, *'Towards a Sustainable European Company Law'* (Kluwer Law International, Alphen aan den Rijn 2009)

Snell J, '*Goods and Services in EU Law – A Study of the Relationship between the Freedoms*' (Oxford University Press, Oxford 2002)

Sommermann K P, '*Staatsziele und Staatszielbestimmungen*' (Mohr Siebeck, Tübingen 1997)

Sornarajah M, '*The International Law of Foreign Investment*' 2nd edition (Cambridge University Press 2004)

Sovacool B K (ed.), *'The Routledge Handbook of Energy Security'* (Routledge Ltd., London, New York 2011)

Sparks D L, '*Environmental Soil Chemistry*' 2nd edition (Academic Press, San Diego 2003)

Sparwasser R, Engel R and Voßkuhle A, '*Umweltrecht: Grundzüge des öffentlichen Umweltschutzrechts*' 5th edition (C F Müller Ltd, Heidelberg 2003)

Squintani L, '*Gold-Plating of European Environmental Law*' (University of Groningen, Proefschrift 2013)

Steinberg R, Müller H and Wickel M, '*Fachplanung*' 4th edition (Nomos, Baden-Baden 2012)

Steiner J and Woods L, *'EU Law'* (Oxford University Press, Oxford 2009)

Stern K, '*Das Staatsrecht der Bundesrepublik Deutschland Band I*' 2nd edition (Beck, München 1984)

Stern K, '*Das Staatsrecht der Bundesrepublik Deutschland Band III/1*' (Beck, München 1994)

Stern K, '*Das Staatsrecht der Bundesrepublik Deutschland Band III/2*' (Beck, München 1994)

Stern P C and Aronson E, '*Energy Use: The Human Dimension*' (Freeman and Company, New York 1984)

Stoneley R, '*An Introduction to Petroleum Exploration for Non-Geologists*' (Oxford University Press, Oxford, 1995)

Streinz R et al., '*EUV/EGV*' (Beck, München 2003)

Streinz R et al., '*EUV/AEUV*' 2nd edition (Beck, München 2012)

Sunstein C, '*The Laws of Fear: Beyond the Precautionary Principle*' (Cambridge University Press, Cambridge 2005)

Talus K, *'EU Energy Law and Policy'* (Oxford University Press, 2013)

Tan K T, *'Principles of Soil Chemistry'* 4th edition (CRC Press, Boca Raton 2011)

Thieme W, '*Umweltschutz im Recht*' (Duncker & Humblot, Berlin 1988)

Tickner J, Raffensperger C and Myers N, '*The precautionary principle in action*' (2000) available at: http://www.biotech-info.net/handbook.pdf [accessed 11 May 2012]

Trouwborst A, '*Evolution and status of the precautionary principle in international law*' (Kluwer Law International, The Hague 2002)

Trouwborst A, '*Precautionary rights and duties of states*' (Martinus Nijhoff Publishers, Leiden 2006)

Tsai T-J, '*Die verfassungsrechtliche Umweltschutzpflicht des Staates*' (Duncker & Humblot, Berlin 1996)

Vedder C, Heintschel von Heinegg W and Folz H-P,'*Europäisches Unionsrecht Handkommentar*' (Nomos Verlagsgesellschaft, Baden-Baden 2012)

Vedder H H B, '*Competition Law and Environmental Protection in Europe – Towards Sustainability?*' (Europa Law Publishing, Groningen 2003)

von Bogdany A and Bast J, '*Europäisches Verfassungsrecht*' 2nd edition (Springer, Berlin 2009)

von der Groeben H and Schwarze J, '*Vertrag über die Europäische Union und Vertrag zur Gründung der Europäischen Gemeinschaft Kommentar Band 3*' 6th edition (Nomos, Baden-Baden 2004)

von Mangoldt H, Klein F and Starck C (eds),'*Das Bonner Grundgesetz Kommentar Band 1*' 6th edition (Verlag Vahlen, München 2010)

von Mangoldt H, Klein F and Starck C (eds), '*Das Bonner Grundgesetz Kommentar Band 2: Artikel 20 bis 78*' 4th edition (Verlag Franz Vahlen, München 2000)

von Münch I and Kunig P, '*Grundgesetz Komentar Band 1: Präambel bis Art. 69*' 6th edition (Beck, München 2012)

von Mutius A, '*Das Widerspruchsverfahren der VwGO als Verwaltungsverfahren und Prozessvoraussetzung*' (Duncker & Humblot, Berlin 1969)

von Savigny F C, '*System des heutigen römischen Rechts Band I*' (Veit, Berlin 1840)

Wagner M,'*Das Konzept der Mindestharmonisierung*' (Duncker & Humblot, Berlin 2000)

Weale A et al., '*Environmental Governance in Europe*' (Oxford University Press 2000)

Weatherill S, *'Cases and Materials on EU Law'* 9th edition (Oxford University Press, Oxford 2010)

Weller H and Kullmann U, *'Bundesberggesetz'* (Beck, München 2012)

Wieberneit B, '*Europarechtlicher Ordnungsrahmen für Umweltsubventionen*' (Duncker & Humblot, Berlin 1997)

Wilms H, '*Staatsrecht I*' (Kohlhammer GmbH, Stuttgart 2007)

Winpenny J T, '*Value for the Environment: A Guide to Economic Appraisal*' 2nd edition (HMSO, London 1993)
Winter G, '*European Environmental Law. A Comparative Perspective*' (Dartmouth, Dordrecht 1996)
Wissinsk B and Bourma J (eds), '*Perspectieven op milieurisico's*' (Wetenschappelijke Raad voor het Regeringsbeleid, The Hague 2002)
Wolf S and Stanley N, '*Wolf and Stanley on Environmental Law*' 5th edition (Routledge, Oxford 2011)
Yergin D, '*The Quest: Energy Security and the Remaking of the Modern World*' (Penguin Ltd., London 2011)
Yew C H, '*Mechanics of hydraulic fracturing*' (Gulf Publications Co., Houston 1997)
Zweigert K and Kötz H, '*Einführung in die Rechtsvergleichung*' (Mohr Siebeck, Tübingen 1971)
Zweigert K and Kötz H and Weir T, '*Introduction to comparative law*' 3rd edition (Clarendon, Oxford 1998)

CHAPTERS IN BOOKS

Ashford N A, 'A Conceptual Framework for the Use of the Precautionary Principle in Law' in Raffensperger C and Tickner J A (eds) '*Protecting Public Health and the Environment: Implementing the Precautionary Principle*' (Island Press Ltd., Washington 1999) 198–206
Barrett K and Raffensperger C, 'Precautionary Science' in Raffensperger C and Tickner J (eds), '*Protecting Public Health and the Environment: Implementing the Precautionary Principle*' (Island Press Ltd., Washington 1999) 106–22
Basedow J, 'Zielkonflikte und Zielhierarchien im Vertrag über die Europäische Gemeinschaft' in Due O (ed.) '*Festschrift für Ulrich Everling*' (Nomos, Baden-Baden 1995) 49–68
Beckerman W, 'The precautionary principle and our obligations to future generations' in Morris J, '*Rethinking Risk and the Precautionary Principle*' (Butterworth-Heinemann, Oxford 2000) 46–60
Bergkamp L and Park D, 'The Organizational and Administrative Structures' in Bergkamp L (ed.), '*The European Union REACH Regulation for Chemicals Law and Practice*' (Oxford University Press, 2013) 23–39
Bleckmann A, 'Handlungspflichten und Zustandssicherung im Europäischen Gemeinschaftsrecht' in Börner B (ed.) '*Einigkeit und Recht und Freiheit – Festschrift für Karl Carstens Band 1*' (Heymann, Köln 1984) 43–61

Bodansky D, 'Deconstructing the Precautionary Principle' in Caron D D and Scheiber H N (eds) *'Bringing New Law to Ocean Waters'* (Konisklijke Brill NV, Leiden 2004) 381–91

Boehmer-Christiansen S, 'The Precautionary Principle in Germany – enabling government' in O'Riordan T and Cameron J '*Interpreting the Precautionary Principle*' (Earthscan Publications Ltd., London 1994) 31–61

Boersma T and Johnson C, 'Twenty Years of US Experience – Lessons Learnt For Europe' in Musialski C et al. (eds) '*Shale Gas in Europe*' (Claeys & Casteels, Deventer 2013) 9–34

Boisson de Chazournes L, 'The Precautionary Principle' in UNEP '*Precaution from Rio to Johannesburg: Proceedings of a Geneva Environment Network Roundtable*' (International Environment House, Geneva 2002) 10–12

Bosecke T, 'Stand der Gesetzgebung zum CCS-Gesetz' in Jaeckel L and Janssen G (eds) *'Riskiodogmatik im Umwelt- und Technikrecht'* (Mohr Siebeck, Tübingen 2012) 127–50

Broomfield M, Johnson M and Dönszelmann E, 'Risks Associated With Shale Gas Extraction and Evidences Supporting the Need for a Specific EU Regulatory Framework' in Musialski C et al. (eds) '*Shale Gas in Europe*' (Claeys & Casteels, Deventer 2013) 335–60

Brown M A and Dworkin M, 'The Environmental Dimension of Energy Security' in Sovacool B K (ed.) '*The Routledge Handbook of Energy Security*' (Routledge Ltd, London, New York 2011) 176–90

Brunnee J, 'A Conceptual Framework for an International Forests Convention: Customary Law and Emerging Principles' in Canadian Council on International Law (ed.) '*Global Forests and International Environmental Law*' (Kluwer Law International, London 1996) 41–77

Calliess C, 'Ökologisierung des EG Vertrages in Baumeister H '*Wege zum ökologischen Rechtsstaat*' (Blottner, Taunusstein 1994) 71–100

Calliess C, 'EU-Umweltrecht' in Hansmann K and Sellner D '*Grundzüge des Umweltrechts*' 4th edition (Erich Schmidt Verlag, Berlin 2012) 55–134

Cameron J, 'The Status of the Precautionary Principle in International Law' in Cameron J and O'Riordan T (eds) '*Interpreting the Precautionary Principle*' (Earthscan Publications Ltd., London 1994) 262–91

Cameron J and Abouchar J, 'The Status of the Precautionary Principle in International Law' in Freestone D and Hey E (eds) '*The Precautionary Principle and International Law: the Challenge of Implementation*' (Kluwer Law International, The Hague 1996) 29–52

Cameron J, Wade-Gery W and Abouchar J, 'Precautionary Principle and

Future Generations' in Agius E '*Future Generations and International Law*' (Earthscan Publications Ltd., London 1998) 93–113

Carpenter R A, 'Limitations in Measuring Ecological Sustainability' in Trzyna T (ed.) '*A Sustainable World: Defining and Measuring Sustainable Development*' (Earthscan Publications Ltd, London 1995) 175–97

Caspar J, '§2 Europäisches und nationales Umweltverfassungsrecht' in Hans-Joachim Koch '*Umweltrecht*' 3rd edition (Franz Vahlen, München 2010) 39–73

Corbeau A S, 'The Introduction of unconventional gas in Europe: Opportunities and Challenges' in Roggenkamp M and Woolley O (eds) '*European Energy Law Report IX*' (Intersentia Publishing Ltd, Cambridge 2012) 191–204

De George R T, 'The environment, rights and future generations' in Partridge E (ed.), '*Responsibilities to Future Generations*' (Prometheus Books Ltd., New York 1981) 157–66

del Guayo I, 'Energy Law in Spain' in Roggenkamp M M et al. (eds) '*Energy Law in Europe*' 3rd edition (Oxford University Press, Oxford 2016) 973–1052

Di Fabio U, 'Voraussetzungen und Grenzen des umweltrechtlichen Vorsorgeprinzips' in Kley M et al., '*FS für Ritter*' (Verlag Dr. Otto Schmidt, Köln 1997) 807–38

Doppelhammer M 'The CCS Directive, its Implementation and the Co-financing of CCS and RES Demonstration Projects under the Emissions Trading System (NER 300 Process)' in Havercroft I, Macrory R and Stewart R (eds) '*Carbon Capture and Storage: Emerging Legal and Regulatory Issues*' (Hart Publishing, Oxford and Portland 2011) 93–103

Elkind J, 'Energy Security Call for a Broader Agenda' in Pascual C and Elkind J (eds) '*Energy Security Economics, Politics, Strategies and Implications*' (Brookings Institution Press, Washington DC 2010)

Feinberger J, 'The rights of animals and unborn generations' in Ernest Partridge (ed.), '*Responsibilities to Future Generations*' (Prometheus Books Ltd, New York 1981) 139–50

Ferrari G F, 'Energy Law in Italy' in Roggenkamp M M et al. (eds) '*Energy Law in Europe*' 3rd edition (Oxford University Press, Oxford 2016) 643–724

Fischer W, 'No CCS in Germany Despite the CCS Act?' in Kuckshinrichs W and Hake J F (eds), '*Carbon Capture, Storage and Use: Technical, Economic, Environmental and Societal Perspectives*' (Springer International, Cham 2015) 255–86

Fisher E C and Harding R, 'The precautionary principle and administrative constitutionalism: the development of frameworks for applying the precautionary principle' in Elisabeth C Fisher et al. (eds) '*Implementing the Precautionary Principle: Perspectives and Prospects*' (Edward Elgar Publications, Cheltenham and Northampton 2006) 113–36

Florini A, 'Global Governance and Energy' in Pascual C and Elkind J (eds), '*Energy Security Economics, Politics, Strategies and Implications*' (Brookings Institution Press, Washington DC 2010) 149–81

Fouillac C, 'CO_2 Capture, Transport and Storage, a Promising Technology for Limiting Climate Change' in Saulnier J B and Varella M (eds) '*Global Change, Energy Issues and Regulation Policies*' (Springer International, Cham 2013) 121–41

Freestone D, 'The Precautionary Principle' in Churchill R R and Freestone D (eds), '*International Law and Global Climate Change*' (Kluwer Law International, London/Dordrecht 1991) 21–39

Freestone D, 'International Fisheries Law Since Rio: The Continued Rise of the Precautionary Principle', in Boyle A and Freestone D (eds) '*International Law and Sustainable Development: Past Achievements and Future Challenges*' (Oxford University Press, Oxford 1999) 135–64

Freestone D and Somsen H, 'The Impact of Subsidiarity' in Jane Holder *'The Impact of EC Environmental Law in the UK'* (Chichester, Wiley & Sons 1997) 87–99

Gilland T, 'Precaution, GM crops and farmland birds' in Morris J '*Rethinking Risk and the Precautionary Principl*e' (Butterworth-Heinemann, Oxford 2000) 60–83

Gordley J, 'The functional method' in Monateri P G (ed.) '*Methods of comparative law*' (Edward Elgar, Cheltenham 2012) 107–19

Gordon G, 'Petroleum Licensing' in Gordon G, Paterson J and Üsenmez E, '*Oil and Gas Law: current practice and emerging trends*' 2nd edition (Dundee University Press, Dundee 2011)

Gordon G, McHarg A and Paterson J, 'Energy Law in the United Kingdom' in Roggenkamp M M et al. (eds) '*Energy Law in Europe*' 3rd edition (Oxford University Press, Oxford 2016) 1053–136

Greinacher D and Helmes S, 'Revising the Environmental Impact Assessment Thresholds: The Case of Germany' in Musialski C et al. (eds) '*Shale Gas in Europe*' (Claeys & Casteels, Deventer 2013) 501–16

Haigh N, 'The Introduction of the Precautionary Principle into the UK' in O'Riordan T and Cameron J (eds), '*Interpreting the Precautionary Principle*' (Earthscan Publications Ltd, London 1994) 229–51

Hancher L and Janssen S, 'Shared Competences and Multi-Faceted Concepts – European Legal Framework for Security of Supply' in Barton B et al. (eds) '*Energy Security: managing risk in a dynamic legal and regulatory environment*' (Oxford University Press, Oxford 2004) 85–120

Hansen B, 'Background and Structure of REACH' in Lucas Bergkamp (ed.) *'The European Union REACH Regulation for Chemicals Law and Practice'* (Oxford University Press, 2013) 17–22

Haszeldine S, 'Geological Factors in Framing Legislation to Enable and Regulate Storage of Carbon Dioxide Deep in the Ground' in Havercroft I, Macrory R and Stewart R (eds) '*Carbon Capture and Storage: Emerging Legal and Regulatory Issues*' (Hart Publishing, Oxford and Portland 2011) 7–23

Hébert R L and Ledésert B, 'Calcimetry at Soultz-Sous-Forets enhanced geothermal system: Relationships with fracture zones, flow pathways and reservoir chemical stimulation results' in Jianwen Yang '*Geothermal energy, technology and geology*' (Nova Science Publishers, New York 2012) 93–114

Herbatschek N, Bergkamp L and Mihova M, 'The REACH Programmes and Procedures' in Lucas Bergkamp (ed.) '*The European Union REACH Regulation for Chemicals Law and Practice*' (Oxford University Press, 2013) 82–172

Heselhaus S, 'Verfassungsrechtliche Grundlagen des Umweltschutzes' in Klaus Hansmann and Dieter Sellner '*Grundzüge des Umweltrechts*' 4th edition (Erich Schmidt Verlag, Berlin 2012) 3–54

Holloway S, Bentham M and Kirk K, 'Underground Storage of Carbon Dioxide' in Shackley S and Gough C (eds) *'Carbon Capture and its Storage'* (Ashgate Publishing Ltd., Hampshire 2006) 15–42

Hossain K, 'Introduction: Permanent Sovereignty over Natural Resources' in Kamal Hossain (ed.) '*Legal Aspects of the new International Economic Order*' (Nichols, New York 1980) 33–44

Hunter T, Usenmez E and Paterson J, 'Future Trends in Shale Gas Law and Policy in the United Kingdom' in Hunter T (ed.) *'Handbook of Shale Gas Law and Policy'* (Intersentia, Cambridge 2015) 383–94

Jasanoff S and Wynne B, 'Science and decision-making' in Rayner S and Malone E L (eds), *'Human Choice and Climate Change Volume 1'* (Battelle Press, Columbus 1998) 1–88

Jimenez Beltran D, 'Preface: To Know and Not to Know, To Act or Not to Act?' in European Environment Agency '*Late Lessons from Early Warnings: The Precautionary Principle 1896–2000*' Environmental Issue Report No. 22 (EEA, Copenhagen 2001) 3–5

Jordan A and O'Riordan T, 'The Precautionary Principle in Contemporary Environmental Policy and Politics' in Raffensperger C and Tickner J A (eds) '*Protecting Public Health and the Environment: Implementing the Precautionary Principle*' (Island Press Ltd., Washington 1999) 15–35

Krämer L, 'Polluter -Pays -Principle in Community Law: The Interpretation of Article 103r of the EEC Treaty' in '*Focus on European Environmental Law*' (Sweet & Maxwell, London 1992) 244–64

Kristl K T, 'Public participation and sustainability: how Pennsylvania's shale gas program thwarts sustainable outcomes' in Dernbach J C and May J R (eds) '*Shale Gas and the Future of Energy Law and Policy for Sustainability*' (Edward Elgar Publishing, Cheltenham, 2016) 125–42

Kruyt B et al., 'Indicators for Energy Security' in Sovacool B K (ed.) '*The Routledge Handbook of Energy Security*' (Routledge Ltd., London, New York 2011) 291–312

Kühne G, 'Energy Security and Conflict with other Values: The Case of Germany' in Barton B et al. (eds) '*Energy Security: managing risk in a dynamic legal and regulatory environment*' (Oxford University Press, Oxford 2004) 337–54

Kutscheidt E, 'Schädliche Umwelteinwirkungen' in Czajka D, Hansmann K and Rebentisch M '*Immissionsschutzrecht in der Bewährung: Festschrift für Gerhard Feldhaus*' (C.F. Müller, Heidelberg/ Karlsruhe1999) 1–19

Lang A and Tosun J, 'The Politics of Hydraulic Fracturing in Germany: An Analysis of Discourse Networks' in Christoph Weible C, Ingold K, Fischer M and Heikkila T (eds) '*Mapping the Political Landscapes of Hydraulic Fracturing*' (Palgrave Macmillan, Basingstoke 2016)

Lauriol T, 'Energy Law in France' in Roggenkamp M M et al. (eds) '*Energy Law in Europe*' 3rd edition (Oxford University Press, Oxford 2016) 481–558

Lemke-Schulte E-M, 'Zwischen Anspruch und Wirklichkeit eines europäischen Umweltrechts' in Calliess C, '*Europäisches Umweltrecht als Chance*' (Blottner, Berlin 1992) 65–8

Lerche P, 'Die Verfassung als Quelle von Optimierungsgeboten?' in Burmeister J et al. '*Verfassungsstaatlichkeit. Festschrift für Klaus Stern*' (Beck, München 1997) 197–210

Lewis M and Westaway N, 'Public Participation in UK CCS Planning and Consent Procedures' in Havercroft I, Macrory R and Stewart R (eds) '*Carbon Capture and Storage: Emerging Legal and Regulatory Issues*' (Hart Publishing, Oxford and Portland) 277–96

Lowe V, 'Sustainable Development and Unsustainable Arguments' in Boyle A and Freestone D (eds) '*International Law and Sustainable Development: Past Achievements and Future Challenges*' (Oxford University Press, Oxford 1999) 19–37

Lübbe-Wolff G, 'Präventiver Umweltschutz' in Bizer J, Denninger E and Koch H J (eds) '*Sicherheit, Vielfalt, Solidarität*' (Nomos Verlag, Baden Baden 1998) 47–74

Luft G, Corin A and Gupta E, 'Energy Security and Climate Change – A tenuous link' in Sovacool B K (ed.) '*The Routledge Handbook of Energy Security*' (Routledge Ltd., London, New York 2011) 43–55

Macklin R, 'Can future generations correctly be said to have rights?' in Partridge E (ed.), *'Responsibilities to Future Generations'* (Prometheus Books Ltd., New York 1981) 151–6

MacLeod N, 'Unitisation' in Gordon G, Paterson J and Üsenmez E, '*Oil and Gas Law: current practice and emerging trends*' 2nd edition (Dundee University Press, Dundee 2011) 411–42

Maichel G, 'Das Energiekapitel in der europäischen Verfassung – mehr Integration oder mehr Zentralismus für die leitungsgebundene Energiewirtschaft Europas' in Hendler R, Ibler M and Martinez Soria J '*"Für Sicherheit, für Europa" Festschrift für Volkmar Götz zum 70. Geburtstag*' (Vandenhoeck & Ruprecht, Göttingen 2005) 55–72

Marburger P, 'EG-Umwelthaftungsrichtlinie und zivilrechtliche Umwelthaftung' in Führ M, Wahl R and von Wilinowsky P '*Umweltrecht und Umweltwissenschaft Festschrift für Eckard Rehbinder*' (Erich Schmidt Verlag, Berlin 2007) 237–52

Maurin C and Vivoda V, 'Shale Gas and the Energy Policy "Trilemma" in Hunter T (ed.) *'Handbook of Shale Gas Law and Policy'* (Intersentia, Cambridge 2016) 369–81

May J R and Dernbach J C, 'Introduction' in Dernbach J C and May J R (eds) '*Shale Gas and the Future of Energy Law and Policy for Sustainability*' (Edward Elgar Publishing, Cheltenham, 2016) 1–17

McHarg A and Poustie M, 'Risk, Regulation, and Carbon Capture and Storage: The United Kingdom Experience' in Zillman D N et al. (eds) '*The Law of Energy Underground: Understanding New Developments in Subsurface Production, Transmission and Storage*' (Oxford University Press, 2014) 249–74

Miller H I and Conko G, 'Genetically modified fear and the international regulation of biotechnology' in Morris J '*Rethinking Risk and the Precautionary Principle*' (Butterworth-Heinemann, Oxford 2000) 84–104

Morgan J, 'Sustainability and stakeholder participation: shale gas extraction in the United Kingdom' in Dernbach J C and May J R (eds) '*Shale Gas and the Future of Energy Law and Policy for Sustainability*' (Edward Elgar Publishing, Cheltenham, 2016) 143–64

Morrissey J and Schumacher J, 'Water quality, water use and wastewater issues related to hydraulic fracturing' in Vivek Bakshi (ed.) '*Shale Gas: A Practitioner's Guide to Shale Gas and Other Unconventional Resources*' (Globe Law Publishing, London 2012) 77–92

Musialski C, 'The EU Legal and Regulatory Framework as Currently Applicable to Shale Gas Extraction and Commercialisation' in Musialski C et al. (eds) '*Shale Gas in Europe – A Multidisciplinary Analysis with a Focus on European Specificities*' (Claes & Casteels, Deventer 2013) 62–76

Musialski C, 'An External Comment on the UK/England & Wales:

Towards an even more refined legal & regulatory framework by imposing A.O. additional and specific requirements for mitigating the risk of induced seismicity?' in Cecile Musialski et al. (eds) '*Shale Gas in Europe*' (Claeys & Casteels, Deventer 2013) 517–36

Musialski C and Altmann M, 'General Conclusions' in Musialski C et al. (eds) '*Shale Gas in Europe*' (Claeys & Casteels, Deventer 2013) 537–53

Nivola P S and Carter E E R, 'Making Sense of "Energy Independence"' in Pascual C and Elkind J (eds) '*Energy Security Economics, Politics, Strategies and Implications*' (Brookings Institution Press, Washington DC 2010) 105–18

Noe S Y and Pring G R, 'The "Fear Factor": Why We Should Not Allow Energy Security Rhetoric to Trump Sustainable Development' in Barton B et al. (eds) '*Energy Security: managing risk in a dynamic legal and regulatory environment*' (Oxford University Press, Oxford 2004) 431–56

Nollkaemper A, 'What You Risk Reveals What You Value', and Other Dilemmas Encountered in Legal Assaults on Risks', in Freestone D and Hey E (eds) '*The Precautionary Principle and International Law*' (Kluwer Law International, The Hague 1996) 73–94

Norton P M, 'Back to the Future: Expropriation and the Energy Charter Treaty' in Wälde T W '*The Energy Charter Treaty – An East West Gateway for Investment and Trade*' (Kluwer Law International, London 1996) 365–85

Pallemaerts M, 'Climate Change, Natural Gas and the Rebirth of EU Energy Policy' in Wylie L and Winand P (eds) '*Energy and the Environmental Challenge: Lessons from the European Union and Australia*' (Peter Lang, Brussels 2011) 37–61

Parker C and Braithwaite J, 'Regulation' in Cane P and Tushnet M (eds) '*The Oxford Handbook of Legal Studies*' (Oxford University Press, 2003) 119–46

Paterson J, 'Decommissioning of Offshore Oil and Gas Installations' in Gordon G, Paterson J and Üsenmez E, '*Oil and Gas Law: current practice and emerging trends*' 2nd edition (Dundee University Press, Dundee 2011) 285–330

Paterson J, 'The Precautionary Principle: practical reason, regulatory decision making and judicial review in the context of functional differentiation' in Philippopoulos-Mihalopoulos A (ed.) '*Law and Ecology: New Environmental Legal Foundations*' (Routledge-Cavendish, London 2011) 83–103

Pearce D, 'The Precautionary Principle and Economic Analysis' in O'Riordan T and Cameron J '*Interpreting the Precautionary Principle*' (Earthscan Publications Ltd., London 1994) 132–51

Perrings C, 'Reserved Rationality and the Precautionary Principle: Technological Change, Time and Uncertainty in Environmental Decision Making' in Costanza R et al. (eds) '*Ecological Economics: The Science and Management of Sustainability*' (Columbia University Press, New York 1991) 153–66

Philippe Q, 'Europe's comparative disadvantage in energy intensive industries: a comparison of shale gas production costs and break-even prices in Europe and the US' in Cecile Musialski et al. (eds) '*Shale Gas in Europe*' (Claeys & Casteels, Deventer 2013) 241–72

Purdy R, 'Geological Carbon Dioxide Storage and the Law' in Shackley S and Gough C (eds) '*Carbon Capture and its Storage*' (Ashgate Publishing Ltd, Hampshire 2006) 87–140

Raszewski S, 'Shale Gas and Energy Security' in Hunter T (ed.) '*Handbook of Shale Gas Law and Policy*' (Intersentia, Cambridge 2016) 123–36

Rehbinder E, 'Vorsorgeprinzip im Umweltrecht und präventive Umweltpolitik' in Simonis U E (ed.) '*Präventive Umweltpolitik*' (Campus Verlag, Frankfurt/Main 1988) 129–41

Rehbinder E, 'Precaution and Sustainability: Two Sides of the Same Coin?' in Kiss A C and Burhenne-Guilmin F '*A Law for the Environment: Essays in Honour of Wolfgang E. Burhenne*' (IUCN, Glend and Cambridge 1994) 91–101

Rehbinder E, 'The Precautionary Principle in an International Perspective' in Basse E M (ed.) '*Miljørettens Grundspørgsmål*' (GadJura, Copenhagen 1994) 93–101

Rehbinder E, 'Ziele, Grundsätze, Strategien und Instrumente' in Hansmann K and Sellner D '*Grundzüge des Umweltrechts*' 4th edition (Erich Schmidt Verlag, Berlin 2012) 135–300

Richardson B J and Razzaque J, 'Public Participation in Environmental Decision-Making' in Richardson B J and Wood S, '*Environmental Law for Sustainability: A Reader*' (Hart Publishing, Portland 2006) 165–94

Robeck M R and Bennett M, 'Shale Gas in the United States: An Institutional Comparison' in Musialski C et al. (eds) '*Shale Gas in Europe*' (Claeys & Casteels, Deventer 2013) 35–60

Roggenkamp M M, 'Regulating Underground Storage of CO_2' in Roggenkamp M M and Woerdman E, '*Legal Design of Carbon Capture and Storage Developments in the Netherlands from an International and EU Perspective*' (Intersentia, Antwerp 2009) 205–27

Roggenkamp M M, 'Protecting Energy Infrastructure in the EU – The Impact of External Damages on Supply Security' in Roggenkamp M M et al. (eds) '*Energy Networks and the Law*' (Oxford University Press, Oxford 2012) 118–40

Roggenkamp M M, 'Energy Law in the Netherlands' in Roggenkamp M

M et al. (eds) '*Energy Law in Europe*' 3rd edition (Oxford University Press, Oxford 2016) 725–806

Ronne A, 'Smart Grids and Intelligent Energy Systems: A European Perspective' in Roggenkamp M M et al. (eds) '*Energy Networks and the Law*' (Oxford University Press, Oxford 2012) 141–60

Ronne A, 'Energy Law in Denmark' in Roggenkamp M M et al. (eds) '*Energy Law in Europe*' 3rd edition (Oxford University Press, Oxford 2016) 403–80

Rose G and Paleokrassis G, 'Compliance with International Environmental Obligations: A Case Study of the International Whaling Commission' in Cameron J, Werksman J and Roderick P (eds) '*Improving Compliance with International Environmental Law*' (Earthscan Publications Ltd, London 1996) 148–75

Rossi L S, 'A New Revision of the EU Treaties After Lisbon?' in Rossi L S and Casolari F '*The EU after Lisbon – Amending or Coping with the Existing Treaties?*' (Springer International, Cham 2014) 3–21

Sacerdoti G, 'Expropriation and Compensation' in Sacerdoti G, '*Bilateral treaties and multilateral instruments on investment protection*' Recueil des Cours Vol. 269 (Martinus Nijhoff, Leiden 1997) 379–411

Sauer H, 'Juristische Methodenlehre' in Julian Krüper (ed.) '*Grundlagen des Rechts*' 2nd edition (Nomos, Baden-Baden 2013) 168–86

Scheuner U, 'Staatszielbestimmungen' in Roman Schnurd (ed.) '*Festschrift für Ernst Forsthoff zum 70. Geburtstag*' (Beck, München 1972) 325–46

Scotford E, 'Mapping the Article 174(2) Case Law: A First Step to Analysing Community Environmental Principles' in Etty T and Somsen H (eds), *The yearbook of European environmental law* Vol. 8 (2008, Oxford University Press) 1–47

Scott J and Voss E, 'The Juridification of Uncertainty: Observations on the Ambivalence of the Precautionary Principle within the EU and the WTO' in Joerges C and Dehousse R (eds) '*Good Governance in Europe's Integrated Market*' (Oxford University Press, 2002) 253–88

Singh N, 'Foreword' in Munro R D and Lammers J G '*Environmental Protection and Sustainable Development; Legal Principles and Recommendations Adopted by the Experts Group on Environmental Law of the World Commission on Environment and Development*' (Graham & Trotman, London 1987)

Stairs K and Taylor P, 'Non-Governmental Organizations and the Legal Protection of the Oceans: A Case Study' in Hurrell A and Kingsbury B (eds) '*The International Politics of the Environment*' (Clarendon Press Ltd, Oxford 1992) 110–41

Stares D, McElfish J and Ubinger Jr J 'Sustainability and community responses to local impacts' in John C Dernbach and James R May (eds)

'Shale Gas and the Future of Energy Law and Policy for Sustainability' (Edward Elgar Publishing, Cheltenham, 2016) 101–21

Stein T, 'Subsidiarität als Rechtsprinzip?' in Merten D (ed.) *'Die Subsidiarität Europas'* 2nd edition (Duncker & Humblot, Berlin 1994) 23–40

Steiner H, 'The rights of future generations' in Maclean D and Brown P G *'The Rights of future generations'* (Rowman and Littlechild Ltd; New Jersey 1983) 151–65

Stirling A, 'On Precautionary and Science Based Approaches to Risk Assessment and Environmental Appraisal: Field Study' in European Science and Technology Observatory *'On Science and Precaution in the Management of Technological Risk Volume II Case studies'* (Joint Research Centre of the European Commission, 2001) available at: http://ftp.jrc.es/EURdoc/EURdoc/eur19056IIen.pdf#page=38 [accessed 21 March 2013] 37–93

Stirling A, 'Risk, Uncertainty and Precaution: Some Instrumental Implications from the Social Science' in Berkhout F *'Negotiating Environmental Change: New Perspectives from Social Science'* (Edward Elgar Publishing, Cheltenham 2003) 33–76

Talus K, 'Security of Supply – An Increasingly Political Notion' in Delvaux B and Hunt M and Talus K *'EU Energy Law and Policy Issues'* (Euroconfidentiel, Rixensart 2008) 100–144

Teitgen F, 'Le principe de proportionnalité en droit français' in Kutscher H, Ress G and Francis Teitgen *'Der Grundsatz der Verhältnismäßigkeit in europäischen Rechtsordnungen'* (C F Müller, Heidelberg 1985) 53–62

Teubner G, 'Juridification: Concepts, Aspects, Limits, Solutions' in Teubner G (ed.) *'Juridification of social spheres'* (Walter de Gruyter & Co, Berlin 1987) 3–48

Tickner J A, 'A Map Toward Precautionary Decision Making' in Raffensperger C and Tickner J A (eds) *'Protecting Public Health and the Environment: Implementing the Precautionary Principle'* (Island Press Ltd., Washington 1999) 162–86

Trudeau N and Taylor P G, 'The Energy Efficiency Dimension of Energy Security' in Benjamin K Sovacool (ed.) *'The Routledge Handbook of Energy Security'* (Routledge Ltd, London, New York 2011) 218–38

Valentine S V, 'The Fuzzy Nature of Energy Security' in Sovacool B K (ed.) *'The Routledge Handbook of Energy Security'* (Routledge Ltd, London, New York 2011) 56–73

Van Gerven W, 'The effect of proportionality on actions of the member states of the European Community: national viewpoints from continental Europe' in Ellis E (ed.) *'The principle of proportionality in the laws of Europe'* (Hart Publishing Ltd; Oxford 1999) 37–64

Van Holten S and van Rijswick M, 'The governance approach in European Union environmental directives and its consequences for flexibility, effectiveness and legitimacy' in Peeters M and Uylenburg R (eds) '*EU Environmental Legislation*' (Edward Elgar, Cheltenham 2014) 13–47

Van Rijswick M,'The Water Framework Directive' in Van Rijswick M (ed.) '*The Water Framework Directive: Implementation into German and Dutch Law*' (Centrum voor Omgevingsrecht en Beleid, Utrecht 2003) 1–30

Vedder H et al., 'EU Energy Law' in Roggenkamp M M et al. (eds) '*Energy Law in Europe*' 3rd edition (Oxford University Press, Oxford 2016) 187–366

von Hippel D F et al., 'Evaluating the Energy Security Impacts of Energy Security' in Sovacool B K (ed.) '*The Routledge Handbook of Energy Security*' (Routledge Ltd, London, New York 2011) 74–95

von Moltke K, 'The Relationship between Policy, Science, Technology, Economics and Law in the Implementation of the Precautionary Principle' in Freestone D and Hey E (eds) '*The Precautionary Principle and International Law*' (Kluwer Law International, The Hague 1996) 97–108

Wiener J B, 'Precaution' in Bodansky D et al., '*The Oxford Handbook of International Environmental Law*' (Oxford University Press, Oxford 2008) 598–612

Wildavsky A, 'Trial and error versus trial without error' in Morris J '*Rethinking Risk and the Precautionary Principle*' (Butterworth-Heinemann, Oxford 2000) 22–45

Wyciszkiewicz E, 'European Institutions in the Debate on Shale Gas' in Musialski C et al. (eds) '*Shale Gas in Europe*' (Claeys & Casteels, Deventer 2013) 111–30

ARTICLES

Adler J H, 'More Sorry than Safe: Assessing the Precautionary Principle and the Proposed International Biosafety Protocol' (2000) 35 *Texas International Law Journal* 173–206

Adler M D and Posner E D, 'Rethinking CBA' (1999) 2 *Yale Law Journal* 165–245

Alexandrov S A, 'Breaches of Contract and Breaches of Treaty – The Jurisdiction of Treaty-based Arbitration Tribunals to Decide Breach of Contract Claims in SGS v. Pakistan and SGS v. Philippines' (2006) 5 *TDM* 555–77

A-Khavari A and Rothwell D R, 'The ICJ and the Danube Dam Case: A missed Opportunity for International Environmental Law?' (1998) 22 *Melbourne University Law Review* 507–36

Andersen I E and Jaeger B, 'Scenario workshops and consensus conferences: Towards more democratic decision-making' (1999) Vol 26 No 5 *Science and Public Policy* 331–40

Antonopoulos I, 'Case Comment Responding to claims of property interests on maritime public property interfering with environmental protection' (2011) 13 (4) *Env. L. Rev*. 318–23

Appel I, 'Europas Sorge um die Vorsorge. Zur Mitteilung der Europäischen Kommission über die Anwendbarkeit des Vorsorgeprinzips' (2001) 4 *NVwZ* 395–8

Applegate J S, 'The Precautionary Preference: An American Perspective on the Precautionary Principle' (2000) Vol 6 No 3, *Human and Ecological Risk Assessment* 413–43

Arbeitskreis Europäische Umweltunion, 'Umweltpolitische Ziele und Grundsätze für die Europäische Union' (1994) 7 *Natur und Recht* 346–51

Argetsinger B, 'The Marcellus Shale: Bridge to a Clean Energy Future or Bridge to Nowhere? Environmental, Energy and Climate Policy Considerations for Shale Gas Development 'in New York State' (2011) 29 *Pace Envtl. L. Rev*. 321–43

Armeni C, 'Case studies on the implementation of Directive 2009/31/EC on the geological storage of carbon dioxide. United Kingdom.' (2011) available at: http://blogs.ucl.ac.uk/law-environment/files/2012/11/Chiara-Armeni-CCLP-EU-Case-Studies-UK-2011.pdf [accessed 25 November 2016] 1–39

Baranzelli C et al., 'Scenarios for shale gas development and their related land use impacts in the Baltic Basin, Northern Poland' (2015) 84 *Energy Policy* 80–95

Barton C, 'Status of the Precautionary Principle in Australia: Its Emergence in Legislation and as a Common Law Doctrine' (1998) Vol. 22 Issue 2 *Harvard Environmental Law Review* 509–58

Battis U, 'Vereinheitlichung des Umweltrechts im europäischen Binnenmarkt?' (1989) 9 *Natur und Recht* 365–9

Baughen S, 'Expropriation and Environmental Regulation: the Lessons of NAFTA Chapter Eleven' (2006) Vol 18 No 2 *Journal of Environmental Law* 207–28

Becker B, 'Das neue Umweltschadensgesetz und das Artikelgesetz zur Umsetzung der Richtlinie über die Umwelthaftung zur Vermeidung und Sanierung von Umweltschäden' (2007) 10 *NVwZ* 1105–13

Beckmann M, 'Anmerkung' (1989) *DVBl* 669–72

Beeckman K, 'Transboundary Damage to the Environment Per Se: Remedial Measures and Standing' (1996) 29 *REDI* 453–92

Bennie L G, 'Brent Spar, Atlantic Oil and Greenpeace' (1998) Vol 51 No 3 *Parliamentary Affairs* 397–410

Bentham R W, 'United Kingdom Offshore Safety Regime: Before and after Piper Alpha' (1991) 9 *Journal of Energy & Natural Resources Law* 273–80

Bernsdorff N, 'Positivierung des Umweltschutzes im Grundgesetz (Art. 20a GG)' (1997) 7 *Natur und Recht* 328–34

Beyerlin U, 'Die "neue" Umweltpolitik der Europäischen Gemeinschaften' (1989) 9 *Umwelt-und Planungsrecht* 361–4

Blohm A et al., 'The significance of regulation and land use patterns on natural gas resource estimates in the Marcellus shale' (2012) Vol 50 *Energy Policy* 358–69

Bodansky D, 'Scientific Uncertainty and the Precautionary Principle' (1991) 33 *Environment* 4–5 and 43–4

Boute A, 'Combating Climate Change and Securing Electricity Supply: The Role of Investment Protection Law' (2007) Vol 16 No 8 *EELR* 227–48

Bradbrook A J, 'Energy Law as an Academic Discipline' (1996) Vol 14 No 2 *Journal of Energy & Natural Resources Law* 193–217

Brandt A R et al., 'Methane Leaks from North American Natural Gas Systems' (2014) Vol. 343 No. 6172 *Science* 733–5

Brombacher M W, 'The precautionary principle threatens to replace science' [1999] *Pollution Engineering* 32–4

Büdenbender U, 'Umweltschutz in der Novelle des Energiewirtschafts-gesetzes' (2005) *DVBl* 1163–74

Burgi M, 'Das Schutz- und Ursprungsprinzip im europäischen Umwelt-recht' (1995) 17 *Natur und Recht* 11–15

Byrne J P, 'Ten Arguments for the Abolition of the Regulatory Takings Doctrine' (1995) 22 *Ecol. L.Q* 89–142

Charnley G and Elliott E D, 'Risk versus precaution: environmental law and public health protection' (2002) 32 *Environmental Law Reporter* 10363–6

Charny D, 'The German Corporate Governance System' (1998) Columbia Business Review 145–66

Chopyak J and Levesque P, 'Public participation in science and technology decision making: trends for the future' (2002) Vol 24 No 1–2 *Technology of Society* 155–66

Christensen M,'The Precautionary Principle and GMOs: an Australian Perspective' (2001) *IUCN-Environmental Law Programme Newsletter* 5–6

Christie G C, 'What Constitutes a Taking of Property Under International Law?' (1962) 38 *Brit. Y.B. Int'l L.* 307–38

Ciuta F, 'Conceptual Notes on Energy Security: Total or Banal Security?' (2010) Vol 41 No 2 *Security Dialogue* 123–44

Colyvan M, 'Is Probability the Only Coherent Approach to Uncertainty?' (2008) Vol 28 No 3 *Risk Analysis* 645–52

Cranor C F, 'Learning for the law to address uncertainty in the precautionary principle' (2001) 7 *Science and Engineering Ethics* 313–26

Cross F B, 'Paradoxical Perils of the Precautionary Principle' (1996) 53 *Washington and Lee Law Review* 851–925

Curtis J B, 'Fractured shale-gas systems' (2002) Vol 86 No 11 *American Association of Petroleum Geologists Bulletin* 1921–38

Czybulka D, 'Naturschutz und Verfassungsrecht' (1999) 2 *Potchefstroom Electronic Law Journal* 1–30

Dąborowski T and Groszkowski J, 'Shale Gas in Bulgaria, the Czech Republic and Romania political context – legal status – outlook' (Centre for Eastern Studies, Warsaw 2012) 10 available at: http://www.osw.waw.pl/sites/default/files/shale_gas_in_bulgaria_the_czech_republic_and_romania_net_0.pdf [accessed 23 May 2014]

DeFur P L and Kaszuba M, 'Implementing the Precautionary Principle' (2002) 288 *The Science of the Total Environment* 155–65

Delfs S, 'Anmerkung' (1999) 10 *Zeitschrift für Umweltrecht* 322–4

Dernbach J C, 'Sustainable Development: Now More than Ever' (2002) 32 *Envtl. Law Institute* 10003–15

Desgagné R, 'Integrating Environmental Values into the ECHR' (1995) 89 *AJIL* 263–95

Diestelmeier L and Kuiken D, 'Sustaining Universal Service Conditions in Smart Electricity Systems' (2016) Vol 18 No 3 *Network Industries Quarterly* 7–10

Dietrich L and Elgeti T, 'Rechtliche Implikationen der Aufsuchung und Förderung von unkonventionellem Erdgas' (2011) 127 (7–8) *Erdöl Erdgas Kohle* 311–15

Dixon T et al., 'Legal and Regulatory Developments on CCS' (2015) Vol 40 *International Journal of Greenhouse Gas Control* 431–48

Doyle A and Carney T, 'Precaution and Prevention: Giving Effect to Article 130r Without Direct Effect' (1999) 8 *EELR* 44–7

Dreier H, 'Die drei Staatsgewalten im Zeichen von Europäisierung und Privatisierung' [2002] *Die öffentliche Verwaltung* (DöV) 537–47

Dzidzornu D, 'Four Principles in Marine Environment Protection: A Comparative Analysis' (1998) 29 *Ocean Development and International Law* 91–123

Edwards V, 'Significant EU Environmental Cases: 2011' (2012) 24 *Journal of Environmental Law* 155–70

Ehricke U and Hackländer D, 'Europäische Energiepolitik auf der Grundlage der neuen Bestimmungen des Vertrags von Lissabon' (2008) 4 *Zeitschrift für Europarechtliche Studien* 579–600

Emiliou N, 'Subsidiarity: an effective barrier against the "enterprise of ambition"?' (1992) 17 *European Law Review* 383–407

Epiney A, 'Umweltrechtliche Querschnittsklausel und freier Warenverkehr' (1995) 10/11 *NuR* 497–504

Fleming R C and Reins L, 'Shale gas extraction, precaution and prevention: a conversation on regulatory responses' (2016) 20 *Energy Research and Social Science* 131–41

Finkel M L and Law A, 'The Rush to Drill for Natural Gas: A Public Health Cautionary Tale' (2011) Vol 101 No 5 *American Journal of Public Health* 784–5

Fischer-Hüftle P, 'Bergbauberechtigungen und naturschutz' 'Bergbauberechtigungen und naturschutzrechtliche Verordnungen' (1989) 11 *Natur und Recht* 106–13

Foster C E, 'The "Real Dispute" in the Southern Bluefin Tuna Case: A Scientific Dispute?' (2001) Vol 16 No 4 *International Journal of Marine and Coastal Law*, 571–601

Frantz J H and Jochen V, 'Shale Gas – When your gas reservoir is unconventional so is our solution' (2005) White Paper Schlumberger Ltd. available at: http://www.pe.tamu.edu/wattenbarger/public_html/Selected_papers/--Shale%20Gas/shale_gas-%20schlumberger.pdf (accessed 20 March 2012)

Freestone D, 'International Environmental Law after the Earth Summit' (1994) 6 *Journal of Environmental Law* 193–218

Frenz W, 'Berg- und Umweltrecht I 1. Teil: Bergrecht' available at: http://www.bur.rwth-aachen.de/download/Skript_Bergrecht.pdf [accessed 12 December 2012]

Frenz W, 'Wirtschaftskrise und nachhaltiger Umweltschutz' (2009) 2 *Umwelt- und Planungsrecht* 48–51

Frenz W, 'Moratorium für Fracking?' (2013) 4 *ZNER* 344–7

George L, 'Expropriation' (2002) 3 *International Energy Law & Taxation Review* 46–52

Glaesner H-J, 'Die Einheitliche Europäische Akte' (1986) 2 *Europarecht* 119–52

Glenn P S, 'Control and Modeling of Fluid Leakoff during Hydraulic Fracturing' (1985) Vol 37 No 6 *Journal of Petroleum Technology* 1071–81

Goerlich H, 'Erfordernisse rationaler Gesetzgebung nach Maßstäben des Bundesverfassungsgerichts' (1977) *JR* 89–95

Goetz M et al., 'Renewable Power-to-Gas: A technological and economic review' (2016) Vol 85 *Renewable Energy* 1371–90

Goldstein B D, 'The Precautionary Principle Also Applies to Public Health Actions' [2001] *American Journal for Public Health* 1358–61

Goldstein B D, Ferrell Bjerke E and Kriesky J, 'Challenges of unconventional shale gas development: so what's the rush?' (2013) 27 *Notre Dame J.L. Ethics & Pub. Pol'y* 149–86

Grabitz E, 'Der Grundsatz der Verhältnismäßigkeit in der Rechtsprechung des Bundesverfassungsgerichts' (1973) 98 *AöR* 568–616

Gray D et al., 'Estimation of stress and geomechanical properties using 3D seismic data' (2012) 30(3) *First Break Magazine* 59–68

Grigas A, 'The Gas Relationship between the Baltic States and Russia' (Oxford Institute for Energy Studies, Oxford 2012) 2 available at: https://www.oxfordenergy.org/wpcms/wp-content/uploads/2012/10/NG_67.pdf [accessed 26 June 2016]

Gullett W, 'Environmental protection and the precautionary principle: a response to scientific uncertainty in environmental management' (1997) 14(1) *Environmental and Planning Law Journal* 52–69

Gunningham N, 'Managing the energy trilemma: The case of Indonesia' (2013) Volume 54 *Energy Policy* 184–93

Gündisch J, 'Preisgarantie für Strom aus Windkraftanlagen keine Beihilfe' (2001) 50 *NJW* 3686–8

Gündling L, 'The Status in International Law of the Precautionary Principle' (1990) 5 *International Journal of Estuarine and Coastal Law* 23–30

Haghighi S S, 'Energy Security and the Division of Competences between the European Community and its Member States' (2008) 14 *European Law Journal* 461–82

Hammit J K et al., 'Precautionary Regulation in Europe and the United States: A Quantitative Comparison' (2005) 25 *Risk Analysis* 1215–28

Haneklaus W, 'Zur Verankerung umweltpolitischer Ziele im EWG-Vertrag' (1990) 21 *Deutsches Verwaltungsblatt* 1135–41

Heffron R J, McCauley D and Sovacool B K, 'Resolving society's energy trilemma through the Energy Justice Metric' (2015) 87 *Energy Policy* 168–76

Hess G, 'Drilling process draws scrutiny' (2010) Vol 88 No 22 *Chemical & Engineering News* 42–5

Heyvaert V, 'Facing the consequences of the precautionary principle in European Community law (2006) 31 *European Law Review* 185–206

Hoppe W, 'Die Einschränkung bergbaurechtlicher Berechtigungen durch eine Nationalparkverordnung – am Beispiel des niedersächsischen Wattenmeeres' (1987) *DVBl* 757–65

Howarth R W, Santoro R and Ingraffea A, 'Methane and the greenhouse-gas footprint of natural gas from shale formations' (2011) Vol 106 No 4 *Climatic Change* 679–90

Htun N,'EIA and Sustainable Development' (1990) Vol 8 No 1–2 *Impact Assessment* 15–23

Jacobs F, 'The Role of the European Court of Justice in the Protection of the Environment' (2006) 18 *JEL* 185–205

Jackson R B et al., 'Increased stray gas abundance in a subset of drinking water wells near Marcellus shale gas extraction' (2013) Vol 110 No 28 *Proceedings of the National Academy of Science of the United States of America (PNAS)* 11250–55

Jaeckel L, 'Die Simulation des Urknalls vor dem Bundesverfassungsgericht' (2011) *DVBl* 13–20

Jahns-Böhm J and Breier S, 'Die Umweltrechtliche Querschnittsklausel des Art. 130r II 2 EWGV' (1992) 2 *Europäische Zeitschrift für Wirtschaftsrecht* 49–55

Jans J H, 'Minimum Harmonisation and the Role of the Principle of Proportionality' available at: www.user.uni-bremen.de/avosetta/janjans article1.pdf [accessed 1 June 2016]

Jans J H, 'Proportionality Revisited' (2000) 27 *LIEI* 239–65

Jimenez A S, 'Ecological Catastrophes in Light of the Rio Agreements' (1996) 39 *German Yearbook of International Law (GYIL)* 388–408

Johnson C and Boersma T, 'Energy (in)security in Poland the case of shale gas' (2013) 53 *Energy Policy* 389–99

Johnson D L et al., 'Meanings of Environmental Terms' (1997) Vol 26 No 3 *Journal of Environmental Quality* 581–9

Johnston A et al., 'The Proposed New EU Renewables Directive: Interpretation, Problems and Prospects' (2008) 17 *EEELR* 126–45

Jones J, 'Regulatory Design for Scientific Uncertainty' (2007) 19 *Journal of Environmental Law* 347–65

Jones P B C, 'Implementing the Precautionary Principle' (2000) available at: http://www.biotech-info.net/BPCJ_viewpoint.html (accessed 27 March 2012)

Kahl W, 'Die Kompetenz der EU in der Energiepolitik nach Lissabon' (2009) 5 *Europarecht* 601–21

Kaiser M, 'Fish-Farming and the Precautionary Principle: Context and Values in Environmental Science for Policy' (1997) 2 *Foundations of Science* 307–41

Klatt M, 'Semantic normativity and the objectivity of legal argumentation' (2004) 90 *Archiv für Rechts-und Sozialphilosophie* 51–65

Klinke A and Renn O, 'Prometheus Unbound Challenges of Risk Evaluation, Risk Classification, and Risk Management' (1999) *Working*

Paper for Akademie für Technikfolgenabschätzung in Baden-Württemberg available at: http://elib.uni-stuttgart.de/opus/volltexte/2004/1712/index.html (accessed 28 March 2012)

Klinke A and Renn O, 'Precautionary principle and discursive strategies: classifying and managing risks' (2001) 4 *Journal of Risk Research* 159–73

Klinke A and Renn O, 'A New Approach to Risk Evaluation and Management: Risk-based, Precaution-based and Discourse-based Strategies' (2002) 22 *Risk Analysis* 1071–94

Koenig C, 'Möglichkeiten und Grenzen von Zertifikatmärkten als Steuerungsmedien im Umweltrecht' (1996) 22 *Die Öffentliche Verwaltung* 943–50

Kolonko B, 'Naturschutz und Bergrecht – zwei unvereinbare Materien?' (1995) 3 *Zeitschrift für Umweltrecht* 126–34

Konoplyanik A and Wälde T, 'The Energy Charter Treaty and its Role in International Energy' (2006) 24 Issue 4 *Journal of Energy and Natural Resources Law* 523–58

Kotsakis A, 'The Regulation of the Technical, Environmental and Health Aspects of Current Exploratory Shale Gas Extraction in the United Kingdom: Initial Lessons for the Future of European Union Energy Policy' (2012) 21 (3) *RECIEL* 282–90

Krämer L, 'Das Verursacherprinzip im Gemeinschaftsrecht / Zur Auslegung von Art. 130r EWG-Vertrag' (1989) *EuGRZ* 353–61

Krämer L, 'Case studies on the implementation of Directive 2009/31/EC on the geological storage of carbon dioxide. Germany.' (2011) available at: http://blogs.ucl.ac.uk/law-environment/files/2012/11/Ludwig-Kraemer-CCLP-EU-Case-Studies-Germany-2011.pdf [accessed 25 November 2016] 1–31

Krimsky S, 'The Funding Effect in Science and Its Implications for the Judiciary' (2005) 13 *Journal of Law and Policy* 43–68

Kruyt B et al., 'Indicators for energy security' (2009) 37 *Energy Policy* 2166–81

Kühn W M, 'Die Entwicklung des Vorsorgeprinzips im Europarecht' (2006) 4 *Zeitschrift für europarechtliche Studien* 487–520

Kühne G, 'Zulassung und Ausübung des Bergbaus bei Kollision mit anderen öffentlichen Interessen – zugleich ein Beitrag zu § 47 Regierungsentwurf BBergG' (1980) 121 *Zeitschrift für Betriebswirtschaft* 58–72

Kühne G, 'Bergrechtliche Aspekte der Endlagerung radioaktiver Stoffe' (1985) *DVBl* 207–11

Kuxenko M, 'Der Gesetzeszweck einer umweltverträglichen Energieversorgung im Planfeststellungsverfahren für Energieleitungen' (2003) 25 *Natur und Recht* 332–9

Kysar D A, 'It Might Have Been: Risk, Precaution and Opportunity Costs'

(2006) Paper 50 *Cornell Law Faculty Publications* available at: http://scholarship.law.cornell.edu/lsrp_papers/50 [accessed 6 August 2013] 1–48

LaFave W R, '"Street Encounters"' and the Constitution: Terry, Sibron, Peters and Beyond' (1968/1969) 67 *Michigan Law Review* 39–126

Lafferty W and Hovden E, 'Environmental policy integration: towards an analytical framework' (2003) 12 *Environmental Politics* 1–22

Lemons J et al., 'The Precautionary Principle: Scientific Uncertainty and Type I and Type II Errors' (1997) 2 *Foundations of Science* 207–36

Little M G and Jackson R B, 'Potential Impacts of Leakage from Deep CO_2 Geosequestration on Overlying Freshwater Aquifers' (2010) 44 *Environmental Science & Technology* 9225–32

Loucaides L, 'Environmental Protection through the Jurisprudence of the ECHR' (2004) 75 *BYBIL* 249–67

Louis H W, 'Die naturschutzrechtliche Befreiung' (1995) 17 *Natur und Recht* 62–71

Lübbe-Wolff G, 'IVU-Richtlinie und Europaisches Vorsorgeprinzip' (1998) 8 *Neue Zeitschrift fur Verwaltungsrecht (NVwZ)* 777–85

Mackenzie C, 'Shale gas: the fugitive methane problem' (2012) Scottish Widows Investment Partnership Limited, Sustainability Research Series available at: http://www.swip.com/sites/docs/SiteCollectionDocuments/SWIP%20Document%20Test%20Folder/May-2012-Sustainability-Research-Note.pdf [accessed 4 June 2012]

Majone G, 'What price safety? The precautionary principle and its policy implications' (2002) 40 *Journal of Common Market Studies* 89–109

Manson N, 'The precautionary principle, the catastrophe argument and Pascal's wager' available at: http://www.abdn.ac.uk/philosophy/endsandmeans/vol4no1/manson.shtml [accessed 21 May 2012]

Marr S, 'The Southern Bluefin Tuna Cases' (2000) 11 *European Journal of International Law* (EJIL) 815–31

Martin P H, 'If You Don't Know How to Fix it Please Stop Breaking it!': The Precautionary Principle and Climate Change' (1992) 2 *Future of Science* 263–92

Materka E, 'End of Transition? Expropriation, Resource Nationalism, Fuzzy Research, and Corruption of Environmental Institutions in the Making of the Shale Gas Revolution in Northern Poland' (2011) Vol 19 Issue 3 *Journal of Contemporary Central and Eastern Europe* 599–631

Mathieu M, Spencer T and Sartor O, 'Economic analysis of the US unconventional oil and gas revolution' *VOX CEPR's Policy Portal* available at: http://voxeu.org/article/limited-economic-impact-us-shale-gas-boom [accessed 26 June 2016]

Matthee M, 'The International Integration of European Precautionary Measures on Biosafety' (2001) 10 *EELR* 183–93

Matthee M and Vermersch D, 'Are the Precautionary Principle and the International Trade of Genetically Modified Organisms Reconcilable?' (2000) 12 *Journal of Agriculture and Environmental Ethics* 59–70

McIntyre O and Mosedale T, 'The Precautionary Principle as a Norm of Customary International Law' (1997) 9 *JEL* 221–4

Merten D, 'Über Staatsziele' (1993) *DÖV* 368–77

Milligan B, 'Planning for offshore CO_2 storage: Law and policy in the United Kingdom' (2014) 48 *Marine Policy* 167

Molkenbur G, 'Umweltschutz in der Europäischen Gemeinschaft' (1990) 13 *Deutsches Verwaltungsblatt* 677–84

Mortelmans K, 'Case Notion on Campus Oil' (1984) *CMLR* 687–713

Murswiek D, 'Staatsziel Umweltschutz (Art. 20a GG)' (1996) 3 *NVwZ* 222–31

Muter Goldberg S, 'Security of supply in the context of European energy market liberalisation: a brief overview' (2011) 4 *International Business Law Journal* 433–62

Myhr A I and Traavik T, 'The Precautionary Principle: Scientific Uncertainty and Omitted Research in the Context of GMO Use and Release' (2002) Vol 15 No 1 *Journal of Agriculture and Environmental Ethics* 73–86

Nicholson B R and Dillard S C, 'Analysis of Litigation Involving Shale and Hydraulic Fracturing: Part 1' (2013) 2 *IELR* 50–66

Nowak C, 'Die Grundfreiheiten des EG-Vertrags und der Umweltschutz' (2002) 93 *Verwaltungsarchiv* 368–93

Oliver P, 'A Review of the Case Law of the Court of Justice on Articles 30 to 36 EEC in 1984' (1985) *CMLR* 301–28

O'Riordan T and Jordan A, 'The Precautionary Principle in Contemporary Environmental Politics' (1995) 4 *Environmental Values* 191–212

Osborn S G et al., 'Methane contamination of drinking water accompanying gas-well drilling and hydraulic fracturing' (2011) Vol 108 No 20 *Proceedings of the National Academy of Science of the United States of America (PNAS)* 8172–6

Pacala S and Socolow R, 'Stabilization Wedges: Solving the Climate Problem for the Next 50 Years with Current Technologies' available at: http://www.princeton.edu/mae/people/faculty/socolow/Science-2004-SW-1100103-PAPER-AND-SOM.pdf [accessed 26 February 2014]

Pache E, 'Der Grundsatz der Verhältnismäßigkeit in der Rechtsprechung der Gerichte der Europäischen Gemeinschaften' (1999) 10 *NVwZ* 1033–40

Paterson J, 'Sustainable development, sustainable decisions and the precautionary principle' (2007) 42 *Natural Hazards* 515–28

Perry S L, 'Development, Land Use, and Collective Trauma: The Marcellus Shale Gas Boom in Rural Pennsylvania' (2012) Vol 34 Issue 1 *Culture, Agriculture, Food and Environment* 81–92

Philippopoulos-Mihalopoulos A,'The Silence of the Sirens: Environmental Risk and the Precautionary Principle' (1999) Vol. 10 No. 2 *Law and Critique* 175–97

Pieper S U, 'Subsidiaritätsprinzip – Strukturprinzip der Europäischen Union' (1993) *DVBl* 705–12

Pieterman R and Hanekamp J C, 'The Cautious Society? An Essay on the Rise of the Precautionary Culture' (2002) Heidelberg Appeal Netherlands report available at: http://www.degroenerekenkamer.nl/grk files/images/The_Cautious_Society__HAN_.pdf [accessed 26 March 2012]

Pirani S, Stern J and Yafimava K,'The Russo-Ukrainian gas dispute of January 2009: a comprehensive assessment' (Oxford Institute for Energy Studies, 2008) available at https://www.oxfordenergy.org/wpcms/wp-content/uploads/2010/11/NG27-TheRussoUkrainianGasDisputeofJanuary2009AComprehensiveAssessment-JonathanSternSimonPiraniKatjaYafimava-2009.pdf

Pravdic V, 'Environmental Capacity: Is a New Scientific Concept Acceptable as a Strategy to Combat Marine Pollution?' (1985) 16 *Marine Pollution Bulletin* 295–96

Rahm B G and Riha S J, 'Toward strategic management of shale gas development: Regional, collective impacts on water resources' (2011) 17 *Environmental Science & Policy* 12–23

Rawlins B L, 'Prioritizing Stakeholders for Public Relations' (2006) available at: http://www.instituteforpr.org/wp-content/uploads/2006_Stakeholders_1.pdf [accessed 28 June 2012]

Raymond L, 'The Ethics of Compensation: Takings, Utility and Justice' (1996) 23 *Ecol. LQ* 577–621

Reins L,'European minimum principles for shale gas: preliminary insights with reference to the precautionary principle' (2014) 22 (1) *Env. Liability* 16–27

Reins L, 'In Search of the Legal Basis for Environmental and Energy Regulation at the EU Level: The Case of Unconventional Gas Extraction' (2014) 23 (1) *RECIEL* 125–33

Reins L, 'What's in a name? When is a technology "new" and does it matter for its regulation.' Presentation at the *19th Ius Commune Conference*, 28th November 2014, Edinburgh, page 7 et sqq. of the paper accompanying the presentation available at http://www.iuscommune.eu/html/

activities/2014/2014-11-27/workshop_3b_Reins.pdf [accessed 25 February 2015]

Rengeling H-W and Heinz K, 'Die dänische Pfandflaschenregelung' (1990) 8 *Juristische Schulung* 613–17

Resnik D, 'Is the precautionary principle unscientific?' (2003) 34 *Studies in History and Philosophy of Biological and Biomedical Sciences* 329–44

Rickman R et al., 'A practical use of shale petrophysics for stimulation design optimization: All Shale plays are not clones of the Barnett Shale' (2008) Society of Petroleum Engineers SPE 115258 available at: http://www.onepetro.org/mslib/app/Preview.do?paperNumber=SPE-115258-MS&societyCode=SPE (accessed 20 March 2012)

Robinson C, 'The Economics of Energy Security: Is Import Dependency a Problem?' (2007) 8 *CRN* 425–51

Rogers M D, 'Scientific and technological uncertainty, the precautionary principle, scenarios and risk management' (2001) 4 *Journal of Risk Research* 1–15

Sampliner G H, 'Arbitration of Expropriation Cases Under US Investment Treaties: A Threat to Democracy or the Dog that didn't Bark?' (2003) 18 *ICSID Review – Foreign Investment Law Journal* 1–43

Sanden J, 'Das Vorsorgeprinzip im europäischen und deutschen Umweltrecht – Weiterentwicklung und Impulse für das internationale Recht' (2005) 53 *Osaka University Law Review* 243–70

Sandin P, 'Dimensions of the precautionary principle' (1999) 5 *Human and Ecological Risk Assessment* 889–907

Sandin P et al., 'Five charges against the precautionary principle' (2002) 5:4 *Journal of Risk Research* 287–99

Scheidler A, 'Umweltschutz durch Umweltverantwortung' (2007) 10 *NVwZ* 1113–19

Schemmel M L and de Regt B, 'The European Court of Justice and the Environmental Protection Policy of the European Community' (1994) Vol 17 Issue 1 *Boston College International and Comparative Law Review* 53–83

Scheuing D H, 'Umweltschutz auf der Grundlage der EEA' (1989) 2 *Europarecht* 152–92

Schink A, 'Umweltschutz als Staatsziel' (1997) 6 *Die Öffentliche Verwaltung* 221–9

Schmidt R, 'Der Staat der Umweltvorsorge' (1994) 47 *Die Öffentliche Verwaltung* 749–56

Schomerus T and Tolkmitt U, 'Die Umweltinformationsgesetze der Länder im Vergleich' (2007) 10 *Neue Zeitschrift fur Verwaltungsrecht (NVwZ)* 1119–25

Schrenk G, 'Mitgliedsstaatliche und gemeinschaftliche Handlungsebene' (1990) 9 *NuR* 391–5
Schreuer C, 'The Concept of Expropriation under the ETC and other Investment Protection Treaties' (Universität Wien 2005) available at: http://www.univie.ac.at/intlaw/pdf/csunpublpaper_3.pdf [accessed 20 November 2012]
Schröder M, 'Die Je-desto-Formel des Bundesverfassungsgerichts in der Espa-Entscheidung und ihre Bedeutung für Grundrechtsabwägungen (2008) *DVBl* 146–50
Schultz H M et al., 'Shale gas in Europe: a regional overview and current research activities' (2010) 7 *Geological Society Petroleum Geology Conference Series* 1079–85
Schwartz I, 'EG-Kompetenz für das Verbot der Tabakwerbung?' (1998) 6 *Zeitschrift für Medien- und Kommunikationsrecht AfP* 553–64
Scott J, 'Flexibility in the Implementation of EC Environmental Law' (2000) 1 *YEEL* 37–60
Sedemund J and Montag F, 'Die Entwicklung des Europäischen Gemeinschaftsrechts' (1989) 22 *Neue Juristische Wochenschrift* 1409–15
Sendler H, 'Die Bedeutung des Abwägungsgebots in § 1 Abs. 6 BauGB für die Belange des Umweltschutzes in der Bauleitplanung' (1995) 2 *Umwelt- und Planungsrecht* 41–9
Sen-Hu L et al., 'Status quo of tight oil exploitation in the United States and its implications' (2011) Vol 23 Issue 4 *Lithologic Reservoirs* 25–30
Shogenova A et al., 'CCS Directive transposition into national laws in Europe: progress and problems by the end of 2011' (2013) Vol 37 *Energy Procedia* 7723–31
Shogenova A et al., 'Implementation of the EU CCS Directive in Europe: results and development in 2013' (2014) Vol 63 *Energy Procedia* 6662–70
Siegel D I et al., 'Methane Concentrations in Water Wells Unrelated to Proximity to Existing Oil and Gas Wells in Northeastern Pennsylvania' (2015) *Environmental Science & Technology* available at http://pubs.acs.org/doi/pdf/10.1021/es505775c [accessed 1 April 2015]
Slot P J,'Harmonisation' (1996) 21 *ELRev* 378–97
Smithson M, 'Ignorance and Science' (1993) 15 *Science Communication* 133–56
Soell H, 'Überlegungen zum europäischen Umweltrecht' (1990) 4 *NuR* 155–61
Soljan L, 'The General Obligation to Prevent Transboundary Harm and its Relation to Four Key Environmental Principles' (1998) 3 *ARIEL* 209–32
Soloway J A, 'Environmental Regulation as expropriation: the case of NAFTA's chapter 11' (2000) 33 *Can. Bus. L.J.* 92–127

Steele K, 'The Precautionary Principle: a new approach to public decision making?' (2006) 5 *Law, Probability and Risk* 19–31

Stern J, 'The Russian-Ukraine Gas Crisis of January 2006' (Oxford Institute for Energy Studies, 2006) available at https://www.oxfordenergy.org/wpcms/wp-content/uploads/2011/01/Jan2006-RussiaUkraineGasCrisis-JonathanStern.pdf

Sterzel D, 'Staatsziele und soziale Grundrechte: zur Steuerungsfunktion eines Staatszieles Umwelt und soziale Grundrechte' (1993) *ZRP* 13–17

Stevens P, 'The "Shale Gas Revolution": Hype and Reality' (Chatham House report 2010) available at: http://www.chathamhouse.org/sites/default/files/public/Research/Energy,%20Environment%20and%20Development/r_0910stevens.pdfhttp://www.chathamhouse.org/sites/default/files/public/Research/Energy,%20Environment%20and%20Development/r_0910stevens.pdf [accessed 8 June 2012]

Stokes, E, 'New EU policy on shale gas' (2014) Volume 16 Issue 1 *Environmental Law Review* 42–9

Stone C D, 'Is There a Precautionary Principle?' (2001) 31 *Environmental Law Reporter* 10790–800

Sumi L, 'Shale Gas: Focus on the Marcellus Shale.' Oil & Gas Accountability Project / Earthworks' available at: http://www.earthworksaction.org/files/publications/OGAPMarcellusShaleReport-6-12-08.pdf?pubs/OGAPMarcellusShaleReport-6-12-08.pdf [accessed 24 April 2012]

Tai S, 'Uncertainty about Uncertainty: The Impact of Judicial Decisions on Assessing Scientific Uncertainty' (2008) 11 *University of Pennsylvania Journal of Constitutional Law* 671–727

Tarka M, 'The legal consequences of European Commission recommendations on minimum principles for shale gas in Poland' available at: http://www.shale-gas-information-platform.org/categories/legislation/expert-articles/tarka-article-poland.html [accessed 18 December 2014]

Toman M A, 'International Oil Security: Problems and Policies' (2002) Vol 20 No 2 *The Brookings Review* 20–23

Tosun J and Knill C, 'The Extraction of Unconventional Gas in France and Germany: Explaining Cross-Country Variation in Regulation' (2012) Presentation in the workshop '*Intra-institutional Politics of EU Energy Policy-Making: Landscape-Level Shifts, Discourses and Organizational Cultures*', Robinson College, Cambridge, England, 27–28 April 2012

Trouwborst A, 'Prevention, Precaution, Logic and Law: The Relationship between the Precautionary Principle and the Preventative Principle in International Law and Associated Questions' (2009) 2 *Erasmus Law Review* 105–27

Turksen U and Wojcik J, 'The European Union and Russia energy trade – thickening of legality and solidarity?' (2012) 1 *IELR* 21–36
Van Calster G, 'Risk Regulation, EU Law and Emerging Technologies: Smother or Smooth?' (2008) 2 *NanoEthics* 61–71
van Ooik R, 'Cross-Pillar Litigation before the ECJ: Demarcation of Community and Union Competences' (2008) Vol 4 Issue 3 *European Constitutional Law Review* 399–419
Veatch R W Jr, 'Overview of Current Hydraulic Fracturing Design and Treatment Technology – Part 1' [1983] *Journal of Petroleum Technology* 677–87
Veatch R W Jr, 'Overview of Current Hydraulic Fracturing Design and Treatment Technology – Part 2' [1983] *Journal of Petroleum Technology* 853–64
Vincent M C, 'The next opportunity to improve hydraulic-fracture stimulation' [2012] *Journal of Petroleum Technology* 118–27
Vogel D, 'The Hare and the Tortoise Revisited: The New Politics of Consumer and Environmental Regulation in Europe' (2003) 33 *British Journal of Political Science* 557–80
von Arnauld A, 'Theorie und Methode des Grundrechtsschutzeses in Europa – am Beispiel des Grundsatzes der Verhältnismäßigkeit' (2008) Beiheft 1 *Europarecht* 41–64
von Borries R, 'Das Subsidiaritätsprinzip im Recht der Europäischen Union' (1994) 29 *Europarecht* 263–300
Wagner W E, 'The Science Charade in Toxic Risk Regulation' (1995) 95 *Columbia Law Review* 1613–1723
Wahl R,'Der Vorrang der Verfassung' (1981) 20 *Der Staat* 485–516
Wälde T and Hober K, 'The First Energy Charter Treaty Arbitral Award' (2005) 22 *Journal of International Arbitration* 83–104
Wälde T and Kolo A, 'Environmental Regulation, Investment Protection and "Regulatory Taking" in International Law' (2001) Volume 50 Issue 4 *International and Comparative Law Quarterly* 811–48
Walker V R, 'The Siren Songs of Science' (1991) 23 *Connecticut Law Review* 567–626
Walker V R, 'The Myth of Science as a "Neutral Arbiter" Triggering Precaution' (2006) 26 *Boston College International and Comparative Law Review* 197–228
Wang Y et al., 'Study of borehole stability of Marcellus shale wells in longwall mining areas' (2014) 4 *Journal of Petroleum Exploration and Production Technology* 59–71
Weber C L and Clavin C, 'Life Cycle Carbon Footprint of Shale Gas: Review of Evidence and Implications' (2012) Vol 46 No 11 *Environmental Science & Technology* 5688–95

Weber M, Hey C and Faulstich M, 'Energiewende – a pricey challenge' (2012) 10 (3) *CESifo DICE Repo*rt 16–23
Weiss C, 'Expressing Scientific Uncertainty' (2003) 2 *Law Probability and Risk* 25–46
Wernsmann R, 'Wer bestimmt den Zweck einer grundrechtseinschränkenden Norm-BVerfG oder Gesetzgeber?' (2000) *NVwZ* 1360–64
Wiener J B and Rogers M D, 'Comparing Precaution in the US and Europe' (2002) 5 *Journal of Risk Research* 317–49
Wilkins H, 'The need for subjectivity in EIA: discourse as a tool for sustainable development' (2003) Vol 23 No 4 *Environmental Impact Assessment Review* 401–14
Winter G, 'Subsidiarität und Deregulierung im Gemeinschaftsrecht (1996) 3 *Zeitschrift für Europarecht* 247–69
Winter G, 'Die Steuerung grenzüberschreitender Abfallströme' (2000) Vol 115 *DVBl* 657–69
Winzer C, 'Conceptualizing Energy Security' (2011) No 1123 *University of Cambridge Electricity Policy Research Group Working Paper* 1–35
Wittig P, 'Zum Standort des Verhältnissmäßigkeitsgrundsatzes im System des Grundgesetzes' (1968) *DÖV* 817–25
Wolf M, 'Der Bergbau und die naturschutzrechtliche Kompensationspflicht' (2006) *Zeitschrift für Umweltrecht* 524–31
Wynne B, 'Uncertainty and Environmental Learning' (1992) 2 *Global Environmental Change* 111–27
Yergin D, 'Ensuring Energy Security' (2006) Vol 85 No 2 *Foreign Affairs* 69–82
Zeuschner R, 'United Kingdom: The Development of the Consenting Regime for Offshore Renewable Energy Projects in Scotland' (2011) 4 *International Energy Law Review* 150–59
Zuleeg M, 'Vorbehaltene Kompetenzen der Mitgliedsstaaten der EG' (1987) 4 *NVwZ* 280–86

WEBPAGES

Agence France Press via phys.org, *'Brussels says no plans for EU-wide shale gas ban'* of 16th July 2013 available at: http://phys.org/news/2013-07-brussels-eu-wide-shale-gas.html [accessed 24 April 2014]
Artus P, 'US Reindustrialisation Poses Challenge for Eurozone' London: FTSE Global Markets, London 2012) available at: http://www.ftseglobalmarkets.com/blog/european-review/us-industrialisation-poses-challengefor-eurozone.html [accessed 17 July 2014]
Baerbock A, 'Fracking: Keine Entwarnung' available at: http://www.

annalena-baerbock.de/pmfracking-keine-entwarnung/ [accessed 12 July 2016]

Balser M, Süddeutsche Zeitung, 'Giftspritze für den Boden' available at: http://www.sueddeutsche.de/wirtschaft/umstrittenes-fracking-gesetz-gescheitert-giftspritze-fuer-den-boden-1.1688444 [accessed 10 September 2013]

Banerjee N, *Los Angeles Times*, 'Mining sand for fracking causes friction in Wisconsin' available at: http://articles.latimes.com/2012/nov/19/nation/la-na-wisconsin-sand-20121119-1 [accessed 25 March 2013]

BBC, '*Blackpool shale gas drilling suspended after quake*' available at: http://www.bbc.co.uk/news/uk-england-lancashire-13599161 [accessed 7 April 2014]

BBC, Dieter Helm of Oxford University, interviewed by Ben King for *BBC News* 'Shale gas pioneer plans world's first offshore wells in Irish Sea' available at: http://www.bbc.com/news/business-26157228 [accessed 9 December 2016]

BBC, '*Fracking tests near Blackpool "likely cause" of tremors*' available at: http://www.bbc.co.uk/news/uk-england-lancashire-15550458 [accessed 4 April 2014]

Brandenburgian Ministry for Economy and European Affairs, '*Brandenburg begrüßt Kompromisse zu CCS und EEG*', available at: http://www.mwe.brandenburg.de/sixcms/detail.php/bb1.c.298597.de [accessed 15 October 2012]

Buckley N, *Financial Times*, 'Romania and Lithuania back fracking' available at: http://www.ft.com/cms/s/fa2812bc-6fa6-11e2-956b-00144feab49a,Authorised=false.html?_i_location=http%3A%2F%2Fwww.ft.com%2Fcms%2Fs%2F0%2Ffa2812bc-6fa6-11e2-956b-00144feab49a.html%3Fsiteedition%3Duk&siteedition=uk&_i_referer=http%3A%2F%2Fsearch.ft.com%2Fsearch%3FqueryText%3DRomania%2Band%2BLithuania%2Bback%2Bfracking#axzz2xo4pyAZd [accessed 3 April 2014]

Bundesministerium für Wirtschaft und Energie, '*Fracking Gesetzesentwürfe zum Fracking von BMWi und BMUB*' available at: http://www.bmwi.de/DE/Themen/Industrie/Rohstoffe-und-Ressourcen/fracking.html [accessed 21 January 2015]

Bundesministerium für Wirtschaft und Technologie, '*Umsetzung der Richtlinie über die Bewirtschaftung von Abfällen aus der mineralgewinnenden Industrie (Bergbauabfallrichtlinie)*' available at: http://www.lbgr.brandenburg.de/media_fast/4055/VortragBMWiBergbauabfallrichtlinienPr%C3%A4sVersion%20040308.pdf [accessed 9 January 2015]

Bundesrat, '*Geplante "Fracking"-Regelung in Deutschland*' available at:

http://www.bundesrat.de/DE/plenum/themen/fracking/fracking.html [accessed 13 January 2015]

Bundesregierung, '*Kabinettbeschluss Fracking: Mehr Schutz durch strenge Regeln*' available at:
http://www.bundesregierung.de/Content/DE/Artikel/2015/04/2015-04-01-fracking-gesetz-kabinett.html;jsessionid=7198048376FAAF673F3B194C34B7DE85.s2t2 [accessed 7 April 2015]

Bundesregierung, '*Regierungspressekonferenz vom 22. Oktober*' available at: http://www.bundesregierung.de/Content/DE/Mitschrift/Pressekonferenzen/2014/10/2014-10-22-regpk.html [accessed 21 January 2015]

Bundestag, '*Fracking vote by name*' *(Fracking Namentliche Abstimmung)* available at: http://www.bundestag.de/bundestag/plenum/abstimmung/grafik [accessed 29 June 2016]

Bundesumweltministerium, '*Die Aarhus Konvention*' available at: http://www.bmub.bund.de/themen/umweltinformation-bildung/umweltinformation/aarhus-konvention/ [accessed 4 March 2015]

Carrington D, *The Guardian*, 'UK defeats European bid for fracking regulations' of 14th January 2014 available at: http://www.theguardian.com/environment/2014/jan/14/uk-defeats-european-bid-fracking-regulations [accessed 23 April 2014]

Community Energy Scotland, '*Horizontal directional drilling*' available at: http://www.communityenergyscotland.org.uk/news/10-mar-2015-horizontal-directional-drilling.asp [accessed 28 June 2016]

Der Spiegel, '*Debatte um CCS-Technologie: Altmaier gegen CO_2-Speicherung*' available at: http://www.spiegel.de/wissenschaft/technik/ccs-altmaier-sieht-keine-zukunft-fuer-co2-speicherung-in-deutschland-a-845868.html [accessed 18 March 2015]

Der Spiegel, 'Kritik im Bundestag: Dutzende Abgeordnete torpedieren umstrittenes Fracking-Gesetz' available at: http://www.spiegel.de/politik/deutschland/fracking-teile-von-cdu-spd-und-gruenen-gegen-gesetz-a-1026585.html [accessed 7 April 2015]

Der Spiegel, 'Nörgeln ist keine Tugend' available at: http://www.spiegel.de/spiegel/print/d-84789654.html [accessed 15 October 2012]

Die Zeit, '*Fracking-Gesetz scheitert am schwarz-gelben Streit 4. Juni. 2013*' available at: http://www.zeit.de/wirtschaft/2013-06/fracking-gesetz-union-fdp [accessed 16 April 2015]

Duke University Nicholas School of the Environment, '*Leaks from CO_2 Stored Deep Underground Could Contaminate Drinking Water*' available at: http://www.nicholas.duke.edu/news/leaks-from-co2-stored-deep-underground-could-contaminate-drinking-water [accessed 6 March 2013]

DutchNews.nl, 19 September 2013 '*No shale gas decision for 18 months, pending more research*' available at: http://www.dutchnews.nl/news/archives/2013/09/no_shale_gas_decision_for_18_m.php [accessed 14 April 2014]

EFTA, '*The European Free Trade Association*' available at: http://www.efta.int/ [accessed 18 September 2014]

EFTA Court, '*Introduction to the EFTA Court*' available at: http://www.eftacourt.int/the-court/jurisdiction-organisation/introduction [accessed 18 September 2014]

Environmental Ministry of Schleswig Holstein, '*Schleswig-Holstein will CCS-Lager per Gesetz landesweit verhindern*' available at: http://www.schleswigholstein.de/MELUR/DE/Service/Presse/PI2012_neu/0712/MELUR_120717_CCS_Gesetz.html [accessed 15 October 2012]

Environmental Protection Agency (EPA), '*Hydraulic Fracturing Background information*' http://water.epa.gov/type/groundwater/uic/class2/hydraulicfracturing/wells_hydrowhat.cfm [accessed 28 March 2013]

EurActiv, '*Shale gas firms to be brought under "robust" new EU law*' of 21 October 2013 available at: http://www.euractiv.com/energy/shale-gas-firms-brought-robust-n-news-531191 [accessed 23 April 2014]

EurActiv, 'Verzicht auf EU-Millionen: Vattenfall stoppt CCS in Brandenburg' of 6 December 2011 available at: http://www.euractiv.de/energie-und-klimaschutz/artikel/verzicht-auf-eu-millionen-vattenfall-stoppt-ccs-in-brandenburg-005708 [accessed 24 October 2013]

European Chemical Agency ECHA, '*ECHA clarifies how to report substances used in hydraulic fracturing*' available at: http://echa.europa.eu/documents/10162/21779840/annex+to+a+news+item+20150318.pdf [accessed: 2 September 2015)

European Commission, '*Environmental Aspects on Unconventional Fossil Fuels*' available at: http://ec.europa.eu/environment/integration/energy/unconventional_en.htm [accessed 24 April 2014]

European Commission, '*Press Release Energy Roadmap 2050: a secure, competitive and low-carbon energy sector is possible*' available at: http://europa.eu/rapid/press-release_IP-11-1543_en.htm [accessed 26 February 2015]

European Commission, '*Review of the Environmental Impact Assessment (EIA) Directive*' available at: http://ec.europa.eu/environment/eia/review.htm [accessed 12 December 2013]

European Commission, '*Studies & Evaluations Energy Studies*' available at: http://ec.europa.eu/energy/studies/energy_en.htm [accessed 28 August 2014]

European Parliament, '*Legislative Powers*' available at: http://www.europarl.europa.eu/aboutparliament/en/0081f4b3c7/Law-making-procedures-in-detail.html [accessed 20 December 2014]

European Union, '*Committee of the Regions*' available at: http://europa.eu/

about-eu/institutions-bodies/cor/index_en.htm [accessed 5 September 2014]

European Union, '*Get involved in European policy making*' available at: http://europa.eu/eu-law/have-your-say/index_en.htm#toc_2 [accessed 3 March 2015]

European Union Assembly of Regional and Local Representatives, '*Fracking: local and regional authorities call for compulsory environmental impact assessments*' available at: http://cor.europa.eu/en/news/Pages/fracking-environmental-impact.aspx [accessed 4 September 2014]

Eurostat, '*Development of the production of primary energy (by fuel type), EU-28, 2003–13*' available at: http://ec.europa.eu/eurostat/statistics-explained/index . php / File : Development _ of _ the _ production _ of _ primary _ energy _(by _ fuel _ type) , _ EU - 28 , _ 2003%E2%80%9313 _ (2003 _ %3D _ 100 , _ based _on_tonnes_of_oil_equivalent)_YB15.png [accessed 26 June 2016]

Exxon Mobil, '*Dialog Process*' available at: http://dialog-erdgasundfrac.de/ [accessed 21 March 2012]

Exxon Mobil, Europeunconventionalgas.org, '*Tight Gas*' available at: http://www . europeunconventionalgas . org / unconventional - gas / types - of - unconventional-gas/tight-gas [accessed 30 June 2016]

Exxon Mobil, '*Schiefergas-ohne giftige Stoffe*' available at: http://www.erdgassuche - in - deutschland . de / erkundung _ foerderung / frac _ fluessigkeiten/index.html [accessed 12 March 2015]

Federal Economic Ministry, '*Die weitere Entwicklung von CCS-Technologien*' available at: http://www.bmwi.de/DE/Themen/Industrie/Industrie-und-Umwelt/ccs.html [accessed 18 March 2015]

Fleischhauer J, *Spiegel Online*, '*S.P.O.N. – Der Schwarze Kanal: Nach der Atomangst kommt die Gasangst*' available at: http://www.spiegel.de/politik/deutschland / fracking - jan - fleischhauer - ueber - die - deutsche - angst - vor-dem-bohren-a-974313.html [accessed 30 June 2014]

Focus Features, '*Promised Land Movie (2012)*' available at: http://www.focusfeatures.com/promised_land [accessed 22 December 2012]

Frack Free Wales, www.frackfreewales.org [accessed 4 August 2014]

Frack Off, http://frack-off.org.uk/fracking-hell/frontline/ [accessed 4 August 2014]

Fracking freies Hessen, www.frackingfreieshessen.de [accessed 2 August 2014]

Fracturing Law Blog, '*The Minimum Principles Applied in Practice: Temporary Ban on Fracking in the Flemish Region in Belgium*' available at: http://fraclawblog.com/2014/07/05/the-minimum-principles-applyed-in-practice-temporary-ban-on-fracking-in-the-flemish-region-in-belgium/ [accessed 25 September 2015]

Frankfurter Allgemeine Zeitung, '*Gasfördertechnik Fracking in Nordhessen gestrichen*' available at: http://www.faz.net/aktuell/rhein-main/nach-gutachten-gasfoerdertechnik-fracking-in-nordhessen-gestrichen-12145389.html [accessed 9 January 2015]

Friends of the Earth, '*Europe opens doors to dangerous fracking*' available at: http://www.foeeurope.org/shale_gas_framework_220114 [accessed 18 December 2014]

Gas Strategies, 'Shale Gas in Europe. A revolution in the making' (2010) available at: http://www.gasstrategies.com/files/files/euro%20shale%20gas_final.pdf [accessed 18 June 2012]

Gasfrac Energy Systems Inc., '*Realizing the Potential*' available at: http://www.gasfrac.com/assets/docs/PDFS/presentations/Investor%20Presentation%20-%20Realizing%20the%20Potential.pdf [accessed 22 October 2012]

Gasland http://www.gaslandthemovie.com/ [accessed 22 December 2012].

GAVC LAW, '*Preussen Elektra confirmed? Bot AG in Essent*' available at: http://gavclaw.com/2013/05/14/preussen-elektra-confirmed-bot-ag-in-essent/ [accessed 3 February 2014]

Gaydazhieva S, New Europe Online, 'Wave of protests against shale gas in Romania, Bulgaria' of 5 April 2013 available at: http://www.neurope.eu/article/wave-protests-against-shale-gas-romania-bulgaria [accessed 24 April 2014]

Gegen Gasbohren, '*BI "No Fracking" Völkersen*' available at: http://www.gegen-gasbohren.de/initiativen/bi-no-fracking-voelkersen/ [accessed 7 February 2013]

Gegen Gasbohren, '*Informationen und Nachrichten über Erdgassuche, Gasförderung und Hydraulic "Fracking" Fracturing*' available at: www.gegen-gasbohren.de [accessed 12 March 2013]

General Court, '*Presentation*' available at: http://curia.europa.eu/jcms/jcms/Jo2_7033/ [accessed 25 September 2014]

Geo.net, '*Gutachten mit Risikostudie zur Exploration und Gewinnung von Erdgas aus unkonventionellen Lagerstätten*' available at: http://ausschreibungen.geo.net/neue/ausschreibung/geologie-geooekologie/gutachten-mit-risikostudie-zur-exploration-und-gewinnung-von.html [accessed 15 June 2012]

German Ministry of Economic Affairs and Energy, '*Fracking*' available at: http://www.bmwi.de/DE/Themen/Industrie/Rohstoffe-und-Ressourcen/fracking,did=653918.html?view=renderPrint [accessed 30 June 2016]

Germany Travel, 'Mineral and thermal springs – harnessing the healing power of water' available at: http://www.germany.travel/en/leisure-and-recreation/health-wellness/spas-and-health-resorts/mineral-and-thermal-springs/mineral-and-thermal-springs.html [accessed 20 July 2015].

GFZ, '*Pilot Site Ketzin*' available at: http://www.co2ketzin.de/en/pilot-site-ketzin/summary.html [accessed 28 November 2016]

Global CCS Institute, '*5. French CCS legislation*' available at: http://hub.globalccsinstitute.com/publications/dedicated-ccs-legislation-current-and-proposed/french-ccs-legislation [accessed 28 November 2016]

Global CCS Institute, '*6. German CCS legislation*' available at: http://hub.globalccsinstitute.com/publications/dedicated-ccs-legislation-current-and-proposed/german-ccs-legislation [accessed 28 November 2016]

Global CCS Institute, '*9. United Kingdom CCS legislation*' available at: https://hub.globalccsinstitute.com/publications/dedicated-ccs-legislation-current-and-proposed/united-kingdom-ccs-legislation [accessed 28 November 2016]

Guardian, 'EU parliament excludes shale gas from tougher environmental code' of 12 March 2014 available at: http://www.theguardian.com/environment/2014/mar/12/eu-parliament-shale-gas-environmental-code [accessed 22 April 2014]

Guardian, 'France cements fracking ban' of 11 October 2013 available at: http://www.theguardian.com/environment/2013/oct/11/france-fracking-ban-shale-gas [accessed 2 May 2013]

Guardian, '*North Yorkshire Council backs First UK Fracking Test for Five Years*' available at: http://www.theguardian.com/environment/2016/may/23/north-yorkshire-council-backs-first-uk-fracking-tests-for-five-years?CMP=fb_gu [accessed 10 June 2016]

Halliburton Press Release, '*El Paso and Halliburton Pioneer the First Natural Gas Completion using all current Cleansuite "Green Technologies" for Hydraulic Fracturing and Water Treatment*' available at: http://www.halliburton.com/public/news/pubsdata/press_release/2011/corpnws_050211_1.html?SRC=ElPasoandHalliburton [accessed 22 October 2012]

Hamid M F and Sulaiman W R W, 'Fundamentals Of Petroleum Engineering Well Completion and Stimulation' available at: http://ocw.utm.my/file.php/12/Chapter_6-OCW.pdf [accessed 27 February 2014]

Hannoversche Allgemeine, '*Die Fracking-Pause dauert bis 2016*' available at: http://www.haz.de/Nachrichten/Politik/Niedersachsen/Die-Fracking-Pause-dauert-bis-2016-in-Niedersachsen [accessed 9 January 2015]

Hessisches Ministerium für Umwelt, Klimaschutz, Landwirtschaft, '*20.08.2014 Fracking-Klage gegen das Land Hessen zurückgenommen*' available at: https://www.hessen.de/presse/pressemitteilung/fracking-klage-gegen-das-land-hessen-zurueckgenommen-0 [accessed 9 January 2015]

HLN.be, '*Vlaamse regering legt tijdelijk verbod op fracking op*' (*Flemish government imposes temporary ban on fracking*) available at: http://www.hln.

be/hln/nl/2764/milieu/article/detail/1936420/2014/07/04/Vlaamse-regering-legt-tijdelijk-verbod-op-fracking-op.dhtml [accessed 19 September 2015]
International Energy Agency (IEA), '*What is Energy Security*' available at: http://www.iea.org/topics/energysecurity/subtopics/whatisenergysecurity/ [accessed 22 September 2015]
International WOW Company, '*Gasland*' available at: http://www.gaslandthemovie.com/ [accessed 22 December 2012]
Johnston I, *NBC News*, 'Molotov cocktails, civil disobedience and middle-class furor: Fracking protests hit Europe' of 24 August 2013 available at: http://www.nbcnews.com/news/other/molotov-cocktails-civil-disobedience-middle-class-furor-fracking-protests-hit-f8C10995311 [accessed 24 April 2014]
Keep Tap Water Safe, '*List of bans worldwide*' available at: http://keeptapwatersafe.org/global-bans-on-fracking/ [accessed 17 December 2013]
Kein Fracking, www.kein-fracking.de [accessed 4 August 2014]
Knuf T, *Frankfurter Rundschau*, 'Proteste gegen Fracking in Deutschland' of 4 April 2014 available at: http://www.fr-online.de/energie/fracking-proteste-gegen-fracking-in-deutschland,1473634,26840316.html [accessed 24 April 2014]
Kruk M, *The Wall Street Journal*, 'Poland Cuts Estimate Of Shale Gas Reserves' available at: http://www.wsj.com/articles/SB100014240527023038129045772957904428444470 [accessed 25 June 2016
Krukowska E and Bakhsh N, *Bloomberg*, 'U.K. Lobbied in Brussels to Halt EU Legislation on Shale Gas' of 15 January 2014 available at: http://www.bloomberg.com/news/2014-01-15/u-k-lobbied-in-brussels-to-prevent-eu-regulation-on-shale-gas.html [accessed 23 April 2014]
Kurtz J, *E&E Publishing*, '"Drill, baby, drill!"' almost didn't happened' available at: http://www.eenews.net/stories/1059969331 [accessed 23 March 2015]
Lain S, *The Guardian*, 'Russia's gas deal with China underlines the risks to Europe's energy security' available at: http://www.theguardian.com/commentisfree/2014/may/26/russia-gas-deal-china-europe-energy-security-danger [accessed 18 June 2014]
Le Blonde J, '*Financial Times*,' Coal resurgence darkens Germans' green image' 13 October 2015 available at: http://www.ft.com/cms/s/0/719ea15e-68fa-11e5-a57f-21b88f7d973f.html#axzz49r8oaw7P [accessed 26 June 2016]
Legifrance.gouv.fr, 'Article L110-1' available at: http://www.legifrance.gouv.fr/affichCodeArticle.do?idArticle=LEGIARTI000022494168&cidTexte=LEGITEXT000006074220 [accessed 22 May 2014]

Leibniz-Institut für Angewandte Geophysik, '*Geothermisches Informations system für Deutschland*' available at: http://www.geotis.de/ [accessed 20 October 2012]

Lower Saxony State Agency for Mining, Energy and Geology, '*Hydraulische Bohrlochbehandlung*' available at: http://www.lbeg.niedersachsen.de/bergbau/genehmigungsverfahren/hydraulische_bohrlochbehandlung/hydraulische-bohrlochbehandlung-110656.html [accessed 17 July 2014]

Mattfeldt A, '*Bundestag Aktuell 24.06.2016*' available at: http://www.andreas-mattfeldt.de/images/newsletter/2016_06_24_Bundestag_Aktuell.pdf [accessed 12 July 2016]

McAleer P, 'Frack Nation' available at: http://fracknation.com/ [accessed 22 December 2012]

Ministerium für Klimaschutz, Umwelt, Landwirtschaft, Natur- und Verbraucherschutz des Landes Nordrhein-Westfalen, '*Pressemitteilung Umweltministerium und Wirtschaftsministerium legen Risikogutachten zu Fracking vor*' 7 September 2012 available at: http://www.umwelt.nrw.de/ministerium/service_kontakt/archiv/presse2012/presse120907_a.php [accessed 15 April 2014]

Mint Press News, '*EPA Pushes Back Fracking Impact Study To 2016*' available at: http://www.mintpressnews.com/epa-pushes-back-fracking-impact-study-to-2016/163927/ [accessed 25 February 2015]

Molnar S, Natural Gas Europe, 'Shale Gas Exploitation in Romania can be Postponed at Least Another Two Years' available at: http://www.naturalgaseurope.com/shale-gas-exploitation-in-romania-postponed [accessed 3 April 2014]

MP Leopold Sulovsky, 'Hlasování v Senátu' available at: http://www.leopoldsulovsky.cz/index.php?mact=News,cntnt01,detail,0&cntnt01articleid=17&cntnt01returnid=58 [accessed 27 March 2014]

Niedersächsisches Ministerium für Umwelt, Energie und Klimaschutz, '*Pressemitteilung Minister Wenzel fordert strenge Auflagen bei der Ergasförderung*' 3 March 2014 available at: http://www.umwelt.niedersachsen.de/aktuelles/pressemitteilungen/strenge-auflagen-bei-der-ergasfoerderung-122495.html [accessed 14 April 2014]

New Brunswick Anti-Shale Gas Alliance, http://www.noshalegasnb.ca/ [accessed 28 November 2016]

NN, '*Bulgarian Parliament to review shale gas moratorium*' available at: http://sofiaecho.com/2012/04/02/1800346_bulgarian-parliament-to-review-shale-gas-moratorium [accessed 5 June 2012]

Pakosta P, 'Břidlicový plyn: Vše, co bychom měli vědět, v otázkách a odpovědích' available at: http://www.pakosta.cz/article452.html [accessed 22 May 2014]

Petkova D and Ewropeiska E investor.bg, 'Iskra Michailova: Moratorium on shale gas extraction only a temporary measure' (*Деляна Петкова and Евгения Европейска investor.bg 'Искра Михайлова: Мораториумът за добив на шистов газ е временна мярка*) of 3 June 2013 available at: http://www.investor.bg/ikonomika-i-politika/332/a/iskra-mihailova-moratoriumyt-za-dobiv-na-shistov-gaz-e-vremenna-miarka-,152074/ [accessed 29 April 2014]

PolishShale.pl, '*European Commission is going to propose a directive concerning shale*' of 18 September 2013 available at: http://polishshale.pl/komisja-europejska-zaproponuje-dyrektywe-ws-lupkow/ [accessed 23 April 2014]

Prague Daily Monitor/Czech News Agency, '*PM: Shale gas mining not on the agenda*' available at: http://praguemonitor.com/2013/05/23/pm-shale-gas-mining-not-agenda [accessed 27 March 2014]

Regnum.ru, '*»Сланцевая революция» в Литве на грани срыва – разведку газа предлагают запретить*' available at: http://www.regnum.ru/news/1621957.html [accessed 7 March 2013]

Reuters, '*Spain to challenge Cantabria's fracking ban in court*' of 24 January 2014 available at: http://www.reuters.com/article/2014/01/24/spain-shale-idUSL5N0KY32K20140124 [accessed 29 April 2014]

Revkin A, *The New York Times*, 'Damning Review of Gas Study Prompts a Shakeup at the University of Texas' available at: http://dotearth.blogs.nytimes.com/2012/12/06/damning-review-of-gas-study-prompts-a-shakeup-at-the-university-of-texas/ [accessed 16 April 2013]

Rijksdienst voor Ondernemend Nederland, '*Schiefergasförderung in den Niederlanden*' available at: http://www.rvo.nl/subsidies-regelingen/thema-schiefergas [accessed 25 September 2015]

Russian Nature, '*Environmental Impact of Oil and Gas Development*' available at: http://www.rusnature.info/env/20.htm [accessed 22 August 2013]

Savu I, *Bloomberg News*, 'Romania Ends Moratorium on Shale Gas Exploration, Premier Says' available at: http://www.bloomberg.com/news/2013-03-19/romania-ends-moratorium-on-shale-gas-exploration-premier-says.html [accessed 1 April 2014]

Savu I and Timu A, 'Romania Shale-Gas, Mine Projects on Hold in 2012, Ponta Says' available at: http://www.businessweek.com/news/2012-06-07/romanian-shale-gas-mining-projects-on-hold-in-2012-ponta-says [accessed 3 April 2014]

Schaliegasvrij Nederland, '*146 gemeenten en acht provincies wijzen schaliegas af*' available at: https://www.schaliegasvrij.nl/2013/11/05/100ste-schalegasvrije-gemeente-hengelo/ [accessed 14 April 2014]

Schavemaker Y, 'Shale Gas in the Netherlands' available at: http://www.

shale-gas-information-platform.org/areas/the-debate/shale-gas-in-the-netherlands.html [accessed 14 April 2014]

Schmidt H, GmbH, 'Horizontalbohrtechnik' available at: http://www.schmidt-rohrleitungsbau.de/leistungen/horizontalbohrtechnik/ [accessed 28 June 2016]

Scottish Government, '*Issued on behalf of the Expert Scientific Panel*' available at: http://news.scotland.gov.uk/News/Issued-on-behalf-of-the-Expert-Scientific-Panel-f2a.aspx [accessed 8 September 2016]

Scottish Government, '*Moratorium called on Fracking*' available at: http://news.scotland.gov.uk/News/Moratorium-called-on-fracking-1555.aspx [accessed 8 September 2016]

Shale Gas Europe, '*Spain, the forgotten shale gas country*' of 10 April 2014 available at: http://shalegas-europe.eu/spain-forgotten-shale-gas-country/ [accessed 28 April 2014]

Sokolski H, *Newsweek*, 'Post Fukushima, Nuclear Power Changes Latitude' available at: http://europe.newsweek.com/post-fukushima-nuclear-power-changes-latitudes-66311?rm=eu [accessed 1 July 2016]

Soraghan M, *New York Times* (24/February/2011), 'Groundtruthing Academy Award Nominee "'Gasland'"' available at: http://www.nytimes.com/gwire/2011/02/24/24greenwire-groundtruthing-academy-award-nominee-gasland-33228.html?pagewanted=all [accessed 22 December 2012]

Statista, '*Bevölkerungsdichte (Einwohner je km²) in Deutschland nach Bundesländern (zum 31. Dezember 2010)*' available at: http://de.statista.com / statistik / daten / studie / 1242 / umfrage / bevoelkerungsdichte - in - deutschland-nach-bundeslaendern/ [accessed 8 June 2012]

Stern, '*Russland vs. Ukraine: Gas-Streit erreicht Deutschland*' available at: http://www.stern.de/wirtschaft/news/unternehmen/russland-vs-ukraine-gas-streit-erreicht-deutschland-650657.html [accessed 18 July 2013]

Stop Fracking Now, http://www.stopfrackingnow.com/ [accessed 28 November 2016]

Tanasuc D, www.vrn.ro, '*UPDATE: Senate today rejected a proposal banning shale drilling by hydraulic fracturing ...*' www.vrn.ro of 6 November 2013 (*UPDATE: Plenul Senatului a respins astăzi propunerea privind interzicerea exploatării gazelor de sist prin fracturare hidraulică...)* available at: http://www.vrn.ro/senatorii-decid-pro-sau-contra-romanilor [accessed 26 May 2014]

Tertzakian P, 'Oil and gas innovation keeps pace with high-tech world' *The Globe and Mail* of 18 June 2013 available at: http://www.theglobeandmail.com/report-on-business/industry-news/energy-and-resources/oil-and-gas-innovation-keeps-pace-with-high-tech-world/article12607183/ [accessed 6 November 2013]

The Telegraph, '*Shale gas drilling suspended after earthquake near Blackpool*' availableat:http://www.telegraph.co.uk/earth/energy/8549340/Shale-gas-drilling-suspended-after-earthquake-near-Blackpool.html [accessed 7 April 2014]

Thompson E, 'Residents reject shale gas drilling' available at: http://www.praguepost.com/business/13127-residents-reject-shale-gas-drilling.html [accessed 15 June 2012]

Tramba D, Civic Democratic Party, 'Tomáš Chalupa: Extraction of shale gas in the Czech Republic? Why not' (*Občanska Demokraticka Strana 'Tomáš Chalupa: Těžba plynu z břidlic v Česku? Proč ne'*) available at: http://www.ods.cz/clanek/1476-tezba-plynu-z-bridlic-v-cesku-proc-ne [accessed 15 June 2012]

Truth Live, '*Prime Minister Victor Ponta: "I would be more popular if I tried to suspend Basescu but the price for Romania is too high"*' (*ADEVĂRUL LIVE Premierul Victor Ponta: "Eu aş fi mai popular dacă m-aş apuca să-l suspend pe Băsescu dar preţul pentru România ar fi prea mare")* http://adevarul.ro/news/politica/live-video-premierul-victor-ponta-direct-studioul-adevarul-ora-1315-1_51483b3d00f5182b85288f32/index.html [accessed 9 June 2014]

Twitter.com, '*Google Trends*' available at: https://twitter.com/googletrends/status/746303118820937728 [accessed 29 June 2016]

UEFA, '*Uefa Euro Championship 2016*' available at: http://www.uefa.com/uefaeuro/index.html [accessed 5 July 2016]

UK Government, '*Local Councils to Receive Millions in Business Rates from Shale Gas Developments*' 13 January 2014 available at: www.gov.uk/government/news/local-councils-to-receive-millions-in-business-rates-from-shale-gas-developments [accessed 10 June 2016]

US Energy Information Administration, '*Energy in Brief Tight Oil*' available at: https://www.eia.gov/energy_in_brief/article/shale_in_the_united_states.cfm [accessed 13 July 2016]

US Environmental Protection Agency, '*EPA's Study of Hydraulic Fracturing and Its Potential Impact on Drinking Water Resources*' available at: http://epa.gov/hfstudy/ [accessed 2 March 2013]

US Environmental Protection Agency, 'Shale Gas Extraction' available at: http://water.epa.gov/scitech/wastetech/guide/shale.cfm [accessed 21 March 2012]

US Environmental Protection Agency, 'Treatment and Disposal of Wastewater from Shale Gas Extraction' available at: http://cfpub.epa.gov/npdes/hydrofracturing.cfm (accessed 20 March 2012)

Vattenfall Press Release of 5 December 2011 available at www.Vattenfall.de [accessed 25 November 2016]

Vereinigung der Initiativen gegen unkontrollierte Erdgassuche, '*Studie sieht Fracking Risiken-Gegner zweifeln an Unabhängigkeit*' available at: http://www.gegen-"Fracking" Fracturing in Deutschland, gasbohren.de/2012/04/25/studie-sieht-fracking-risiken-gegner-zweifeln-an-unabhaengigkeit/ [accessed 27 June 2012]

Vidic R D, 'Sustainable Water Management for Marcellus Shale' available at: http://www.temple.edu/environment/NRDP_pics/shale/presentations_TUsummit/Vidic-Temple-2010.pdf [accessed 7 June 2012]

Vitík M, 'MŽP připravuje moratorium na průzkum těžby břidlicového plynu' available at: http://www.mzp.cz/cz/news_120504_plyn [accessed 2 April 2014]

WAZ, '*NRW-Umweltminister Remmel erteilt Fracking-Pilotprojekt eine Absage*' available at: http://www.derwesten.de/politik/nrw-umweltminister-remmel-erteilt-fracking-pilotprojekt-eine-absage-id6591914.html [accessed 27 June 2012]

Wirtschaftsblatt, '*OMV will Mega-Gasvorrat im Weinviertel ab 2020 fördern*' availableat:http://wirtschaftsblatt.at/home/boerse/wien/1213426/index?_vl_pos=r.1.NT [accessed 22 October 2012]

World Energy Council, '*World Energy Trilemma*' available at: https://www.worldenergy.org/work-programme/strategic-insight/assessment-of-energy-climate-change-policy/ [accessed 22 September 2016].

World Nuclear Association, 'Nuclear Power in Italy' available at: http://www.world-nuclear.org/information-library/country-profiles/countries-g-n/italy.aspx [accessed 1 June 2016]

World Nuclear Association, '*Nuclear Power in Sweden*' available at: http://www.world-nuclear.org/information-library/country-profiles/countries-o-s/sweden.aspx [accessed 1 June 2016]

Ziare.com, 'Gazele de sist - obiectiv prioritar in programul de guvernare al lui Ponta' available at: http://www.ziare.com/victor-ponta/guvern/gazele-de-sist-obiectiv-prioritar-in-programul-de-guvernare-al-lui-ponta-1208368 [accessed 3 April 2014]

Ziare.com, '*Rovana Plumb, despre gazele de sist: Suntem sub moratoriu si ramanem asa*' available at: http://www.ziare.com/rovana-plumb/ministrul-mediului/rovana-plumb-despre-gazele-de-sist-suntem-sub-moratoriu-si-ramanem-asa-1190245 [accessed 3 April 2014]

Zpravy E 15, 'Chalupa chce stop průzkumu plynu z břidlic do roku 2014' available at: http://zpravy.e15.cz/byznys/prumysl-a-energetika/chalupa-chce-stop-pruzkumu-plynu-z-bridlic-do-roku-2014-910421 [accessed 31 March 2014]

Zpravy Rozhlas, '*Petice proti těžbě břidlicového plynu už podepsalo přes 35 tisíc lidí*' available at: http://www.rozhlas.cz/zpravy/regiony/_zprava/1054012 [accessed 15 June 2012]

Index